AS/A Level course structure

This book has been written to support students studying for AQA AS Biology and for students in their first year of studying for AQA A Level Biology. It covers the AS sections from the specification, the content of which will also be examined at A Level. The sections covered are shown in the contents list, which also shows you the page numbers for the main topics within each section. There is also an index at the back to help you find what you are looking for. If you are studying for AS Biology, you will only need to know the content in the blue box.

AS exam

Year 1 content

1 Biological molecules
2 Cells
3 Organisms exchange substances with their environment
4 Genetic information, variation, and relationships between organisms

Year 2 content

5 Energy transfers in and between organisms
6 Organisms respond to changes in their internal and external environment
7 Genetics, populations, evolution, and ecosystems
8 The control of gene expression

A level exam

A Level exams will cover content from Year 1 and Year 2 and will be at a higher demand. You will also carry out practical activities throughout your course. There are **twelve** required practicals: six from the AS and six A-Level.

Contents

HAMPTON SCHOOL
BIOLOGY DEPARTMENT

How to use this book

This book contains many different features. Each feature is designed to support and develop the skills you will need for your examinations, as well as foster and stimulate your interest in biology.

Terms that you will need to be able to define and understand are shown in **bold type** within the text.

Where terms are not explained within the same topic, they are highlighted in **bold orange text**. You can look these words up in the glossary.

Application features

These features contain important and interesting applications of biology in order to emphasise how scientists and engineers have used their scientific knowledge and understanding to develop new applications and technologies. There are also application features to develop your maths skills, with the icon $\sqrt{x}$, and to develop your practical skills, with the icon 🧪.

Extension features

These features contain material that is beyond the specification designed to stretch and provide you with a broader knowledge and understanding and lead the way into the types of thinking and areas you might study in further education. As such, neither the detail nor the depth of questioning will be required for the examinations. But this book is about more than getting through the examinations.

1   Extension and application features have questions that link the material with concepts that are covered in the specification. There are also extension features which develop your maths skills, with the icon $\sqrt{x}$, and to develop your practical skills with the icon 🧪.

Summary questions

1   These are short questions that test your understanding of the topic and allow you to apply the knowledge and skills you have acquired. The questions are ramped in order of difficulty.

2   $\sqrt{x}$ Questions that will test and develop your mathematical and practical skills are labelled with the mathematical symbol ($\sqrt{x}$) and the practical symbol (🧪).

Section 3

...ubstances
...t

7 Mass transport

...on

...organisms exchange material between themselves and
...ent. To enter or leave an organism, substances must
...lasma membrane. Single-celled and small multicellular
organisms can satisfactorily exchange materials over their body
surfaces using diffusion alone, especially if their metabolic rate is low.
As organisms evolved and became larger, their surface area to volume
ratios decreased and specialised respiratory surfaces evolved to meet
the increasing requirement to exchange ever larger quantities of
materials.

Where large size is combined with a high metabolic rate there is
a requirement for a mass transport system to move substances
between the exchange surface and the cells of which the organism
is composed. In animals these systems often involve circulating a
specialised transport medium (blood) through vessels using a
pump (heart).

Plants do not move from place to place and have a relatively low
metabolic rate and consequently reduced demand for oxygen and
glucose. Coupled with their large surface area, essential for obtaining
light for photosynthesis, they have not evolved a pumped circulatory
system. Plants do, however, transport water up from their roots to
the leaves and distribute the products of photosynthesis. Their mass
transport system comprises vessels too – xylem and phloem, but the
movement of fluid within them is largely a passive process.

The internal environment of a cell or organism differs from the
environment around it. The cells of large multicellular animals are
surrounded by tissue fluid, the composition of which is kept within
a suitable metabolic range. In both plants and animals, it is the mass
transport system that maintains the final diffusion gradients which
allows substances to be exchanged across cell-surface membranes.

Working scientifically

Studying exchange between organisms and the environment allows you
to carry out practical work and to develop practical skills. A required
practical activity is the dissection of an animal or plant gas exchange
system or mass transport system or of an organ within such a system.

You will require a range of mathematical skills: in particular the ability
to change the subject of an equation and calculate the surface areas
and volumes of various shapes.

What you already know

The material in this unit is intended to be self-explanatory. However, there is
some knowledge from GCSE that will aid your understanding of this section. This
information includes:

○ The effectiveness of a gas-exchange surface is increased by having a large surface
area, being thin, having an efficient blood supply and being ventilated.

○ In humans the surface area of the lungs is increased by alveoli and that of the
small intestine by villi. The villi provide a large surface area with an extensive
network of capillaries to absorb the products of digestion by diffusion and active
transport.

○ Breathing in involves the ribcage moving out and up and the diaphragm
becoming flatter. Breathing out involves these changes being reversed.

○ In plants, water and mineral ions are absorbed by roots, the surface area of
which is increased by root hairs.

○ Plants have stomata in their leaves through which carbon dioxide and oxygen
are exchanged with the atmosphere by diffusion. The size of stomata is
controlled by guard cells that surround them and help control water loss.

○ In flowering plants, xylem tissue transports water and mineral ions from the
roots to the stem and leaves and phloem tissue carries dissolved sugars from the
leaves to the rest of the plant.

○ In animals a circulatory system transports substances using a heart, which is a
muscular organ with four main chambers – left and right atria and ventricles.

○ Blood flows from the heart to the organs through arteries and returns through
veins. Arteries have thick walls containing muscle and elastic fibres. Veins have
thinner walls and often have valves to prevent back-flow of blood.

○ Blood is a tissue and consists of plasma in which red blood cells, white blood
cells and platelets are suspended.

○ Red blood cells have no nucleus and are packed with haemoglobin. In the lungs
haemoglobin combines with oxygen to form oxyhaemoglobin. In other organs
oxyhaemoglobin splits up into haemoglobin and oxygen.

○ White blood cells have a nucleus and form part of the body's defence system
against microorganisms.

breathing — gas exchange
(topic 6.7)   (topic 6.8)

specific
enzymes

human
lung
(topic 6.6)

single-celled organisms
and insect
tracheal system
(topic 6.2)

digestion of
carbohydrates
lipids
proteins

large surface area
to volume ratio

partially
permeable

examples of
gas exchange

absorption
up the
ileum
(topic 6.10)

short diffusion
distance

exchange
surfaces
(topic 6.1)

digestive
system
(topic 6.9)

monosaccharides

maintaining
concentration
gradient

exchange
(topic 6)

amino acids

organisms exchange
substances with their
environment (section 3)

triglycerides

bohr effect (CO₂)

micelles

oxygen dissociation
curves (topic 7.2)

mass
transport
(topic 7)

water transport

cohesiontension
theory

Hb
(topic 7.1)

xylem (topic 7.7)

Hb in different
organisms

animals

plants

transpiration

closed

double

evidence

potometer

circulation
(topic 7.3)

atria

ventricles

phloem
(topic 7.8)

blood vessels
structure/function
(topic 7.6)

heart
(topic 7.4)

experiments
(topic 7.9)

veins

valves

ringing
and
tracers

arteries

cardiac cycle
(topic 7.5)

transport
of sucrose
and amino
acids

translocation

capillaries

tissue fluid
formation

Practical skills

In this section you have met
the following practical skills:

• Evaluating the results of
scientific experiments

• Using appropriate
apparatus, such
as a potometer, to
obtain quantitative
measurements

• Commenting on
experimental design
and suggesting
improvements.

Maths skills

In this section you have met the following maths skills:

• Calculating surface area to volume ratios

• Changing the subject in pulmonary ventilation and
cardiac output equations

• Using appropriate units in calculations

• Substituting values in, and solving, algebraic
equations

• Interpreting graphs and translating information
between graphical and numerical forms

• Recognising expressions in decimal and standard forms

• Understanding simple probability

• Interpreting bar charts.

Extension tasks

Using the knowledge that you have gained
from this section make a comprehensive list
of each of the following:

a   The general features of all transport systems.

b   The differences between transport
systems in plants and transport systems
in animals.

c   An explanation for each of the differences
you have listed under b).

Figure 1 shows you how to take a person's
pulse at the wrist. Each pulse is equivalent to
a single heartbeat.

You should take each person's pulse for
30 seconds and double the reading to give the
number of heart beats per minute (heart rate).

Find out what is meant by the 'recovery
heart rate' and then design and carry out
an experiment to determine any difference
between the recovery heart rate of people
who exercise frequently and those who do
not. Draw conclusions from your results and
suggest an explanation for them.

Section 5 Skills in A level Biology
Chapter 11 Mathematical skills

Biology students are often less comfortable with the application of mathematics compared with students such as physicists, for whom complex maths is a more obvious everyday tool. Nevertheless, it is important to realise that biology does require competent maths skills in many areas. It is important to practise these skills so you are familiar with them as part of your routine study of the subject.

Confidence with mental arithmetic is very helpful, but among the most important skills is that of taking care and checking calculations. We may not be required to understand the detailed theory of the maths we use, but we do need to be able to apply the skills accurately, whether simply calculating percentages or means, or substituting numbers into complex-looking algebraic equations, such as in statistical tests.

This chapter is designed to help with some of the regularly encountered mathematical problems in biology.

Working with the correct units
In biology it is very important to be secure in the use of correct units. These must always be written clearly in calculations.

Base units
The units we use are from the Système Internationale – the SI units. In biology we most commonly use the SI base units:
- metre (m) for length, height, distance
- kilogram (kg) for mass
- second (s) for time
- mole (mol) for the amount of a substance.

You should develop good habits right from the start, being careful to use the correct abbreviation for each unit used. For example, seconds should be abbreviated to s, not 'sec' or 'S'.

Derived units
Biologists also use SI derived units, such as:
- square metres (m^2) for area
- cubic metre (m^3) for volume
- cubic centimetre (cm^3), also written as millilitre (ml), for volume
- degree Celsius (°C) for temperature
- mole per litre (mol/L, mol dm⁻³) is usually used for concentration of a substance in solutions (although the official SI derived unit is moles per cubic metre)
- joule (J) for energy
- pascal (Pa) for pressure
- volt (V) for electrical potential.

Maths link
MS 0.1

265

> Mathematical section to support and develop your mathematical skills required for your course. Remember, at least 10% of your exam will involve mathematical skills.

Chapter 12 Practical skills

Practical skills are at the heart of Biology. In the AS skills is found only in the written exams and there marks in the AS papers will relate to practical work.

By undertaking the set practical activities in this co manipulative skills with specific apparatus and tech deeper understanding into the processes of scientifi planning, implementing by making and processing evaluating results will be reinforced and enhanced.

It is advantageous for you to answer practical ques practical – any exam questions on practical skills w that you will have carried out the practical activitie helps with the teaching and learning of concepts in experience will be gained if you do more practicals than the following six set practical activities in Table 1. For each activity, Table 1 references the relevant topic(s) in this book.

AS required practical activities

▼ Table 1

	Practical	Topic
1	Investigation into the effect of a named variable on the rate of an enzyme-controlled reaction	1.8 Factors affecting enzyme action
2	Preparation of prepared squashes of cells from plant root tips; set-up and use of an optical microscope to identify the stages of mitosis in these stained squashes and calculation of a mitotic index	3.1 Methods of studying cells 3.7 Mitosis
3	Production of a dilution series of a solute to produce a calibration curve with which to identify the water potential of plant tissue	4.3 Osmosis
4	Investigation into the effect of a named variable on the permeability of cell-surface membranes	4.1 Structure of the cell-surface membrane
5	Dissection of animal or plant gas exchange systems, a mass transport system or of an organ within such a system	6.2 Gas exchange in insects 6.3 Gas exchange in fish 6.4 Gas exchange in the leaf of a plant 6.6 Mammalian lungs
6	Use of aseptic technique to investigate the effect of antimicrobial substances on microbial growth	9 Genetic diversity and adaptation

Practical questions

The following questions are designed to give you some practice at this practical style of question. If you haven't completed the practical yet, just think of similar practicals you have done or when you have used the apparatus and this will help you.

Practical 1 – The effect of pH on catalases

A celery extract was liquidised and prepared by the technician as a source of the enzyme catalase. A burette had been filled up to the 50 cm³ mark with hydrogen peroxide. 10 cm³ of celery extract was added and the height of the upper level of the frothing liquid was recorded. The class was asked to repeat the procedure adding the following to the H_2O_2:

- Add 2 drops HCl / 2 drops distilled water
- Add 4 drops HCl
- Add 2 drops NaOH / 2 drops distilled water
- Add 4 drops NaOH

277

> Practical skills section with questions for each suggested practical on the specification. Remember, at least 15% of your exam will be based on practical skills.

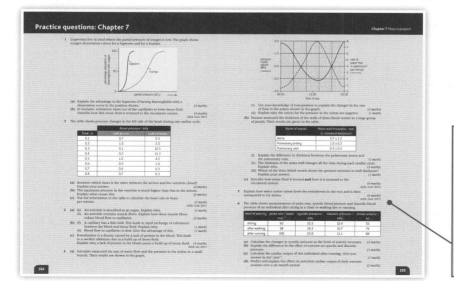

> Practice questions at the end of each chapter and each section, including questions that cover practical and maths skills. There are also additional practice questions at the end of the book.

Kerboodle

This book is supported by next generation Kerboodle, offering unrivalled digital support for independent study, differentiation, assessment, and the new practical endorsement.

If your school subscribes to Kerboodle, you will also find a wealth of additional resources to help you with your studies and with revision:

- Study guides
- Maths skills boosters and calculation worksheets
- On your marks activities to help you achieve your best
- Practicals and follow up activities to support the practical endorsement
- Interactive objective tests that give question-by-question feedback
- Animations and revision podcasts
- Self-assessment checklists.

Revise with ease using the study guides to guide you through each chapter and direct you towards the resources you need.

Transport of water in plants

xylem

cambium

phloem

If you are a teacher reading this, Kerboodle also has plenty of further assessment resources, answers to the questions in the book, and a digital markbook along with full teacher support for practicals and the worksheets, which include suggestions on how to support and stretch your students. All of the resources are pulled together into teacher guides that suggest a route through each chapter.

Section 1
Biological molecules

Introduction

Biology covers a wide field of information over a considerable size range. On the one hand it involves the movement of electrons in photosynthesis and on the other, the migrations of populations around the Earth. In the same way, living organisms have an extremely diverse range of form and function.

This section explores the fundamental building blocks of these organisms – the molecules of which their cells are composed. Their cells are made up of only a few groups of molecules that react chemically with each other in very similar ways. More importantly, these molecules are all based on carbon.

Examples of biologically important carbon-based molecules that will be explored in this section:

- **Carbohydrates** that are a respiratory substrate from which cells release the energy required to carry out their functions. They also have structural roles in cell walls, and form part of glycoproteins and glycolipids which act as recognition sites in plasma membranes.
- **Lipids** form a major component of plasma membranes. They also make up certain hormones and act as respiratory substrates.
- **Proteins** display a very diverse range of structure and therefore function also. They too are found in plasma membranes but perhaps their most important role is as enzymes. In addition they are chemical messengers within and between cells as well as being important components of the blood, for example antibodies.
- **Nucleic acids** such as deoxyribonucleic acid (DNA) carry genetic information that determines the structure of proteins. Others, like ribonucleic acid (RNA) have a role in the synthesis of these proteins.

A review of biologically important molecules would not be complete without a mention of **water**. It is not a carbon-based molecule but, despite its simplicity, serves a wide range of roles in living organisms. It is the most common component of cells and all life as we know it relies on this simple molecule.

Working scientifically

The study of biological molecules provides many opportunities to carry out practical work and to develop practical skills. A required practical activity is an investigation into the effect of a named variable on the rate of an enzyme-controlled reaction. In carrying out this activity you should look to develop practical skills such as:

- using appropriate instrumentation to record quantitative measurements
- using laboratory glassware apparatus and qualitative reagents to identify biological molecules
- identifying variables that must be controlled and calculating the uncertainty of the measurements you make
- considering margins of error, accuracy and precision of data.

You will require a range of mathematical skills, including the ability to use a calculator's logarithmic functions, to plot two variables from experimental data and draw and use the slope of a tangent to a curve as a measure of rate of change.

What you already know

While the material in this unit is intended to be self-explanatory, there is certain information from GCSE that will prove helpful to the understanding of this section. This information includes:

- ◯ The glucose produced by plants during photosynthesis may be converted into insoluble starch for storage
- ◯ During aerobic respiration chemical reactions occur that use glucose and oxygen to release energy
- ◯ Some glucose in plants and algae is used to produce fat for storage, and cellulose which strengthens the cell wall and proteins
- ◯ Protein molecules are made up of long chains of amino acids. These long chains are folded to produce a specific shape that enables other molecules to fit into the protein
- ◯ Proteins act as structural components of tissues such as muscles, hormones, antibodies and catalysts
- ◯ Catalysts increase the rate of chemical reactions – biological catalysts are proteins called enzymes
- ◯ The shape of an enzyme is vital for its function. High temperatures change the shape of the enzyme
- ◯ Different enzymes work best at different pH values
- ◯ Some enzymes work outside the body's cells

Learning objectives

→ Describe what a mole is, and what is meant by a molar solution.

→ Explain bonding and the formation of molecules.

→ Describe polymerisation and state what macromolecules are.

→ Describe condensation and hydrolysis.

→ Describe metabolism.

Specification reference: 3.1.1

Biological molecules are particular groups of chemicals that are found in living organisms. Their study is known as **molecular biology**. All molecules, whether biological or not, are made up of units called atoms.

Bonding and the formation of molecules

Atoms may combine with each other in a number of ways:

- **Covalent bonding** – atoms share a pair of electrons in their outer shells. As a result the outer shell of both atoms is filled and a more stable compound, called a molecule, is formed.

- **Ionic bonding** – ions with opposite charges attract one another. This electrostatic attraction is known as an ionic bond. For example, the positively charged sodium ion Na^+ and negatively charged chloride ion Cl^- form an ionic bond to make sodium chloride. Ionic bonds are weaker than covalent bonds.

- **Hydrogen bonding** – the electrons within a molecule are not evenly distributed but tend to spend more time at one position. This region is more negatively charged than the rest of the molecule. A molecule with an uneven distribution of charge is said to be **polarised**, in other words it is a **polar molecule**. The negative region of one polarised molecule and the positively charged region of another attract each other. A weak electrostatic bond is formed between the two. Although each bond is individually weak, they can collectively form important forces that alter the physical properties of molecules. This is especially true for water.

Polymerisation and the formation of macromolecules

Certain molecules, known as **monomers**, can be linked together to form long chains. These long chains of monomer sub-units are called **polymers** and the process by which they are formed is therefore called **polymerisation**. The monomers of a polymer are usually based on carbon. Many, such as polythene and polyesters, are industrially produced. Others, like polysaccharides, polypeptides and polynucleotides, are made naturally by living organisms. The basic sub-unit of a polysaccharide is a monosaccharide or single sugar (Topic 1.3), for example glucose. Polynucleotides are formed from mononucleotide sub-units. Polypeptides are formed by linking together peptides that have amino acids as their basic sub-unit (Topic 1.6).

Synoptic link

Examples of Polynucleotides are given in Topic 2.1 and of how polypeptides are formed in Topics 8.4 and 8.5.

Condensation and hydrolysis reactions

In the formation of polymers by polymerisation in organisms, each time a new sub-unit is attached a molecule of water is formed. Reactions that produce water in this way are termed **condensation reactions**. Therefore the formation of a polypeptide from amino acids and that of the polysaccharide starch from the monosaccharide glucose are both condensation reactions.

Polymers can be broken down through the addition of water. Water molecules are used when breaking the bonds that link the sub-units of a polymer, thereby splitting the molecule into its constituent parts. This type of reaction is called **hydrolysis** ('hydro' = water; 'lysis' = splitting). Thus polypeptides can be hydrolysed into amino acids, and starch can be hydrolysed into glucose. Figure 1 summarises atomic and molecular organisation.

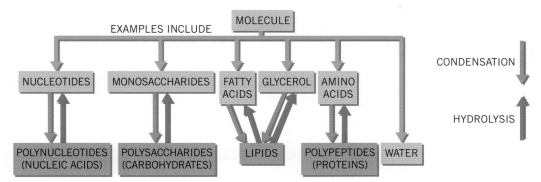

▲ **Figure 1** *Summary of atomic and molecular organisation*

Metabolism

All the chemical processes that take place in living organisms are collectively called metabolism.

The mole and molar solution

The mole is the SI unit for measuring the amount of a substance and is abbreviated to mol.

One mole contains the same number of particles as there are in 12 g of carbon-12 atoms (^{12}C). 12 g of carbon-12 atoms contain 6.022×10^{23} carbon atoms. 6.022×10^{23} is called the Avogadro number or Avogadro constant.

A **molar solution** (M) is a solution that contains one mole of solute in each litre of solution. A mole is the molecular mass (molecular weight) expressed as grams (= one gram molecular mass).

As an example, to make a molar solution of sodium chloride we must first find its molecular mass.

The chemical formula for sodium chloride is NaCl, which means that a molecule of sodium chloride contains one sodium atom and one chlorine atom. The atomic weight of sodium (Na) is 23 while that of chlorine (Cl) is 35.5. The molecular mass of NaCl is therefore Na (23) + Cl (35.5) = NaCl (58.5).

Therefore, a 1 M solution of sodium chloride contains 58.5 grams of sodium chloride in 1 litre of solution.

Atoms, Isotopes and the formation of ions

Atoms

Atoms are the smallest units of a chemical element that can exist independently. An atom comprises a nucleus that contains particles called protons and neutrons (the hydrogen atom is the only exception as it has no neutrons). Tiny particles called electrons orbit the nucleus of the atom. The main features of these sub-atomic particles are:

- **Neutrons** – occur in the nucleus of an atom and have the same mass as protons but no electrical charge.
- **Protons** – occur in the nucleus of an atom and have the same mass as neutrons but do have a positive charge.
- **Electrons** – orbit in shells around the nucleus but a long way from it. They have such a small mass that their contribution to the overall mass of the atom is negligible. They are, however, negatively charged and their number determines the chemical properties of an atom.

In an atom the number of protons and electrons is the same and therefore there is no overall charge.

Two important terms are:

- the **atomic number** – the number of protons in an atom
- the **mass number** – the total number of protons and neutrons in an atom.

The atomic structure of three biological elements is given in Figure 2.

Isotopes

While the number of protons in an element always remains the same, the number of neutrons can vary. The different types of the atom so produced are called **isotopes**. Isotopes of any one element have the same chemical properties but differ in mass. Each type is therefore recognised by its different mass number. Isotopes, especially radioactive ones, are very useful in biology for tracing the route of certain elements in biological processes and for dating fossils.

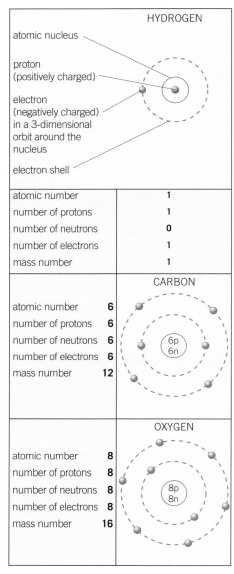

▲ **Figure 2** *Atomic structure of three commonly occurring biological elements*

The formation of ions

If an atom loses or receives an electron it becomes an **ion**.

- The loss of an electron leads to the formation of a positive ion, for example, the loss of an electron from a hydrogen atom produces a positively charged hydrogen ion, written as H^+.

- The receiving of an electron leads to the formation of a negative ion, for example, if a chlorine atom receives an additional electron it becomes a negatively charged chloride ion, written as Cl^-.

More than one electron may be lost or received, for example the loss of two electrons from a calcium atom forms the calcium ion, Ca^{2+}. Ions may be made up of more than one type of atom, for example a sulfate ion is formed when one sulfur atom and four oxygen atoms receive two electrons and form the sulfate ion, SO_4^{2-}.

1 a Name the element that contains one proton and one electron.

   b If an atomic particle with no overall charge is added to this element, state which general term can be used to describe the new form of the element.

   c Determine by what percentage the element's mass number is altered by the addition of this new particle.

   d Determine how the atomic number is affected by the addition of this new particle.

2 a State what is formed if a negatively charged particle is removed from a hydrogen atom.

   b State how the mass number is changed by the removal of this negatively charged particle.

Specification reference: 3.1.2

Learning objectives

→ Describe how carbohydrates are constructed.

→ Describe the structure of monosaccharides.

→ Describe how to carry out the Benedict's test for reducing and non-reducing sugars.

As the word suggests, carbohydrates are carbon molecules (carbo) combined with water (hydrate). Some carbohydrate molecules are small while others are large.

Life based on carbon

Carbon atoms have an unusual feature. They very readily form bonds with other carbon atoms. This allows a sequence of carbon atoms of various lengths to be built up. These form a 'backbone' along which other atoms can be attached. This permits a large number of different types and sizes of molecule, all based on carbon. The variety of life that exists on Earth is a consequence of living organisms being based on the versatile carbon atom. Carbon-containing molecules are known as organic molecules. In living organisms, there are relatively few other atoms that attach to carbon. Life is therefore based on a small number of chemical elements.

Hint

In biology certain prefixes are commonly used to indicate numbers. There are two systems, one based on Latin and the other on Greek. The Greek terms which are used when referring to chemicals are:

- mono – one
- di – two
- tri – three
- tetra – four
- penta – five
- hexa – six
- poly – many

The making of large molecules

Many organic molecules, including carbohydrates, are made up of a chain of individual molecules. Each of the individual molecules that make up these chains is given the general name **monomer**. Examples of monomers include monosaccharides, amino acids and nucleotides. Monomers can join together to form long chains called **polymers**. How this happens is explained in Topic 1.3. Biological molecules like carbohydrates and proteins are often polymers. These polymers are based on a surprisingly small number of chemical elements. Most are made up of just four elements: carbon, hydrogen, oxygen and nitrogen.

In carbohydrates, the basic monomer unit is a sugar, otherwise known as a saccharide. A single monomer is therefore called a **monosaccharide**. A pair of monosaccharides can be combined to form a **disaccharide**. Monosaccharides can also be combined in much larger numbers to form **polysaccharides**.

Monosaccharides

Monosaccharides are sweet-tasting, soluble substances that have the general formula $(CH_2O)_n$, where n can be any number from three to seven.

Examples of monosaccharides include glucose, galactose and fructose. Glucose is a hexose (6-carbon) sugar and has the formula $C_6H_{12}O_6$. However, the atoms of carbon, hydrogen and oxygen can be arranged in many different ways. For example, glucose has two isomers – α-glucose and β-glucose. Their structures are shown in Figure 1.

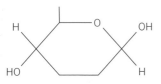

α-glucose

β-glucose

▲ **Figure 1** *Molecular arrangements of α-glucose and β-glucose (five carbon atoms at the intersection of the lines and one at the end of the vertical line at the top have been omitted for simplicity. Each line represents a covalent bond)*

Test for reducing sugars 🧪

All monosaccharides and some disaccharides (e.g., maltose) are reducing sugars. Reduction is a chemical reaction involving the gain of electrons or hydrogen. A reducing sugar is therefore a sugar that can donate electrons to (or reduce) another chemical, in this case Benedict's reagent. The test for a reducing sugar is therefore known as the Benedict's test.

Benedict's reagent is an alkaline solution of copper(II) sulfate. When a reducing sugar is heated with Benedict's reagent it forms an insoluble red precipitate of copper(I) oxide. The test is carried out as follows:

- Add 2 cm³ of the food sample to be tested to a test tube. If the sample is not already in liquid form, first grind it up in water.
- Add an equal volume of Benedict's reagent.
- Heat the mixture in a gently boiling water bath for five minutes.

Study tip

The Benedict's test may be a practical exercise but be certain to learn the *details* of the procedure. 'Add Benedict's and look for a red colour' is not enough.

Food sample dissolved in water.

Equal volume of Benedict's reagent added.

Heated in water bath. If reducing sugar present solution turns orange–brown.

▲ **Figure 2** *The Benedict's test*

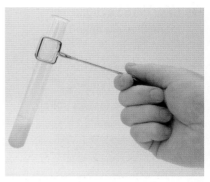

▲ **Figure 3** *If a reducing sugar is present an orange-brown colour is formed*

Summary questions

1 Large molecules often contain carbon. Explain why this is.

2 State the general name for a molecule that is made up of many similar repeating units.

3 Explain why Benedict's reagent turns red when heated with a reducing sugar.

Semi-quantitative use of the Benedict's test 🧪

Table 1 shows the relationship between the concentration of reducing sugar and the colour of the solution and precipitate formed during the Benedict's test. The differences in colour mean that the Benedict's test is semi-quantitative, that is it can be used to estimate the approximate amount of reducing sugar in a sample.

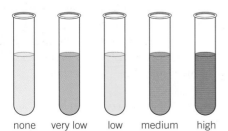

none   very low   low   medium   high

▲ **Figure 4** *Results of Benedict's test according to the concentration of reducing sugar present*

The Benedict's test was carried out on five food samples. The results are shown in Table 1.

▼ **Table 1**

Sample	Colour of solution
A	yellowish brown
B	green
C	red
D	dark brown
E	yellowish green

1 List the letters in sequence of the increasing amount of reducing sugar in each sample.

2 Suggest a way, other than comparing colour changes, in which different concentrations of reducing sugar could be estimated.

3 Explain why it is not possible to distinguish between very concentrated samples, even when their concentrations are different.

Learning objectives

→ Explain how monosaccharides are linked together to form disaccharides.

→ Describe how α-glucose molecules are linked to form starch.

→ Describe the test for non-reducing sugars.

→ Describe the test for starch.

Specification reference: 3.1.2

In Topic 1.2 we saw that in carbohydrates, the monomer unit is called a monosaccharide. Pairs of monosaccharides can be combined to form a **disaccharide**. Monosaccharides can also be combined in much larger numbers to form **polysaccharides**.

Disaccharides

When combined in pairs, monosaccharides form a disaccharide. For example:

- Glucose joined to glucose forms maltose.
- Glucose joined to fructose forms sucrose.
- Glucose joined to galactose forms lactose.

When the monosaccharides join, a molecule of water is removed and the reaction is therefore called a **condensation reaction**. The bond that is formed is called a **glycosidic bond**.

When water is added to a disaccharide under suitable conditions, it breaks the glycosidic bond releasing the constituent monosaccharides. This is called **hydrolysis** (addition of water that causes breakdown).

Figure 1a) illustrates the formation of a glycosidic bond by the removal of water (condensation reaction). Figure 1b) shows the breaking of the glycosidic bond by the addition of water (hydrolysis reaction).

a *Formation of glycosidic bond by removal of water (condensation reaction)*

b *Breaking of glycosidic bond by addition of water (hydrolysis reaction)*

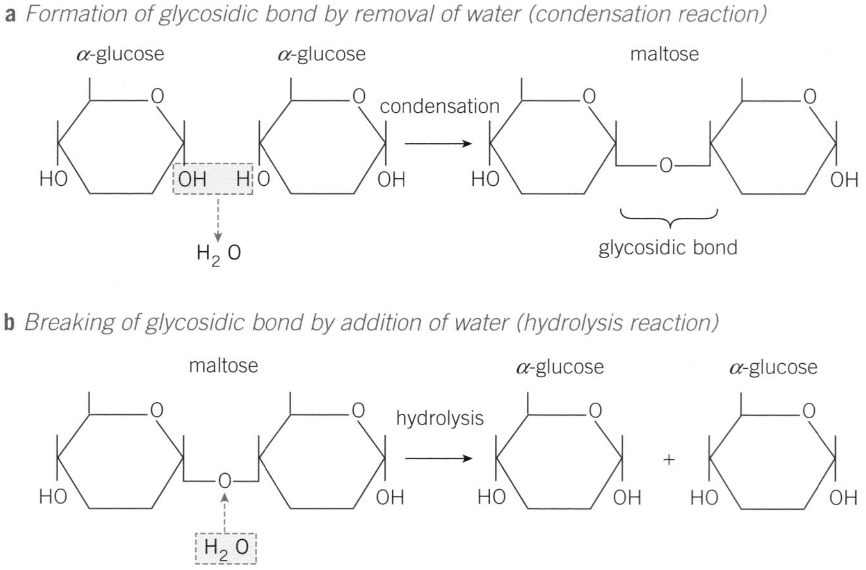

▲ **Figure 1** *Formation and breaking of a glycosidic bond by condensation and hydrolysis*

Test for non-reducing sugars ⚠

Some disaccharides (e.g. maltose) are reducing sugars. To detect these we use the Benedict's test, as described in Topic 1.2, Carbohydrates – monosaccharides. Other disaccharides, such as sucrose, are known as non-reducing sugars because they do not change the colour of Benedict's reagent when they are heated with it. In order to detect a non-reducing sugar it must first be hydrolysed into its monosaccharide components by hydrolysis. The process is carried out as follows:

- If the sample is not already in liquid form, it must first be ground up in water.
- Add 2 cm³ of the food sample being tested to 2 cm³ of Benedict's reagent in a test tube and filter.
- Place the test tube in a gently boiling water bath for 5 minutes. If the Benedict's reagent does not change colour (the solution remains blue), then a reducing sugar is *not* present.
- Add another 2 cm³ of the food sample to 2 cm³ of dilute hydrochloric acid in a test tube and place the test tube in a gently boiling water bath for five minutes. The dilute hydrochloric acid will hydrolyse any disaccharide present into its constituent monosaccharides.
- Slowly add some sodium hydrogencarbonate solution to the test tube in order to neutralise the hydrochloric acid. (Benedict's reagent will not work in acidic conditions.) Test with pH paper to check that the solution is alkaline.
- Re-test the resulting solution by heating it with 2 cm³ of Benedict's reagent in a gently boiling water bath for five minutes.
- If a non-reducing sugar was present in the original sample, the Benedict's reagent will now turn orange–brown. This is due to the reducing sugars that were produced from the hydrolysis of the non-reducing sugar.

Polysaccharides

Polysaccharides are polymers, formed by combining together many monosaccharide molecules. The monosaccharides are joined by glycosidic bonds that were formed by **condensation reactions**. As polysaccharides are very large molecules, they are insoluble. This feature makes them suitable for storage. When they are hydrolysed, polysaccharides break down into disaccharides or monosaccharides (Figure 2). Some polysaccharides, such as cellulose (see Topic 1.4), are not used for storage but give structural support to plant cells.

> ### Hint
>
> Polysaccharides illustrate an important principle: that a few basic monomer units can be combined in a number of different ways to give a large range of different biological molecules.

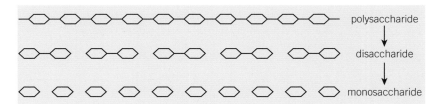

▲ **Figure 2** *The hydrolysis of a polysaccharide into disaccharides and monosaccharides*

Starch is a polysaccharide that is found in many parts of plants in the form of small granules or grains, for example starch grains in chloroplasts. It is formed by the joining of between 200 and 100 000 α-glucose molecules by glycosidic bonds in a series of condensation reactions. More details of starch and its functions are given in Topic 1.4.

1 Two drops of iodine solution added to test solution

2 If starch is present it turns the iodine a blue–black colour

▲ **Figure 3** *Test for starch*

Test for starch

Starch is easily detected by its ability to change the colour of the iodine in potassium iodide solution from yellow to blue–black (Figure 3). The test is carried out at room temperature. The test is carried out as follows:

- Place $2\,cm^3$ of the sample being tested into a test tube (or add two drops of the sample into a depression on a spotting tile).
- Add two drops of iodine solution and shake or stir.
- The presence of starch is indicated by a blue–black coloration.

Summary questions

1 Identify which one, or more, monomer units make up each of the following carbohydrates.

   a lactose

   b sucrose

   c starch

2 Glucose $(C_6H_{12}O_6)$ combines with fructose $(C_6H_{12}O_6)$ to form the disaccharide sucrose. From your knowledge of how disaccharides are formed, deduce the formula of sucrose.

3 🧪 To hydrolyse a disaccharide it can be boiled with hydrochloric acid but if hydrolysis is carried out by an enzyme a much lower temperature $(40\,°C)$ is required. Explain why.

1.4 Starch, glycogen and cellulose

In organisms, a wide range of different molecules with very different properties can be made from a limited range of smaller molecules. What makes the larger molecules different is the various ways in which the smaller molecules are combined to form them and small differences in the monomers used. You will look at some of these larger molecules by considering three important polysaccharides.

Starch

Starch is a polysaccharide that is found in many parts of a plant in the form of small grains. Especially large amounts occur in seeds and storage organs, such as potato tubers. It forms an important component of food and is the major energy source in most diets. Starch is made up of chains of α-glucose monosaccharides linked by glycosidic bonds that are formed by **condensation reactions**. The chains may be branched or unbranched. The unbranched chain is wound into a tight coil that makes the molecule very compact. The structure of a starch molecule is shown in Figure 1.

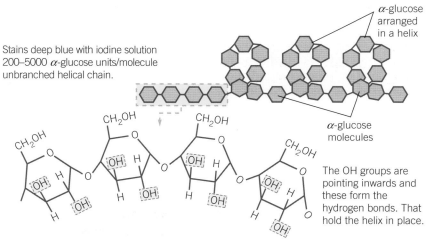

Stains deep blue with iodine solution 200–5000 α-glucose units/molecule unbranched helical chain.

α-glucose arranged in a helix

α-glucose molecules

The OH groups are pointing inwards and these form the hydrogen bonds. That hold the helix in place.

▲ **Figure 1** *Structure of a starch molecule*

The main role of starch is energy storage, something its structure is especially suited for because:

- it is insoluble and therefore doesn't affect water potential, so water is not drawn into the cells by **osmosis**

- being large and insoluble, it does not diffuse out of cells

- it is compact, so a lot of it can be stored in a small space

- when hydrolysed it forms α-glucose, which is both easily transported and readily used in respiration

- the branched form has many ends, each of which can be acted on by enzymes simultaneously meaning that glucose monomers are released very rapidly.

Starch is never found in animal cells. Instead a similar polysaccharide, called glycogen, serves the same role.

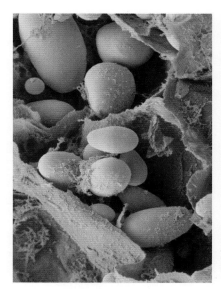

▲ **Figure 2** *False colour scanning electron micrograph (SEM) of starch grains (blue) in the cells of a potato. Starch is a compact storage material.*

Glycogen

Glycogen is found in animals and bacteria but never in plant cells. Glycogen is very similar in structure to starch but has shorter chains and is more highly branched. It is sometimes called 'animal starch' because it is the major carbohydrate storage product of animals. In animals it is stored as small granules mainly in the muscles and the liver. The mass of carbohydrate that is stored is relatively small because fat is the main storage molecule in animals. Its structure suits it for storage because:

- it is insoluble and therefore does not tend to draw water into the cells by osmosis
- being insoluble, it does not diffuse out of cells
- it is compact, so a lot of it can be stored in a small space
- It is more highly branched than starch and so has more ends that can be acted on simultaneously by enzymes. It is therefore more rapidly broken down to form glucose monomers, which are used in respiration. This is important to animals which have a higher metabolic rate and therefore respiratory rate than plants because they are more active.

Cellulose

Cellulose differs from starch and glycogen in one major respect: it is made of monomers of β-glucose rather than α-glucose. This seemingly small variation produces fundamental differences in the structure and function of this polysaccharide.

Rather than forming a coiled chain like starch, cellulose has straight, unbranched chains. These run parallel to one another, allowing hydrogen bonds (Topic 1.6, Proteins) to form cross-linkages between adjacent chains. While each individual hydrogen bond adds very little to the strength of the molecule, the sheer overall number of them makes a considerable contribution to strengthening cellulose, making it the valuable structural material that it is. The arrangement of β-glucose chains in a cellulose molecule is shown in Figure 3.

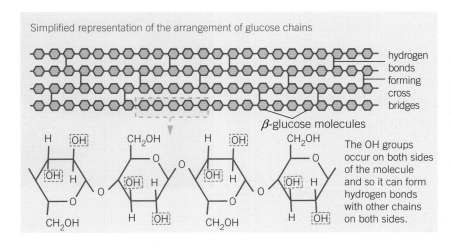

Simplified representation of the arrangement of glucose chains

hydrogen bonds forming cross bridges

β-glucose molecules

H OH CH₂OH H OH CH₂OH

OH H OH H OH H OH H

O O O O

CH₂OH H OH CH₂OH H OH

The OH groups occur on both sides of the molecule and so it can form hydrogen bonds with other chains on both sides.

The cellulose chain, unlike that of starch, has adjacent glucose molecules rotated by 180°. This allows hydrogen bonds to be formed between the hydroxyl (–OH) groups on adjacent parallel chains that help to give cellulose its structural stability.

◀ **Figure 3** *Structure of a cellulose molecule*

The cellulose molecules are grouped together to form microfibrils (Figure 4) which, in turn, are arranged in parallel groups called fibres.

Cellulose is a major component of plant cell walls and provides rigidity to the plant cell. The cellulose cell wall also prevents the cell from bursting as water enters it by osmosis. It does this by exerting an inward pressure that stops any further influx of water. As a result, living plant cells are turgid and push against one another, making non-woody parts of the plant semi-rigid. This is especially important in maintaining stems and leaves in a turgid state so that they can provide the maximum surface area for photosynthesis.

In summary, the structure of cellulose is suited to its function of providing support and rigidity because:

- cellulose molecules are made up of β-glucose and so form long straight, unbranched chains
- these cellulose molecular chains run parallel to each other and are crossed linked by hydrogen bonds which add collective strength
- these molecules are grouped to form microfibrils which in turn are grouped to form fibres all of which provides yet more strength.

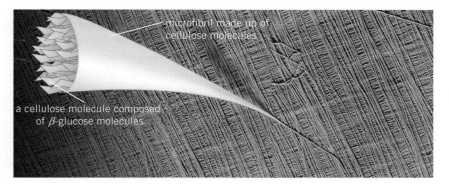

▲ **Figure 4** *Structure of a cellulose microfibril*

Synoptic link

More detail of the cell wall in plants is given in Topic 3.4 and its importance in supporting non-woody plant tissues is discussed in Topic 4.3.

Summary questions

From the following list of carbohydrates choose **one or more** that most closely fit each of the statements below. Each carbohydrate may be used once, more than once, or not at all.

α-glucose   starch   cellulose β-glucose   glycogen

1 Stains deep blue with iodine solution.

2 Is known as 'animal starch'.

3 Found in plants.

4 Are polysaccharides.

5 Monosaccharide found in starch.

6 Has a structural function.

7 Can be hydrolysed.

8 Easily moves in and out of cells by facilitated diffusion.

Learning objectives

→ Describe the structure of triglycerides and how this relates to their function.

→ Describe the roles of lipids.

→ Describe the structure of a phospholipids and how this relates to their function.

→ Describe the test for a lipid.

Specification reference: 3.1.3

Hint

Fats are generally made of saturated fatty acids, while oils are made of unsaturated ones.

Fats are solid at room temperature (10–20 °C), whereas oils are liquid.

Lipids are a varied group of substances that share the following characteristics:

- They contain carbon, hydrogen and oxygen.
- The proportion of oxygen to carbon and hydrogen is smaller than in carbohydrates.
- They are insoluble in water.
- They are soluble in organic solvents such as alcohols and acetone.

The main groups of lipids are **triglycerides** (fats and oils) and phospholipids.

Roles of lipids

Lipids have many roles, one role of lipids is in the **cell membranes** (cell-surface membranes and membranes around organelles). Phospholipids contribute to the flexibility of membranes and the transfer of lipid-soluble substances across them. Other roles of lipids include:

- **source of energy**. When oxidised, lipids provide more than twice the energy as the same mass of carbohydrate and release valuable water.
- **waterproofing**. Lipids are insoluble in water and therefore useful as a waterproofing. Both plants and insects have waxy, lipid cuticles that conserve water, while mammals produce an oily secretion from the sebaceous glands in the skin.
- **insulation**. Fats are slow conductors of heat and when stored beneath the body surface help to retain body heat. They also act as electrical insulators in the myelin sheath around nerve cells.
- **protection**. Fat is often stored around delicate organs, such as the kidney.

Fats are solid at room temperature (10–20 °C) whereas oils are liquid.

Triglycerides

Triglycerides are so called because they have three (tri) fatty acids combined with glycerol (glyceride). Each fatty acid forms an ester bond with glycerol in a **condensation reaction** (Figure 1). **Hydrolysis** of a triglyceride therefore produces glycerol and three fatty acids.

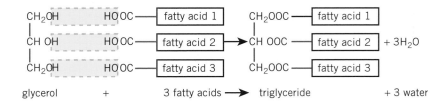

▲ **Figure 1** *The formation of a triglyceride. The three fatty acids may all be the same, thereby forming a simple triglyceride, or they may be different, in which case a mixed triglyceride is produced. In either case it is a condensation reaction*

As the glycerol molecule in all triglycerides is the same, the differences in the properties of different fats and oils come from variations in the fatty acids. There are over 70 different fatty acids and all have a carboxyl (—COOH) group with a hydrocarbon chain attached. If this chain has no carbon–carbon double bonds, the fatty acid is then described as **saturated**, because all the carbon atoms are linked to the maximum possible number of hydrogen atoms, in other words they are saturated with hydrogen atoms. If there is a single double bond, it is **mono-unsaturated** – if more than one double bond is present, it is **polyunsaturated**. These differences are illustrated in Figure 2.

The structure of triglycerides related to their properties

- Triglycerides have a high ratio of energy-storing carbon–hydrogen bonds to carbon atoms and are therefore an excellent source of energy.
- Triglycerides have low mass to energy ratio, making them good storage molecules because much energy can be stored in a small volume. This is especially beneficial to animals as it reduces the mass they have to carry as they move around.
- Being large, non-polar molecules, triglycerides are insoluble in water. As a result their storage does not affect osmosis in cells or the water potential of them.
- As they have a high ratio of hydrogen to oxygen atoms, triglycerides release water when oxidised and therefore provide an important source of water, especially for organisms living in dry deserts.

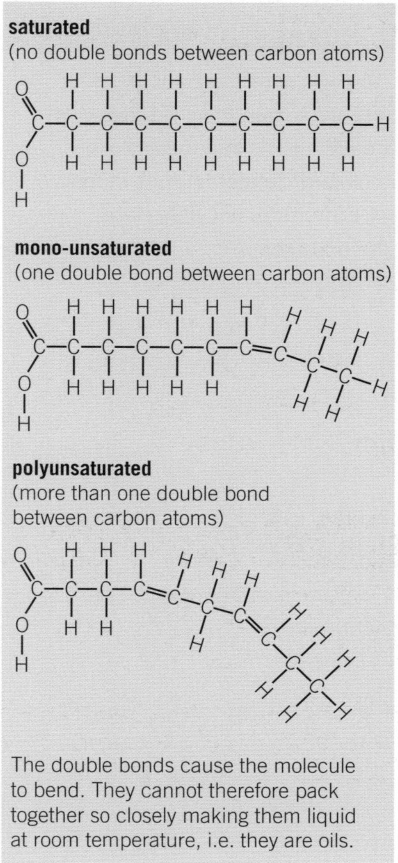

Phospholipids

Phospholipids are similar to lipids except that one of the fatty acid molecules is replaced by a phosphate molecule (Figure 3). Whereas fatty acid molecules repel water (are hydrophobic), phosphate molecules attract water (are hydrophilic). A phospholipid is therefore made up of two parts:

- **a hydrophilic 'head'**, which interacts with water (is attracted to it) but not with fat
- **a hydrophobic 'tail'**, which orients itself away from water but mixes readily with fat.

Molecules that have two ends (poles) that behave differently in this way are said to be **polar**. This means that when these polar phospholipid molecules are placed in water they position themselves so that the hydrophilic heads are as close to the water as possible and the hydrophobic tails are as far away from the water as possible (Figure 4).

saturated
(no double bonds between carbon atoms)

mono-unsaturated
(one double bond between carbon atoms)

polyunsaturated
(more than one double bond between carbon atoms)

The double bonds cause the molecule to bend. They cannot therefore pack together so closely making them liquid at room temperature, i.e. they are oils.

▲ **Figure 2** *Saturated and unsaturated fatty acids*

Study tip

Do not use terms like 'water-loving' and 'water-hating'. Use the correct scientific terms '**hydrophilic**' and '**hydrophobic**'.

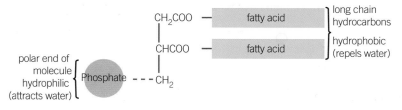

▲ **Figure 3** *Structure of a phospholipid*

The structure of phospholipids related to their properties

- Phospholipids are polar molecules, having a hydrophilic phosphate head and a hydrophobic tail of two fatty acids. This means that in an aqueous environment, phospholipid molecules form a bilayer within cell-surface membranes. As a result, a hydrophobic barrier is formed between the inside and outside of a cell.
- The hydrophilic phosphate 'heads' of phospholipid molecules help to hold at the surface of the cell-surface membrane.
- The phospholipid structure allows them to form glycolipids by combining with carbohydrates within the cell-surface membrane. These glycolipids are important in cell recognition.

Test for lipids 🧪

The test for lipids is known as the emulsion test and is carried out as follows:

1 Take a completely dry and grease-free test tube.
2 To $2\,cm^3$ of the sample being tested, add $5\,cm^3$ of ethanol.
3 Shake the tube thoroughly to dissolve any lipid in the sample.
4 Add $5\,cm^3$ of water and shake gently.
5 A cloudy-white colour indicates the presence of a lipid.
6 As a control, repeat the procedures using water instead of the sample; the final solution should remain clear.

The cloudy colour is due to any lipid in the sample being finely dispersed in the water to form an emulsion. Light passing through this emulsion is refracted as it passes from oil droplets to water droplets, making it appear cloudy.

Summary questions

1 In the following passage state the most suitable word for each of the letters **a** to **e**.

Fats and oils make up a group of lipids called **a** which, when hydrolysed, form **b** and fatty acids. A fatty acid with more than one carbon–carbon double bond is described as **c**. In a phospholipid the number of fatty acids is **d**; these are described as **e** because they repel water.

2 List **two** differences between a triglyceride molecule and a phospholipid molecule.

3 Organisms that move, e.g. animals, and parts of organisms that move, e.g., some plant seeds, use lipids rather than carbohydrates as an energy store. Suggest **one** reason why this is so.

1.6 Proteins

Proteins are usually very large molecules. The types of carbohydrates and lipids in all organisms are relatively few and they are very similar. However, each organism has numerous proteins that differ from species to species. The shape of any one type of protein molecule differs from that of all other types of proteins. Proteins are very important molecules in living organisms. Indeed the word 'protein' is a Greek word meaning 'of first importance'. One group of proteins, enzymes, is involved in almost every living process. There is a vast range of different enzymes that between them perform a very diverse number of functions.

Structure of an amino acid

Amino acids are the basic **monomer** units which combine to make up a **polymer** called a polypeptide. Polypeptides can be combined to form proteins. About 100 amino acids have been identified, of which 20 occur naturally in proteins. The fact that the same 20 amino acids occur in all living organisms provides indirect evidence for evolution.

Every amino acid has a central carbon atom to which are attached four different chemical groups:

- amino group ($—NH_2$) – a basic group from which the amino part of the name amino acid is derived
- carboxyl group (—COOH) – an acidic group which gives the amino acid the acid part of its name
- hydrogen atom (—H)
- R (side) group – a variety of different chemical groups. Each amino acid has a different R group. These 20 naturally occurring amino acids differ only in their R (side) group.

The general structure of an amino acid is shown in Figure 1.

The formation of a peptide bond

In a similar way that monosaccharide monomers combine to form disaccharides (see Topic 1.3), so amino acid monomers can combine to form a dipeptide. The process is essentially the same: namely the removal of a water molecule in a **condensation** reaction. The water is made by combining an —OH from the carboxyl group of one amino acid with an —H from the amino group of another amino acid. The two amino acids then become linked by a new **peptide bond** between the carbon atom of one amino acid and the nitrogen atom of the other. The formation of a peptide bond is illustrated in Figure 3. In a similar way as a glycosidic bond of a disaccharide can be broken by the addition of water (hydrolysis), so the peptide bond of a dipeptide can also be broken by hydrolysis to give its two constituent amino acids.

The primary structure of proteins – polypeptides

Through a series of condensation reactions, many amino acid monomers can be joined together in a process called **polymerisation**. The resulting chain of many hundreds of amino acids is called a **polypeptide**. The sequence of amino acids in a polypeptide chain forms the primary structure of any protein. As we shall see in Topic 8.1, this sequence is

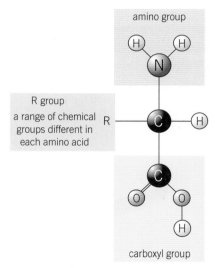

▲ **Figure 1** *The general structure of an amino acid*

$$H_2N—\overset{\overset{\displaystyle R}{|}}{\underset{\underset{\displaystyle H}{|}}{C}}—COOH$$

▲ **Figure 2** *Simplified structural formula of an amino acid*

19

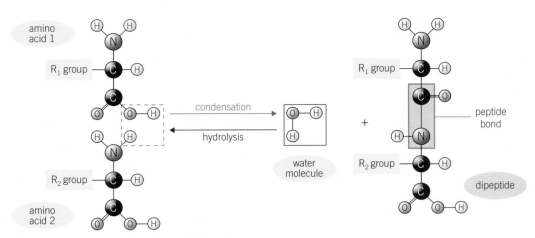

▲ **Figure 3** *The formation of a peptide bond*

Synoptic link

You will learn more about DNA structure in Topic 2.1, and its function in Topic 8.1.

Study tip

Distinguish between **condensation reactions** (molecules combine producing water) and **hydrolysis reactions** (molecules are split up by taking in water).

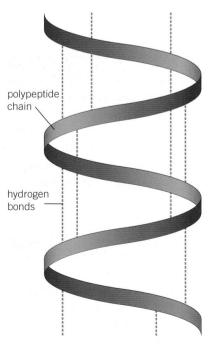

▲ **Figure 4** *Structure of the α-helix*

determined by DNA. As polypeptides have many (usually hundreds) of the 20 naturally occurring amino acids joined in different sequences, it follows that there is an almost limitless number of possible combinations, and therefore types, of primary protein structure.

It is the primary structure of a protein that determines its ultimate shape and hence its function. A change in just a single amino acid in this primary sequence can lead to a change in the shape of the protein and may stop it carrying out its function. In other words, a protein's shape is very specific to its function. Change its shape and it will function less well, or differently.

A simple protein may consist of a single polypeptide chain. More commonly, however, a protein is made up of a number of polypeptide chains.

The secondary structure of proteins

The linked amino acids that make up a polypeptide possess both —NH and —C=O groups on either side of every peptide bond. The hydrogen of the —NH group has an overall positive charge while the O of the —C=O group has an overall negative charge. These two groups therefore readily form weak bonds, called **hydrogen bonds**. This causes the long polypeptide chain to be twisted into a 3-D shape, such as the coil known as an α-helix. Figure 4 illustrates the structure of an α-helix.

Tertiary structure of proteins

The α-helices of the secondary protein structure can be twisted and folded even more to give the complex, and often specific, 3-D structure of each protein (Figure 5). This is known as the tertiary structure. This structure is maintained by a number of different bonds. Where the bonds occur depends on the primary structure of the protein. These bonds include:

*   **disulfide bridges** – which are fairly strong and therefore not easily broken.
*   **ionic bonds** – which are formed between any carboxyl and amino groups that are not involved in forming peptide bonds. They are weaker than disulfide bonds and are easily broken by changes in pH.
*   **hydrogen bonds** – which are numerous but easily broken.

It is the 3-D shape of a protein that is important when it comes to how it functions. It makes each protein distinctive and allows it to recognise, and be recognised by, other molecules. It can then interact with them in a very specific way.

a The primary structure of a protein is the sequence of amino acids found in its polypeptide chains. This sequence determines its properties and shape. Following the elucidation of the amino acid sequence of the hormone insulin by Frederick Sanger in 1954, the primary structure of many other proteins is now known.

b The secondary structure is the shape which the polypeptide chain forms as a result of hydrogen bonding. This is most often a spiral known as the α-helix, although other configurations occur.

c The tertiary structure is due to the bending and twisting of the polypeptide helix into a compact structure. All three types of bond, disulfide, ionic and hydrogen, contribute to the maintenance of the tertiary structure.

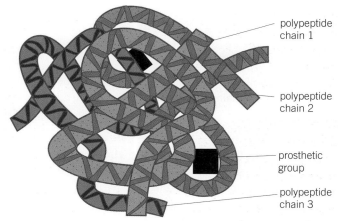

d The quaternary structure arises from the combination of a number of different polypeptide chains and associated non-protein (prosthetic) groups into a large, complex protein molecule, e.g., haemoglobin.

▲ **Figure 5** *Structure of proteins*

Quaternary structure of proteins

Large proteins often form complex molecules containing a number of individual polypeptide chains that are linked in various ways. There may also be non-protein (prosthetic) groups associated with the molecules (Figure 5d), such as the iron-containing haem group in haemoglobin. Remember that, although the 3-D structure is important to how a protein functions, it is the sequence of amino acids (primary structure) that determines the 3-D shape in the first place.

Test for proteins 🧪

The most reliable protein test is the Biuret test, which detects peptide bonds. It is performed as follows:

* Place a sample of the solution to be tested in a test tube and add an equal volume of sodium hydroxide solution at room temperature.
* Add a few drops of very dilute (0.05%) copper(II) sulfate solution and mix gently.
* A purple coloration indicates the presence of peptide bonds and hence a protein. If no protein is present, the solution remains blue.

Hint

Think of the polypeptide chain as a piece of string. In a fibrous protein many pieces of the string are twisted together into a rope, while in a globular protein the pieces of string, usually fewer, are rolled into a ball.

Study tip

You can simply refer to adding **Biuret** reagent to test for protein. A purple colour shows protein is present – a blue colour indicates that protein is absent.

1 Name the type of bond that joins amino acids together.

2 State the type of reaction involved in joining amino acids together.

3 List *four* different components that make up an amino acid.

Protein shape and function

Proteins perform many different roles in living organisms. Their roles depend on their molecular shape, which can be of two basic types.

- Fibrous proteins, such as collagen, have structural functions.
- Globular proteins, such as enzymes and haemoglobin, carry out metabolic functions.

It is the very different structure and shape of each of these types of proteins that enables them to carry out their functions.

Fibrous proteins
Fibrous proteins form long chains which run parallel to one another. These chains are linked by cross-bridges and so form very stable molecules. One example is **collagen**. Its molecular structure is as follows:

- The primary structure is an unbranched polypeptide chain.
- In the secondary structure the polypeptide chain is very tightly wound.
- Lots of the amino acid, glycine helps close packing.
- In the tertiary structure the chain is twisted into a second helix.
- Its quaternary structure is made up of three such polypeptide chains wound together in the same way as individual fibres are wound together in a rope.

Collagen is found in tendons. Tendons join muscles to bones. When a muscle contracts the bone is pulled in the direction of the contraction.

1 Explain why the quaternary structure of collagen makes it a suitable molecule for a tendon.

The individual collagen polypeptide chains in the fibres are held together by bonds between amino acids of adjacent chains.

2 Suggest how the cross-linkages between the amino acids of polypeptide chains increase the strength and stability of a collagen fibre.

The points where one collagen molecule ends and the next begins are spread throughout the fibre rather than all being in the same position along it.

3 Explain why this arrangement of collagen molecules is necessary for the efficient functioning of a tendon.

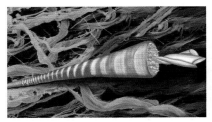

◄**Figure 6** *Fine structure of the fibrous protein collagen*

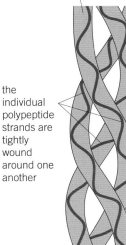

each polypeptide forms a long, unfolded strand

the individual polypeptide strands are tightly wound around one another

long fibrous molecular shape

fibrous protein, e.g., collagen

◄**Figure 7** *Structure of fibrous proteins*

Enzymes are globular proteins that act as catalysts. Catalysts alter the rate of a chemical reaction without undergoing permanent changes themselves. They can be reused repeatedly and are therefore effective in small amounts. Enzymes do not make reactions happen; they speed up reactions that already occur, sometimes by a factor of many millions.

Enzymes as catalysts lowering activation energy

Let us consider a typical chemical reaction:

$$\text{sucrose + water} \longrightarrow \text{glucose + fructose}$$
$$\text{(substrates)} \qquad \text{(products)}$$

For reactions like this to take place naturally a number of conditions must be satisfied:

- The sucrose and water molecules must collide with sufficient energy to alter the arrangement of their atoms to form glucose and fructose.
- The free energy of the products (glucose and fructose) must be less than that of the substrates (sucrose and water).
- Many reactions require an initial amount of energy to start. The minimum amount of energy needed to activate the reaction in this way is called the **activation energy**.

There is an activation energy level, like an energy hill or barrier, which must initially be overcome before the reaction can proceed. Enzymes work by lowering this activation energy level (Figure 1). In this way enzymes allow reactions to take place at a lower temperature than normal. This enables some metabolic processes to occur rapidly at the human body temperature of 37 °C, which is relatively low in terms of chemical reactions. Without enzymes these reactions would proceed too slowly to sustain life as we know it.

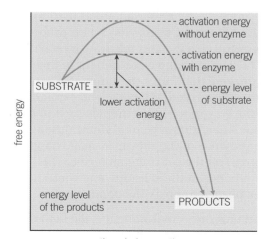

▲ **Figure 1** How enzymes lower activation energy

Learning objectives

→ Explain how enzymes speed up chemical reactions.

→ Describe how the structure of enzyme molecules relates to their function.

→ Explain the lock and key model of enzyme action.

→ Explain the induced-fit model of enzyme action.

Specification reference: 3.1.4.2

Hint

Free energy is the energy of a system that is available to perform work.

Hint

To help you understand the importance of enzymes, it is necessary to appreciate that they catalyse a wide range of reactions both inside the cell (intracellular) and outside the cell (extracellular). In doing so, enzymes determine the structures and functions of all parts of living matter from cells to complete organisms.

Hint

If a stone is lying behind a mound, we need to expend energy to move it down a hillside, either by pushing the stone over the mound or reducing the height of the mound. Once it starts to move, the stone gathers momentum and rolls to the bottom. Hence an initial input of energy (**activation energy**) starts a reaction that then continues of its own accord. Enzymes achieve the equivalent of lowering the mound of earth.

Enzyme structure

From Topic 1.6 you will be aware that enzymes, being globular proteins, have a specific 3-D shape that is the result of their sequence of amino acids (primary protein stucture). A specific region of the enzyme is functional, this is known as the **active site**. The active site is made up of a relatively small number of amino acids. The active site forms a small depression within the much larger enzyme molecule.

The molecule on which the enzyme acts is called the **substrate**. This fits neatly into this depression and forms an **enzyme–substrate complex** (Figure 2). The substrate molecule is held within the active site by bonds that temporarily form between certain amino acids of the active site and groups on the substrate molecule.

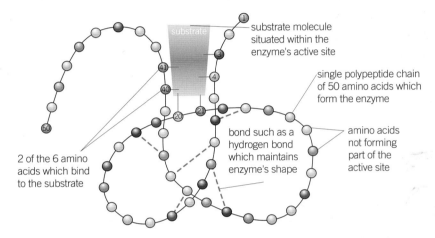

substrate molecule situated within the enzyme's active site

single polypeptide chain of 50 amino acids which form the enzyme

2 of the 6 amino acids which bind to the substrate

bond such as a hydrogen bond which maintains enzyme's shape

amino acids not forming part of the active site

▲ **Figure 2** *Example of an enzyme–substrate complex showing the six out of the 50 amino acids that form the active site*

Induced fit model of enzyme action

Scientists often try to explain their observations by producing a representation of how something works. This is known as a scientific model. Examples include the physical models used to explain enzyme action. The induced fit model of enzyme action proposes that the active site forms as the enzyme and substrate interact. The proximity of the substrate (a change in the environment of the enzyme) leads to a change in the enzyme that forms the functional active site (Figure 3). In other words, the enzyme is flexible and can mould itself around the substrate in the way that a glove moulds itself to the shape of the hand. The enzyme has a certain general shape, just as a glove has, but this alters in the presence of the substrate. As it changes its shape, the enzyme puts a strain on the substrate molecule. This strain distorts a particular bond or bonds in the substrate and consequently lowers the activation energy needed to break the bond.

Any change in an enzyme's environment is likely to change its shape. The very act of colliding with its substrate is a change in its environment and so its shape changes – induced fit.

Study tip

The substrate does *not* have the 'same shape' as the active site. The substrate has a *complementary shape* to the active site.

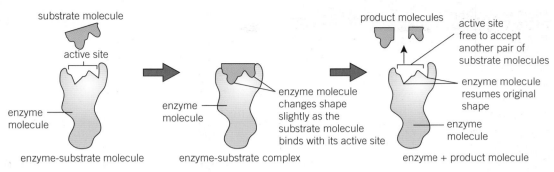

▲ **Figure 3** *Mechanism of enzyme action*

Summary questions

1 Define a catalyst.
2 Explain why enzymes are effective in tiny quantities.
3 Outline why changing one of the amino acids that make up the active site could prevent the enzyme from functioning.
4 Explain why changing certain amino acids that are not part of the active site also prevents the enzyme from functioning.

Lock and key model of enzyme action

One earlier model of enzyme action proposed that enzymes work in the same way as a key operates a lock – each key has a specific shape that fits and operates only a single lock. In a similar way, a substrate will only fit the active site of one particular enzyme. This model was supported by the observation that enzymes are specific in the reactions that they catalyse. The shape of the substrate (key) exactly fits the active site of the enzyme (lock). This is known as the **lock and key model**.

One limitation of this model is that the enzyme, like a lock, is considered to be a rigid structure. However, scientists had observed that other molecules could bind to enzymes at sites other than the active site. In doing so, they altered the activity of the enzyme. This suggested that the enzyme's shape was being altered by the binding molecule. In other words, its structure was not rigid but flexible. In true scientific fashion this led to an alternative model being proposed, one that better fitted the current observations. This was the called the induced fit model as described above. The induced fit model is therefore a modified version of the lock and key model.

1 Explain why the induced fit model is a better explanation of enzyme action than the lock and key model.

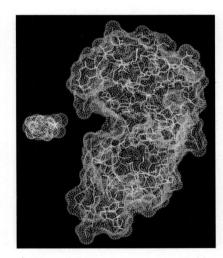

▲ **Figure 4** *Molecular computer graphics image of the enzyme ribonuclease A (right) and its substrate (left) approaching the enzyme's active site*

Hint

Enzymes have an active site but not all proteins are enzymes. Many proteins have binding sites or receptor sites that are not active sites. Some hormones are proteins and these have receptor sites but they are *not* active sites.

Before considering how pH and temperature affect enzymes, it is worth bearing in mind that, for an enzyme to work, it must:

• come into physical contact with its **substrate**,

• have an **active site** which fits the substrate.

Almost all factors that influence the rate at which an enzyme works do so by affecting one or both of the above. In order to investigate how enzymes are affected by various factors we need to be able to measure the rate of the reactions they catalyse.

Measuring enzyme-catalysed reactions

To measure the progress of an enzyme-catalysed reaction we usually measure its time-course, that is how long it takes for a particular event to run its course. The two changes most frequently measured are:

• the formation of the products of the reaction, for example the volume of oxygen produced when the enzyme catalase acts on hydrogen peroxide (Figure 1)

• the disappearance of the substrate, for example the reduction in concentration of starch when it is acted upon by amylase (Figure 2).

Although the graphs in Figures 1 and 2 differ, the explanation for their shapes is the same:

• At first there is a lot of substrate (hydrogen peroxide or starch) but no product (water and oxygen, or maltose).

• It is very easy for substrate molecules to come into contact with the empty active sites on the enzyme molecules.

• All enzyme active sites are filled at any given moment and the substrate is rapidly broken down into its products.

• The amount of substrate decreases as it is broken down, resulting in an increase in the amount of product.

• As the reaction proceeds, there is less and less substrate and more and more product.

• It becomes more difficult for the substrate molecules to come into contact with the enzyme molecules because there are fewer substrate molecules and also the product molecules may 'get in the way' of substrate molecules and prevent them reaching an active site.

• It therefore takes longer for the substrate molecules to be broken down by the enzyme and so its rate of disappearance slows, and consequently the rate of formation of product also slows. Both graphs 'tail off'.

• The rate of reaction continues to slow until there is so little substrate that any further decrease in its concentration cannot be measured.

• The graphs flatten out because all the substrate has been used up and so no new product can be produced.

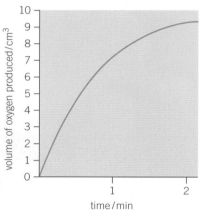

▲ **Figure 1** *Measurement of the formation of oxygen due to the action of catalase on hydrogen peroxide*

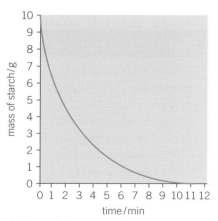

▲ **Figure 2** *Measurement of the disappearance of starch due to the action of amylase*

Measuring rate of change $\sqrt{x}$

We can measure the change in the rate of a reaction at any point on the curve of a graph such as those in Figures 1 and 2. We do so by measuring the gradient at our chosen point. The gradient is equal to the gradient of the tangent to the curve at that point. This tangent is the point at which a straight line touches the curve but without cutting across it (see Maths skills chapter).

Accurately drawing the tangent to a curve is not easy but can be achieved by making use of the normal line. The normal line is a line that passes through a point at a 90° angle.

Let us look at an example. In Figure 3, you see the curve showing the formation of oxygen due to the action of catalase on hydrogen peroxide. Suppose you want to measure the rate of change in this reaction at point X as shown on Figure 3. You draw the tangent to the curve at this point as shown on Figure 3. Using this line you can find the gradient, in your case $= \dfrac{a}{b}$.

This technique is useful in a variety of practical situations, including ones involving measuring the rates of enzyme reactions.

Before we look at the effects of different factors on the rate of enzyme action, it is important to stress the fundamental experimental technique of changing only a single variable in each experiment. When investigating the effect of a named variable on the rate of an enzyme reaction all the other variables must be kept constant. For example, if measuring the effect of temperature, then pH, enzyme concentration and substrate concentration must be kept constant and all possible inhibitors should be absent. Another thing to remember is that the active site and the substrate are not 'the same', any more than a key and a lock are the same – in some senses they are more like opposites. The correct term is **complementary**.

Effect of temperature on enzyme action

A rise in temperature increases the kinetic energy of molecules. As a result, the molecules move around more rapidly and collide with each other more often. In an enzyme-catalysed reaction, this means that the enzyme and substrate molecules come together more often in a given time. There are more effective collisions resulting in more enzyme-substrate complexes being formed and so the rate of reaction increases.

Shown on a graph, this gives a rising curve. However, the temperature rise also begins to cause the hydrogen and other bonds in the enzyme molecule to break. This results in the enzyme, including its active site, changing shape. At first, the substrate fits less easily into this changed active site, slowing the rate of reaction. For many human enzymes this may begin at temperatures of around 45 °C.

At some point, usually around 60 °C, the enzyme is so disrupted that it stops working altogether. It is said to be denatured. **Denaturation** is a permanent change and, once it has occurred, the enzyme does not function again. Shown on a graph, the rate of this reaction follows

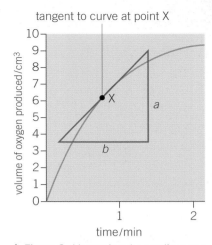

▲ **Figure 3** *Measuring the gradient at a point on a curve*

Study tip

Rate is always expressed per unit time.

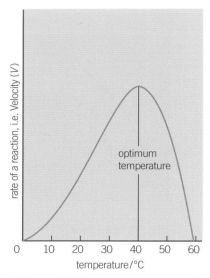

▲ **Figure 4** *Effect of temperature on the rate of an enzyme-controlled reaction*

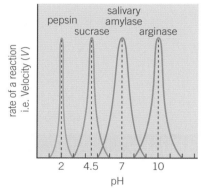

▲ **Figure 5** *Effect of pH on the rate of an enzyme-controlled reaction*

Maths link ✓x̄

MS 0.5, see Chapter 11.

Hint

When considering how factors affect enzyme action, think 'shape change'.

Study tip

Enzymes are not alive and so cannot be 'killed'. Use the correct term: *denatured*.

a falling curve. The actual effect of temperature on the rate of an enzyme reaction is a combination of these two factors (Figure 4). The optimum working temperature differs from enzyme to enzyme. Some work fastest at around 10 °C, while others continue to work rapidly at 80 °C. For example, enzymes used in biological washing powders and in the polymerase chain reaction (Topic 21.3). Many enzymes in the human body have an optimum temperature of about 40 °C. Our body temperatures have, however, evolved to be 37 °C. This may be related to the following:

- Although higher body temperatures would increase the metabolic rate slightly, the advantages are offset by the additional energy (food) that would be needed to maintain the higher temperature.
- Other proteins, apart from enzymes, may be denatured at higher temperatures.
- At higher temperatures, any further rise in temperature, for example, during illness, might denature the enzymes.

Different species of mammals and birds have different body temperatures. many birds, for example, have a normal body temperature of around 40 °C because they have a high metabolic rate for the high energy requirement of flight.

Effect of pH on enzyme action

The pH of a solution is a measure of its hydrogen ion concentration. Each enzyme has an optimum pH, that is a pH at which it works fastest (Figure 5). The pH of a solution is calculated using the formula: $pH = -\log_{10}[H^+]$. A hydrogen ion $[H^+]$ concentration of 1×10^{-9} therefore has a pH of 9. In a similar way to a change in temperature affecting the rate of enzyme action, a change in pH away from the optimum affects the rate of enzyme action. An increase or decrease in pH reduces the rate of enzyme action. If the change in pH is more extreme then, beyond a certain pH, the enzyme becomes denatured.

The pH affects how an enzyme works in the following ways:

- A change in pH alters the charges on the amino acids that make up the active site of the enzyme. As a result, the substrate can no longer become attached to the active site and so the enzyme–substrate complex cannot be formed.
- Depending on how significant the change in pH is, it may cause the bonds maintaining the enzyme's tertiary structure to break. The active site therefore changes shape.

The arrangement of the active site is partly determined by the hydrogen and ionic bonds between —NH_2 and —COOH groups of the polypeptides that make up the enzyme. The change in H^+ ions affects this bonding, causing the active site to change shape.

It is important to note that pH fluctuations inside organisms are usually small, this means they are far more likely to reduce an enzyme's activity than to denature it.

Effect of enzyme concentration on the rate of reaction

Once an **active site** on an enzyme has acted on its substrate, it is free to repeat the procedure on another substrate molecule. This means that enzymes, being catalysts, are not used up in the reaction and therefore work efficiently at very low concentrations. In some cases, a single enzyme molecule can act on millions of substrate molecules in one minute.

As long as there is an excess of substrate, an increase in the amount of enzyme leads to a proportionate increase in the rate of reaction. A graph of the rate of reaction against enzyme concentration will initially show a proportionate increase. This is because there is more substrate than the enzyme's active sites can cope with. If you therefore increase the enzyme concentration, some of the excess substrate can now also be acted upon and the rate of reaction will increase. If, however, the substrate is limiting, in other words there is not sufficient to supply all the enzyme's active sites at one time, then any increase in enzyme concentration will have no effect on the rate of reaction. The rate of reaction will therefore stabilise at a constant level, meaning the graph will level off. This is because the available substrate is already being used as rapidly as it can be by the existing enzyme molecules. These events are summarised in Figure 6.

> ### Practical link
>
> Required practical 1. Investigation into the effect of a named variable on the rate of an enzyme controlled reaction.

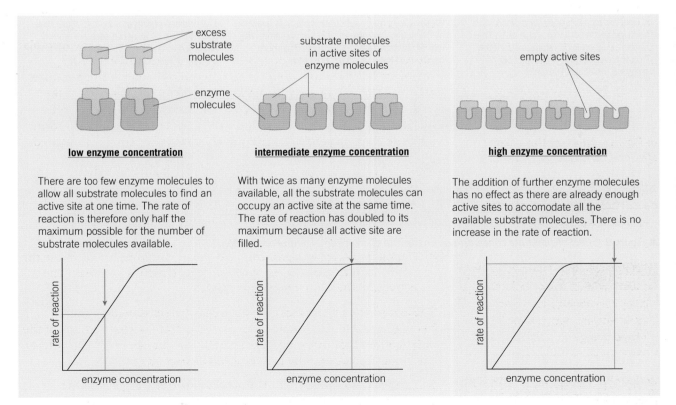

▲ **Figure 6** *Effect of enzyme concentration on the rate of enzyme action*

▲ **Figure 7** *Enzymes in the algae in this hot spring remain functional at temperatures of 80 °C whereas in most organisms they are denatured at temperatures of 40 °C*

Effects of substrate concentration on the rate of enzyme action

If the concentration of enzyme is fixed and substrate concentration is slowly increased, the rate of reaction increases in proportion to the concentration of substrate. This is because, at low substrate concentrations, the enzyme molecules have only a limited number of substrate molecules to collide with, and therefore the active sites of the enzymes are not working to full capacity. As more substrate is added, the active sites gradually become filled, until the point where all of them are working as fast as they can. The rate of reaction is at its maximum (V_{max}). After that, the addition of more substrate will have no effect on the rate of reaction. In other words, when there is an excess of substrate, the rate of reaction levels off. A summary of the effect of substrate concentration on the rate of enzyme action is given in Figure 8.

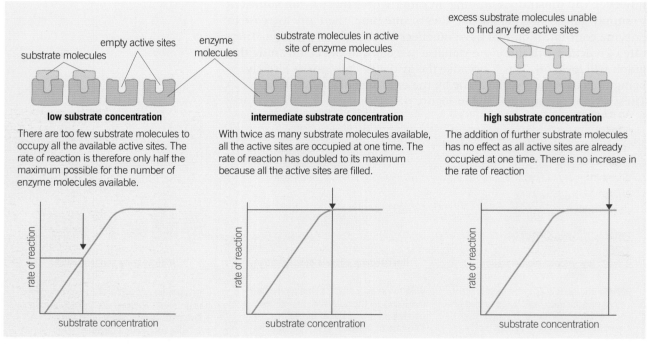

▲ **Figure 8** *Effect of substrate concentration on the rate of an enzyme-controlled reaction*

Enzyme action

Different enzymes can function at a wide range of temperatures. Shrimps that live in Arctic waters have enzymes that function fastest at around 4 °C and are denatured at around 15 °C. By contrast, bacteria that live in hot springs have enzymes that function fastest at 95 °C and continue to operate effectively above 100 °C. These bacteria are called thermophilic (heat-loving) bacteria.

Enzyme X is produced by thermophilic bacteria and hydrolyses many proteins including haemoglobin and egg albumin.

Enzyme Y is found in the stomach of young mammals where it acts on a single soluble protein found in milk, causing it to coagulate (clot).

1 a   From the descriptions, comment on the differences in the specificity of the two enzymes.

   b   Enzymes X and Y are each used for different commercial purposes. Suggest what this might be in each case.

   c   Suggest a possible purpose of enzyme Y in the mammalian stomach.

   d   Use the information about the two enzymes to suggest a possible difference in the type of bonding found in the tertiary structure of each. Explain your reasoning.

An experiment was carried out with enzyme X in which the time taken for it to fully hydrolyse 5 g of its protein substrate was measured at different temperatures. It is important when investigating the effect of a named variable on the rate of an enzyme reaction that the other variables are kept constant. For example, if measuring the effect of temperature, then pH, enzyme concentration, substrate concentration must be kept constant and all possible inhibitors should be absent. The following data were obtained:

Temperature / °C	Time / min for hydrolysis of protein	Rate of reaction μ 1/time
15	5.8	
25	3.4	
35	1.7	
45	0.7	
55	0.6	
65	0.9	
75	7.1	

2 a   Calculate the relative rate of reaction for each temperature.

   b   Plot a graph that shows the effect of temperature on the rate of reaction of enzyme X.

   c   Measure the optimum temperature for the action of enzyme X.

   d   Suggest how you might determine this optimum temperature more precisely.

Maths link

MS 3.2 and 3.4, see Chapter 11.

Summary questions

1   Explain why enzymes function less well at lower temperatures.

2   Explain how high temperatures may completely prevent enzymes from functioning.

3   Enzymes produced by microorganisms are responsible for spoiling food. Using this fact and your knowledge of enzymes, deduce why each of the following procedures are carried out.

   a   Food is heated to a high temperature before being canned.

   b   Some foods, such as onions, are preserved in vinegar.

4   Calculate the pH of a solution that has a hydrogen ion concentration of 0.0001 M

Enzyme inhibitors are substances that directly or indirectly interfere with the functioning of the active site of an enzyme and so reduce its activity. There are a number of types of enzyme inhibitor, two of which are:

- competitive inhibitors – which bind to the active site of the enzyme
- non-competitive inhibitors – which bind to the enzyme at a position other than the active site.

Competitive inhibitors

Competitive inhibitors have a molecular shape similar to that of the substrate. This allows them to occupy the active site of an enzyme. They therefore compete with the substrate for the available active sites (Figure 1). It is the difference between the concentration of the inhibitor and the concentration of the substrate that determines the effect that this has on enzyme activity. If the substrate concentration is increased, the effect of the inhibitor is reduced. The inhibitor is not permanently bound to the active site and so, when it leaves, another molecule can take its place. This could be a substrate or inhibitor molecule, depending on how much of each type is present. Sooner or later, all the substrate molecules will occupy an active site, but the greater the concentration of inhibitor, the longer this will take. An example of competitive inhibition occurs with an important respiratory enzyme that acts on succinate. Another compound, called malonate, can inhibit the enzyme because it has a very similar molecular shape to succinate. It therefore easily combines with the enzyme and blocks succinate from combining with the enzyme's active site. Another example is the inhibition of the enzyme transpeptidase by penicillin.

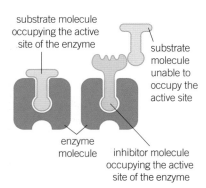

▲ **Figure 1** *Competitive inhibition*

Non-competitive inhibitors

Non-competitive inhibitors attach themselves to the enzyme at a binding site which is not the active site. Upon attaching to the enzyme, the inhibitor alters the shape of the enzyme and thus its active site in such a way that substrate molecules can no longer occupy it, and so the enzyme cannot function (Figure 2). As the substrate and the inhibitor are not competing for the same site, an increase in substrate concentration does not decrease the effect of the inhibitor (Figure 3).

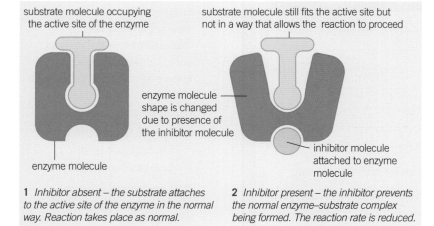

1 *Inhibitor absent – the substrate attaches to the active site of the enzyme in the normal way. Reaction takes place as normal.*

2 *Inhibitor present – the inhibitor prevents the normal enzyme–substrate complex being formed. The reaction rate is reduced.*

▲ **Figure 2** *Non-competitive inhibition*

Control of metabolic pathways

A metabolic pathway is a series of reactions in which each step is catalysed by an enzyme. In the tiny space inside a single cell, there are many hundreds of different metabolic pathways. The pathways are not at all haphazard, but highly structured. The enzymes that control a pathway are often attached to the membrane of a cell organelle in a very precise sequence. Inside each organelle optimum conditions for the functioning of particular enzymes may be provided. To keep a steady concentration of a particular chemical in a cell, the same chemical often acts as an inhibitor of an enzyme at the start of a reaction.

Let us look at the example illustrated in Figure 4. The end product inhibits enzyme A. If for some reason the concentration of end product increases above normal, then there will be greater inhibition of enzyme A. As a result, less end product will be produced and its concentration will return to normal. If the concentration of the end product falls below normal there will be less of it to inhibit enzyme A. Consequently, more end product will be produced and, again, its concentration will return to normal. In this way, the concentration of any chemical can be maintained relatively constant. This is known as **end-product inhibition**. This type of inhibition is usually non-competitive.

1 Different conditions affect how enzymes work. Name one that might vary between one organelle and another.
2 Suggest why enzymes are attached to the inner membrane of an organelle 'in a very precise sequence'.
3 If an end product inhibits enzyme B rather than enzyme A, predict what would be:
   a the initial effect on the concentration of intermediate 1
   b the overall longer term effect on the concentration of the end product.
4 Suggest one advantage of end-product inhibition being non-competitive rather than competitive. Relate your answer to how the two types of inhibition take place.

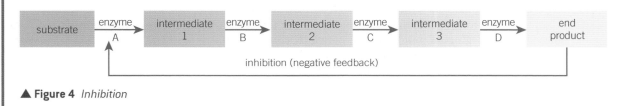

▲ **Figure 4** *Inhibition*

Summary questions

1 Distinguish between a competitive and a non-competitve inhibitor
2 🧪 An enzyme-controlled reaction is inhibited by substance X. Suggest a simple way in which you could tell whether substance X is acting as a competitive or a non-competitive inhibitor.

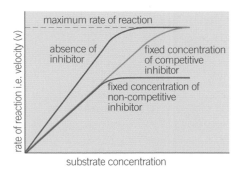

▶ **Figure 3** *Comparison of competitive and non-competitive inhibition on the rate of an enzyme-controlled reaction at different substrate concentrations*

1 (a) The table shows some substances found in cells. Complete the table to show the properties of these substances. Put a tick in the box if the statement is correct.

Statement	Substance			
	Starch	Glycogen	Deoxyribose	DNA helicase
Substance contains only the elements carbon, hydrogen and oxygen				
Substance is made from amino acid monomers				
Substance is found in both animal cells and plant cells				

(4 marks)

(b) The diagram shows two molecules of β-glucose.

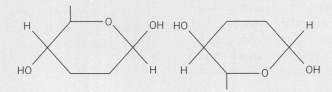

On the diagram, draw a box around the atoms that are removed when the two β-glucose molecules are joined by condensation. (2 marks)

(c) (i) Hydrogen bonds are important in cellulose molecules. Explain why. (2 marks)
(ii) A starch molecule has a spiral shape. Explain why this shape is important to its function in cells. (1 mark)

AQA Jan 2011

2 (a) Omega-3 fatty acids are unsaturated. What is an *unsaturated* fatty acid? (2 marks)
(b) Scientists investigated the relationship between the amount of omega-3 fatty acids eaten per day and the risk of coronary heart disease. The graph shows their results. Do the data show that eating omega-3 fatty acids prevents coronary heart disease? Explain your answer. (3 marks)

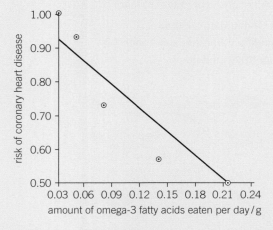

(c) Olestra is an artificial lipid. It is made by attaching fatty acids, by condensation, to a sucrose molecule. The diagram shows the structure of olestra. The letter R shows where a fatty acid molecule has attached.

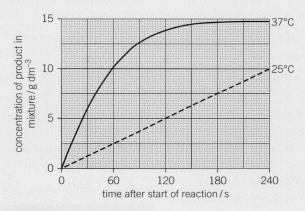

(i)   Name bond **X**. (*1 mark*)

(ii) A triglyceride does **not** contain sucrose or bond **X**. Give **one** other way in which the structure of a triglyceride is different to olestra. (*1 mark*)

(iii) Starting with separate molecules of glucose, fructose and fatty acids, how many molecules of water would be produced when one molecule of olestra is formed? (*1 mark*)

AQA Jan 2011

3   A technician investigated the effect of temperature on the rate of an enzyme-controlled reaction. At each temperature, he started the reaction using the same volume of substrate solution and the same volume of enzyme solution.

▲ **Figure 2** *shows his results*

(a) Give **one** other factor the technician would have controlled. (*1 mark*)

(b) Calculate the rate of reaction at 25°C. (*2 marks*)

(c) Describe and explain the differences between the two curves. (*5 marks*)

AQA SAMS PAPER 1

HAMPTON SCHOOL
BIOLOGY DEPARTMENT

Nucleic acids are a group of the most important molecules of which the best known are **ribonucleic acid** (**RNA**) and **deoxyribonucleic acid** (**DNA**). The double helix structure of deoxyribonucleic acid (DNA) makes it immediately recognisable. DNA carries genetic information. The identification of this extraordinary molecule as the material that passes on the features of organisms from one generation to the next is one of the most remarkable feats of experimental biology. The discovery of the precise molecular arrangement of DNA was no less remarkable. Despite its complex structure, DNA is made up of nucleotides that have just three basic components.

Nucleotide structure

Individual nucleotides are made up of three components:

- a pentose sugar (so called because it has five carbon atoms)
- a phosphate group
- a nitrogen-containing organic base. These are: cytosine **C**, thymine **T**, Uracil **U**, adenine **A** and guanine **G**.

The pentose sugar, phosphate group and organic base are joined, as a result of **condensation reactions**, to form a single nucleotide (**mononucleotide**) as shown in Figure 1. Two mononucleotides may, in turn, be joined as a result of a condensation reaction between the deoxyribose sugar of one mononucleotide and the phosphate group of another. The bond formed between them is called a **phosphodiester bond** (Figure 3). The new structure is called a **dinucleotide**. The continued linking of mononucleotides in this way forms a long chain known as a **polynucleotide**. In addition to DNA and RNA, some other biologically important molecules contain nucleotides. For simplicity the various components of nucleotides are represented by symbols, as shown in Table 1.

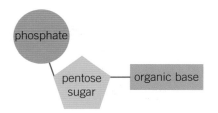

▲ **Figure 1** *Simplified structure of a nucleotide*

▼ **Table 1** *Components of nucleotides.*

Name of molecule	Symbol
phosphate	
pentose sugar	
adenine	adenine
guanine	guanine
cytosine	cytosine
thymine	thymine
uracil	uracil

Ribonucleic acid (RNA) structure

Ribonucleic acid is a polymer made up of nucleotides. It is a single, relatively short, polynucleotide chain in which the pentose sugar is always **ribose** and the organic bases are adenine, guanine, cytosine and **uracil** (Figure 2). One type of RNA transfers genetic information from DNA to the ribosomes. The ribosomes themselves are made up of proteins and another type of RNA. A third type of RNA is involved in protein synthesis.

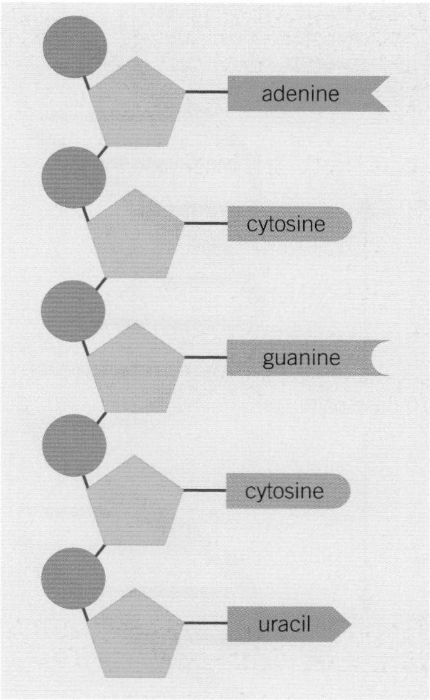

DNA structure

In 1953, James Watson and Francis Crick worked out the structure of DNA, following pioneering work by Rosalind Franklin on the X-ray diffraction patterns of DNA. This opened the door for many of the major developments in biology over the next half-century.

In DNA the pentose sugar is deoxyribose and the organic bases are adenine, thymine, guanine and cytosine. DNA is made up of two strands of nucleotides (polynucleotides). Each of the two strands is extremely long, and they are joined together by **hydrogen bonds** formed between certain bases. In its simplified form, DNA can be thought of as a ladder in which the phosphate and deoxyribose molecules alternate to form the uprights and the organic bases pair together to form the rungs (Figure 4).

▲ **Figure 2** *Section of an RNA molecule*

Base pairing

The bases on the two strands of DNA attach to each other by hydrogen bonds. It is these hydrogen bonds that hold the two strands together. The base pairing is specific:

* Adenine always pairs with thymine
* Guanine always pairs with cytosine

As a result of these pairings, adenine is said to be **complementary** to thymine and guanine is said to be complementary to cytosine.

It follows that the quantities of adenine and thymine in DNA are always the same, and so are the quantities of guanine and cytosine. However, the ratio of adenine and thymine to guanine and cytosine varies from species to species.

The double helix

In order to appreciate the structure of DNA, you need to imagine the ladder-like arrangement of the two polynucleotide chains being twisted. In this way, the uprights of phosphate and deoxyribose wind around one another to form a double helix. They form the structural backbone of the DNA molecule.

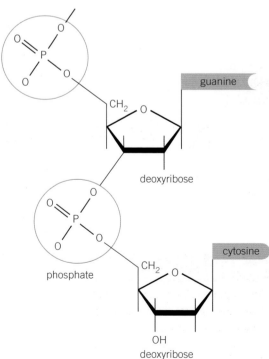

▲ **Figure 3** *The structure of a phosphodiester bond between a guanine nucleotide and a cytosine nucleotide*

The uprights are composed of deoxyribose–phosphate molecules, and the rungs are pairs of bases.

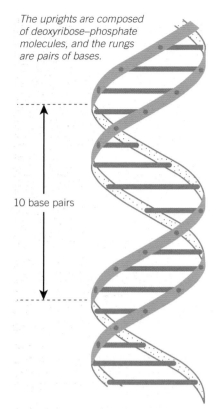

10 base pairs

▲ **Figure 5** *The double helix structure of DNA*

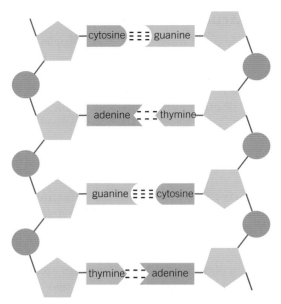

▲ **Figure 4** *Basic structure of DNA*
DNA structure may be likened to a ladder in which alternating phosphate and deoxyribose molecules make up the 'uprights' and pairs of organic bases comprise the 'rungs'. Note the base pairings are always cytosine–guanine and adenine–thymine. This ensures a standard 'rung' length. Note also that the 'uprights' run in the opposite direction to each other (i.e. are antiparallel).

The stability of DNA
DNA is a stable molecule because:

- The phosphodiester backbone protects the more chemically reactive organic bases inside the double helix.
- Hydrogen bonds link the organic base pairs forming bridges (rungs) between the phosphodiester uprights. As there are three hydrogen bonds between cystine and guanine, the higher the proportion of C—G pairings, the more stable the DNA molecule.

There are other interactive forces between the base pairs that hold the molecule together (= base stacking).

Function of DNA
DNA is the hereditary material responsible for passing genetic information from cell to cell and generation to generation. In total, there are around 3.2 billion base pairs in the DNA of a typical mammalian cell. This vast number means that there is an almost infinite variety of sequences of bases along the length of a DNA molecule. It is this variety that provides the genetic diversity within living organisms.

The DNA molecule is adapted to carry out its functions in a number of ways:

- It is a very stable structure which normally passes from generation to generation without change. Only rarely does it mutate.

Hint

In every molecule of DNA, the phosphate group, the deoxyribose and the four bases are always the same. What differs between one DNA molecule and another are the proportions, and more importantly the sequence, of each of the four bases.

Synoptic link

More detail on mutations is given in Topic 9.1, Gene mutations, and about protein synthesis in Topic 8.4, Polypeptide synthesis – transcription and splicing.

- Its two separate strands are joined only with hydrogen bonds, which allow them to separate during DNA replication (Topic 2.2) and protein synthesis.
- It is an extremely large molecule and therefore carries an immense amount of genetic information.
- By having the base pairs within the helical cylinder of the deoxyribose–phosphate backbone, the genetic information is to some extent protected from being corrupted by outside chemical and physical forces.
- Base pairing leads to DNA being able to replicate and to transfer information as mRNA.

The function of the remarkable molecule that is DNA depends on the sequence of base pairs that it possesses. This sequence is important to everything it does and, indeed, to life itself.

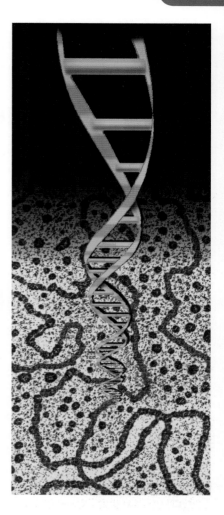

Summary questions

1 List the three basic components of a nucleotide.
2 Suggest why the base pairings of adenine with cytosine and guanine with thymine do not occur.
3 If the bases on one strand of DNA are TGGAGACT, determine the base sequence on the other strand.
4 If 19.9% of the base pairs in human DNA are guanine, calculate what percentage of human DNA is thymine. Show your reasoning.

 ## Unravelling the role of DNA

We now take for granted that DNA is the hereditary material that passes genetic information from cell to cell and generation to generation. This was not always the case because there were other contenders for this role, in particular proteins.

With the knowledge available at the time, scientists thought that proteins were the more likely candidate because of their considerable chemical diversity. DNA was considered to have too few components and to be chemically too simple to fulfil the role. However, not all scientists were convinced and so they set about finding experimental evidence to determine the true nature of hereditary material.

1 Assess the advantages of scientists questioning the validity of a current theory rather than automatically accepting it.

Scientists work by using **observations** and current knowledge to form a **hypothesis**. From this, they make **predictions** about the outcome of a particular **investigation**. By carrying out this investigation a number of times, they collect the experimental evidence that allows them to accept or reject their hypothesis.

2 Explain what is meant by the term 'hypothesis' in the scientific sense.

Investigations were needed to test the hypothesis that DNA was the hereditary material.

One investigation to test the hypothesis that DNA was the hereditary material involved experiments using mice and a bacterium that can cause pneumonia. The bacterium exists in two forms:

- a safe form that does not cause pneumonia, known as the R-strain,

- a harmful form that causes pneumonia, known as the S-strain.

Mice were separately injected with living bacteria of the safe form and dead bacteria from the harmful form.

The group of mice injected with the living safe form of bacteria remained healthy, as did the group injected with the dead harmful form of bacteria. So, when mice were injected with both types together, it would not have been surprising to get a similar result. These mice, however, developed pneumonia. The experiment and the results are summarised in Figure 7.

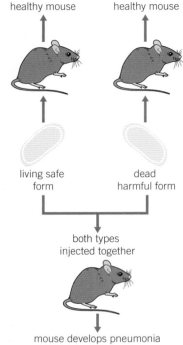

▲ **Figure 7** *Summary of an experiment to determine the nature of hereditary material in an organism*

Living bacteria of the harmful form were isolated from the mice with pneumonia. There are three possible explanations for this:

- Experimental error, for example, the harmful forms in the mixture were not all killed.

- The living safe form had mutated into the harmful form. This is possible but extremely unlikely, especially given that the experiment was repeated many times with the same result.

- Pneumonia is caused by a toxin. The harmful form of the bacterium has the information on how to make the toxin but, being dead, cannot do so. The safe form has the means of making the toxin but lacks the information on how to do so. The information on how to make the toxin may have been transferred from the harmful form to the safe form, which then produced it.

3 ⚗ State what simple procedure could be carried out to discount the first explanation.

4 Mutations happen very rarely. Explain why this helps to discount the second explanation.

The third explanation was considered worthy of further investigation and so a series of experiments was designed and carried out as follows:

- The living harmful bacteria that were found in the mice with pneumonia, were collected.
- Various substances were isolated from these bacteria and purified.
- Each substance was added to suspensions of living safe bacteria to see whether it would transform them into the harmful form.
- The only substance that produced this transformation was purified DNA.
- When an enzyme that breaks down DNA was added, the ability to carry out the transformation ceased.

Other experiments provided further proof that DNA was the hereditary material and also suggested a mechanism by which it could be transferred from one bacterial cell to another.

- It had been observed that viruses infect bacteria, causing the bacteria to make more viruses.
- As the virus is made up of just protein and DNA, one or the other must possess the instructions that the bacteria use to make new viruses.
- The protein and DNA in the viruses were each labelled with a different radioactive element.
- One sample of bacteria was infected by viruses with radioactive protein while another sample was infected by viruses with radioactive DNA.
- In a later stage, the viruses and bacteria in both samples were separated from one another.
- Only the sample with bacteria that had been infected by viruses labelled with radioactive DNA showed signs of radioactivity.

This was evidence that DNA was the material that had provided the bacteria with the genetic information needed to make the viruses. It also showed how DNA can be passed from one bacterium to another, for example, by viruses.

5 A new scientific discovery often presents moral, economic and ethical issues. Justify why it is necessary for society to analyse the risks and benefits of these discoveries before they are developed.

A prime location

In order to understand how nucleotides are arranged in nucleic acids, it is necessary to know how the carbon atoms in the pentose molecule are numbered. Of particular importance is the numbering of the 3′ (3-prime) and 5′ (5-prime) carbon atoms. The 5′ carbon has an attached phosphate group, while the 3′ has a hydroxyl group.

Figure 8 shows a nucleotide with the 3′ and 5′ carbon atoms marked on its pentose sugar.

When nucleotides are organised into the double strands of a DNA molecule, one strand runs in the 5′ to 3′ direction while the other runs the opposite way – in the 3′ to 5′ direction. The two strands are therefore said to be antiparallel.

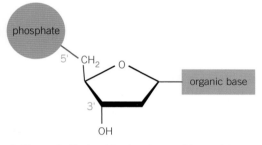

▲ **Figure 8** *Nucleotide showing positions of the 3-prime (3′) and 5-prime (5′) carbon atoms on the pentose sugar*

Nucleic acids can only be synthesised 'in vivo' in the 5′-to-3′ direction. This is because the enzyme DNA polymerase that assembles nucleotides into a DNA molecule can only attach nucleotides to the hydroxyl (OH) group on the 3′ carbon molecule.

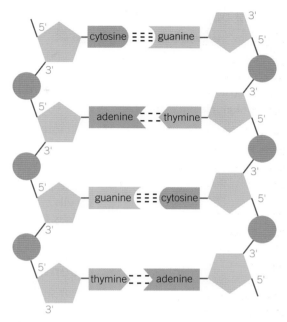

▲ **Figure 9** *DNA molecule showing the 3-prime and 5-prime carbon atoms labelled. Notice that one strand runs 5 to 3 while the other runs 3 to 5 . They are **antiparallel***

1 Suggest what the term 'in vivo' means in the context of synthesising DNA.
2 From your knowledge of the way enzymes work, explain why DNA polymersae can only attach nucleotides to the hydroxyl (OH) group on the 3 carbon molecule.

2.2 DNA replication

The cells that make up organisms are always derived from existing cells by the process of division. Cell division occurs in two main stages:

- **Nuclear division** is the process by which the nucleus divides. There are two types of nuclear division, mitosis and **meiosis**.
- **Cytokinesis** follows nuclear division and is the process by which the whole cell divides.

Before a nucleus divides its DNA must be replicated (copied). This is to ensure that all the daughter cells have the genetic information to produce the enzymes and other proteins that they need.

The process of DNA replication is clearly very precise because all the new cells are more or less genetically identical to the original one. How then does DNA replication take place? It is the semi-conservative model that is universally accepted.

Semi-conservative replication

For semi-conservative replication to take place there are four requirements:

- The four types of nucleotide, each with their bases of adenine, guanine, cytosine or thymine, must be present.
- Both strands of the DNA molecule act as a template for the attachment of these nucleotides.
- The enzyme DNA polymerase.
- A source of chemical energy is required to drive the process.

The process of semi-conservative replication is illustrated in Figure 1. It takes place as follows:

- The enzyme **DNA helicase** breaks the hydrogen bonds linking the base pairs of DNA.
- As a result the double helix separates into its two strands and unwinds.
- Each exposed polynucleotide strand then acts as a template to which complementary free nucleotides bind by specific base pairing
- Nucleotides are joined together in a condensation reaction by the enzyme **DNA polymerase** to form the 'missing' polynucleotide strand on each of the two original polynucleotide strands of DNA.
- Each of the new DNA molecules contains one of the original DNA strands, that is, half the original DNA has been saved and built into each of the new DNA molecules (Figure 2). The process is termed 'semi-conservative replication'.

> **Study tip**
>
> Remember that DNA replication uses complementary base pairings to produce two identical copies.

a A representative portion of DNA, which is about to undergo replication.

b An enzyme, DNA helicase, causes the two strands of the DNA to separate by breaking the hydrogen bonds that join the complementary bases together.

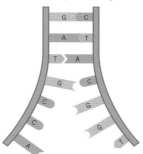

c DNA helicase completes the separation of the strand. Meanwhile, free nucleotides that have been activated bind specifically to their complementary bases.

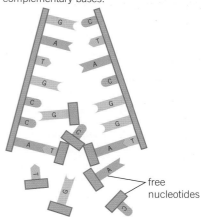

free nucleotides

d Once the activated nucleotides are bound, they are joined together by DNA polymerase which makes phosphodiester bonds (bottom three nucleotides). The remaining unpaired bases continue to attract their complementary nucleotides.

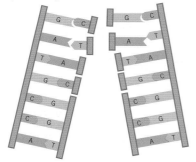

e Finally, all the nucleotides are joined to form a complete polynucleotide chain using DNA polymerase. In this way, two identical molecules of DNA are formed. As each molecule retains half of the original DNA material, t method of replication is called the semi-conservative method.

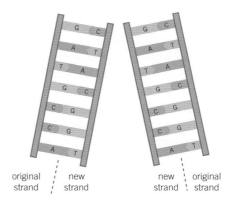

original   new   new   original
strand   strand   strand   strand

▲ **Figure 1** *The semi-conservative replication of DNA*

Summary questions

1   If the bases on a portion of the original strand of DNA are ATGCTACG, determine the equivalent sequence of bases on the newly formed strand.

2   Explain why the process of DNA replication is described as semi-conservative.

3   If an inhibitor of DNA polymerase were introduced into a cell, explain what the effect would be on DNA replication.

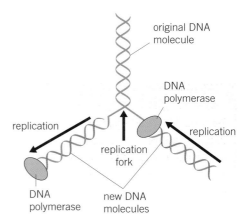

original DNA molecule

DNA polymerase

replication   replication

replication fork

DNA polymerase   new DNA molecules

▲ **Figure 2** *Role of DNA polymerase in the semi-conservative replication of DNA*

Evidence for semi-conservative replication

This account illustrates how scientists use theories and models to attempt to explain observations. Scientific progress is made when experimental evidence is produced that supports a new theory or model.

When James Watson and Francis Crick worked out the structure of DNA in 1953, with the help of Rosalind Franklin's X-ray diffraction studies, they remarked in their paper:

> It has not escaped our notice that the specific pairing we have postulated immediately suggests a possible copying mechanism for the genetic material.

Their idea, namely the semi-conservative method, was, however, only one of two possible mechanisms. Both needed to be scientifically tested before a definite conclusion could be drawn. The two hypotheses were:

- **The conservative model** suggested that the original DNA molecule remained intact and that a separate daughter DNA copy was built up from new molecules of deoxyribose, phosphate and organic bases. Of the two molecules produced, one would be made of entirely new material while the other would be entirely original material (Figure 3).

- **The semi-conservative model** proposed that the original DNA molecule split into two separate strands, each of which then replicated its mirror image (i.e. the missing half). Each of the two new molecules would therefore have one strand of new material and one strand of original material (Figure 3).

If we look at Figure 3, we can see that the distribution of the strands from the original DNA molecule after replication is different in each model. To find out which mechanism was correct was therefore easy, at least in theory – simply label the original DNA in some way and then look at how it was distributed after replication. The next stage was to design an experiment to test which hypothesis was correct. Two scientists, Meselsohn and Stahl, achieved this in a neat and elegant experiment.

They based their work on three facts:

- All the bases in DNA contain nitrogen.
- Nitrogen has two forms: the lighter nitrogen ^{14}N and the **isotope** ^{15}N, which is heavier.
- Bacteria will incorporate nitrogen from their growing medium into any new DNA that they make.

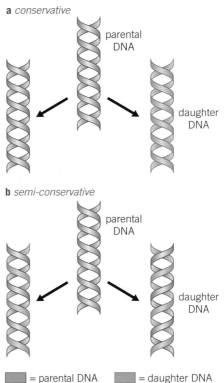

a *conservative*

parental DNA

daughter DNA

b *semi-conservative*

parental DNA

daughter DNA

■ = parental DNA ■ = daughter DNA

▲ **Figure 3** *Different models of DNA replication*

They reasoned that bacteria grown on a medium containing ^{14}N would have DNA that was lighter than bacteria grown on a medium containing ^{15}N. They labelled the original DNA of bacteria by growing them on a medium of ^{15}N. They then transferred the bacteria to a medium of ^{14}N for a single generation to allow it to replicate once. The mass of each 'new' DNA molecule would depend upon which method of replication had taken place (Figure 3). To separate out the different DNA types, they centrifuged the extracted DNA in a special solution. The lighter the DNA, the nearer the top of the centrifuge tube it collected. The heavier the DNA, the nearer the bottom of the tube it collected (see Topic 3.1). They also analysed the DNA after two, then three, generations. By interpreting the results they could determine which hypothesis was correct. Their work is summarised in Figure 4.

1 Name the part of the DNA molecule that contains nitrogen.
2 Explain why, after one generation, all the DNA is made up of an equal mixture of ^{14}N and ^{15}N.

3 🧪 Suppose DNA were replicated by the conservative model. Sketch a tube showing the position of DNA after one generation.

4 🧪 From Figure 4, copy the chart for tube 4. Draw bars on the chart to show the percentage of each of the three possible types of DNA.

5 🧪 √x After three generations (tube 5), calculate what percentage of the DNA will be made up of ^{14}N only.

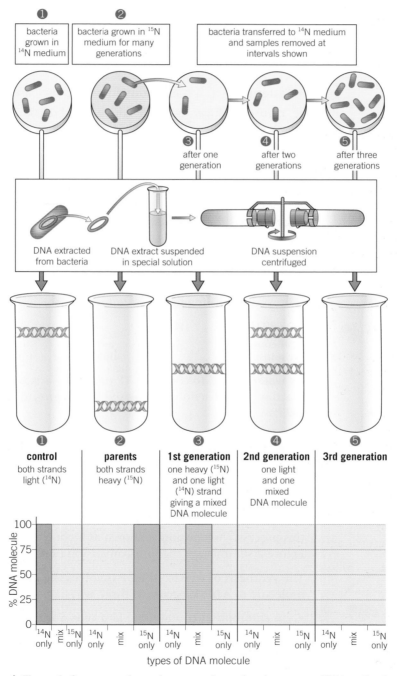

▲ **Figure 4** *Summary of experiments to determine the nature of DNA replication*

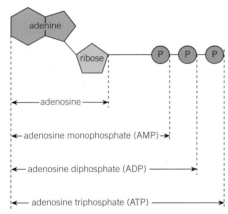

▲ **Figure 1** *Structure of ATP*

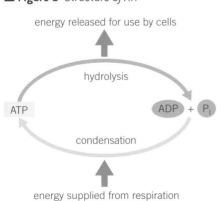

▲ **Figure 2** *Interconversion of ATP and ADP*

All living organisms require energy in order to remain alive. This energy comes initially from the Sun. Plants use solar energy to combine water and carbon dioxide into complex organic molecules by the process of photosynthesis. Both plants and animals then oxidise these organic molecules to make adenosine triphosphate (ATP), which is used as the main energy source to carry out processes within cells.

Structure of ATP

The ATP molecule (Figure 1) is a phosphorylated macromolecule. It has three parts:

- **adenine** – a nitrogen-containing organic base
- **ribose** – a sugar molecule with a 5-carbon ring structure (pentose sugar) that acts as the backbone to which the other parts are attached
- **phosphates** – a chain of three phosphate groups.

How ATP stores energy

Adenosine triphosphate (ATP) is a nucleotide and as the name suggests, has three phosphate groups. These are the key to how ATP stores energy. The bonds between these phosphate groups are unstable and so have a low **activation energy**, which means they are easily broken. When they do break they release a considerable amount of energy. Usually in living cells it is only the terminal phosphate that is removed, according to the equation:

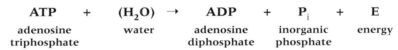

$$\textbf{ATP} + \textbf{(H}_2\textbf{O)} \rightarrow \textbf{ADP} + \textbf{P}_i + \textbf{E}$$

| adenosine triphosphate | water | adenosine diphosphate | inorganic phosphate | energy |

As water is used to convert ATP to ADP, this is known as a hydrolysis reaction. The reaction is catalysed by the enzyme **ATP hydrolase** (ATPase).

Synthesis of ATP

The conversion of ATP to ADP is a reversible reaction and therefore energy can be used to add an inorganic phosphate to ADP to re-form ATP according to the reverse of the equation above. This reaction is catalysed by the enzyme **ATP synthase**. As water is removed in this process, the reaction is known as a condensation reaction. Figure 2 summarises the interconversion of ATP and ADP.

The synthesis of ATP from ADP involves the addition of a phosphate molecule to ADP. It occurs in three ways:

- in chlorophyll-containing plant cells during photosynthesis (photophosphorylation)
- in plant and animal cells during respiration (oxidative phosphorylation)
- in plant and animal cells when phosphate groups are transferred from donor molecules to ADP (substrate-level phosphorylation).

Roles of ATP

The same feature that makes ATP a good energy donor, namely the instability of its phosphate bonds, is also a reason why it is not a good long-term energy store. Fats, and carbohydrates such as glycogen, serve this purpose far better. ATP is therefore the **immediate energy source** of a cell. As a result, cells do not store large quantities of ATP, but rather just maintain a few seconds' supply. This is not a problem, as ATP is rapidly re-formed from ADP and inorganic phosphate (P_i) and so a little goes a long way. ATP is a better immediate energy source than glucose for the following reasons:

- Each ATP molecule releases less energy than each glucose molecule. The energy for reactions is therefore released in smaller, more manageable quantities rather than the much greater, and therefore less manageable, release of energy from a glucose molecule.

- The hydrolysis of ATP to ADP is a single reaction that releases immediate energy. The breakdown of glucose is a long series of reactions and therefore the energy release takes longer.

ATP cannot be stored and so has to be continuously made within the mitochondria of cells that need it. Cells, such as muscle fibres and the epithelium of the small intestine, which require energy for movement and active transport respectively, possess many large mitochondria.

ATP is used in energy-requiring processes in cells including:

- **metabolic processes**. ATP provides the energy needed to build up macromolecules from their basic units. For example, making starch from glucose or polypeptides from amino acids.

- **movement**. ATP provides the energy for muscle contraction. In muscle contraction, ATP provides the energy for the filaments of muscle to slide past one another and therefore shorten the overall length of a muscle fibre.

- **active transport**. ATP provides the energy to change the shape of carrier proteins in plasma membranes. This allows molecules or ions to be moved against a concentration gradient.

- **secretion**. ATP is needed to form the lysosomes necessary for the secretion of cell products.

- **activation of molecules**. The inorganic phosphate released during the hydrolysis of ATP can be used to phosphorylate other compounds in order to make them more reactive, thus lowering the activation energy in enzyme-catalysed reactions. For example – the addition of phosphate to glucose molecules at the start of glycolysis.

Hint

Think of the unstable bonds that link the phosphates in ATP as coiled springs. Due to these spring-like bonds the end phosphate is straining to break away from its nearest partner. Any small addition of energy and the end phosphate springs away, releasing all the energy that is stored in the 'spring', that is, stored in the bond.

Hint

ATP is synthesised during reactions that *release* energy and it is hydrolysed to provide energy for reactions that *require* it.

Study tip

Don't think about ATP as a 'high-energy' substance. ATP is an 'intermediate energy' substance that is used to transfer energy.

Synoptic link

Concentration gradients are looked at in more detail in Topic 4.4, while the role of lysosomes in secretion is covered in Topic 3.4.

Summary questions

1 ATP is sometimes referred to as 'an immediate energy source'. Explain why.

2 Explain how ATP can make an enzyme-catalysed reaction take place more readily.

3 State **three** roles of ATP in plant cells.

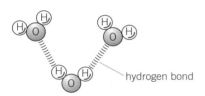

▲ **Figure 1** *Water molecules showing hydrogen bonding*

▲ **Figure 2** *Due to surface tension, pond skaters walk on water*

Water is a major component of cells. Although water is the most abundant liquid on Earth, it is certainly no ordinary molecule. Its unusual properties are due to its dipolar nature and the subsequent hydrogen bonding that this allows.

The dipolar water molecule

A water molecule is made up of two atoms of hydrogen and one of oxygen as shown in Figure 1. Although the molecule has no overall charge, the oxygen atom has a slight negative charge, while the hydrogen atoms have a slight positive one. In other words, the water molecule has both positive and negative poles and is therefore described as **dipolar**.

Water and hydrogen bonding

Different poles attract, and therefore the positive pole of one water molecule will be attracted to the negative pole of another water molecule. The attractive force between these opposite charges is called a hydrogen bond (Figure 2). Although each bond is fairly weak (about one-tenth as strong as a **covalent bond**), together they form important forces that cause the water molecules to stick together, giving water its unusual properties.

Specific heat capacity of water

Because water molecules stick together, it takes more energy (heat) to separate them than would be needed if they did not bond to one another. For this reason the boiling point of water is higher than expected. Without its hydrogen bonding, water would be a gas (water vapour) at the temperatvures commonly found on Earth and life as we know it would not exist. For the same reason, it takes more energy to heat a given mass of water, that is water has a high specific heat capacity. Water therefore acts as a buffer against sudden temperature variations, making the aquatic environment a temperature-stable one. As organisms are mostly water, it also buffers them against sudden temperature changes especially in terrestrial environments.

Latent heat of vaporisation of water

Hydrogen bonding between water molecules means that it requires a lot of energy to evaporate 1 gram of water. This energy is called the **latent heat of vaporisation**. Evaporation of water such as sweat in mammals is therefore a very effective means of cooling because body heat is used to evaporate the water.

Cohesion and surface tension in water

The tendency of molecules to stick together is known as cohesion. With its hydrogen bonding, water has large cohesive forces and these allow it to be pulled up through a tube, such as a **xylem vessel** in plants. In the same way, where water molecules meet air they tend to be pulled back into the body of water rather than escaping from it. This force is called surface tension and means that the water surface acts like a skin and is strong enough to support small organisms such as pond skaters (Figure 3).

The importance of water to living organisms

Water is the main constituent of all organisms – up to 98% of a jellyfish is water and mammals are typically 65% water. Water is also where life on Earth arose and it is the environment in which many species still live. It is important for other reasons too.

Water in metabolism

- Water is used to break down many complex molecules by hydrolysis, for example, proteins to amino acids. Water is also produced in condensation reactions.
- Chemical reactions take place in an aqueous medium.
- Water is a major raw material in photosynthesis.

Water as a solvent

Water readily dissolves other substances:

- gases such as oxygen and carbon dioxide
- wastes such as ammonia and urea
- inorganic ions and small hydrophilic molecules such as amino acids, monosaccharides and ATP
- enzymes, whose reactions take place in solution.

Other important features of water

- Its evaporation cools organisms and allows them to control their temperature.
- It is not easily compressed and therefore provides support, for example the hydrostatic skeleton of animals such as the earthworm and turgor pressure in herbaceous plants.
- It is transparent and therefore aquatic plants can photosynthesise and also light rays can penetrate the jelly-like fluid that fills the eye and so reach the retina.

Inorganic ions

Inorganic ions are found in organisms where they occur in solution in the cytoplasm of cells and in body fluids and as well as part of larger molecules. They may be in concentrations that range from very high to very low.

Inorganic ions perform a range of functions. The specific function a particular ion per forms is related to its properties. For example, as we saw in Topic 1.6, iron ions are found in haemoglobin where they play a role in the transport of oxygen. Other examples we have looked at include the phosphate ions that form a structural role in DNA molecules (Topic 2.1) and a role in storing energy in ATP molecules (Topic 2.3). Hydrogen ions are important in determining the pH of solutions and therefore the functioning of enzymes (Topic 1.8). We shall see in Topic 4.5 that sodium ions are important in the transport of glucose and amino acids across plasma membranes.

▲ **Figure 3** *Evaporation of water during sweating helps to maintain body temperature*

Synoptic link

Xylem vessels will be covered in Topic 7.7 and condensation and hydrolysis were covered in Topic 1.3.

Summary questions

In the following passage, state the missing word indicated by the letters **a–f**.

A water molecule is said to be **a** because it has a positive and a negative pole as a result of the uneven distribution of **b** within it. This creates attractive forces called **c** between water molecules, causing them to stick together. This stickiness of water means that its molecules are pulled inwards at its surface. This force is called **d**. Water is able to split large molecules into smaller ones by a process known as **e**. Water is the raw material for the process of **f** in green plants.

1 (a) **Figure 1** shows one base pair of a DNA molecule.

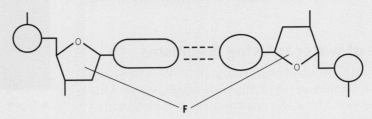

▲ **Figure 1**

(i) Name part **F** of each nucleotide. (*1 mark*)

(ii) Scientists determined that a sample of DNA contained 18% adenine.

What were the percentages of thymine and guanine in this sample of DNA? (*2 marks*)

AQA SAMS PAPER 1

(b) Organisms vary widely in the number of genes they have. Figure 2 shows the total length of DNA in six organisms plotted against the number of functional genes. The length of DNA is measured in numbers of base pairs.

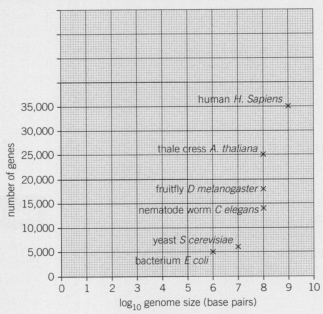

(i) A double-stranded DNA molecule is 2 µm long for every thousand base pairs. Use Figure 2 to calculate the total length of DNA in a human cell in metres. Show your working. (*2 marks*)

(ii) Calculate the ratio of DNA length to number of genes in humans and in *Escherichia coli* and suggest why there is this difference. (*2 marks*)

2 **Figure 3** shows a short section of a DNA molecule.

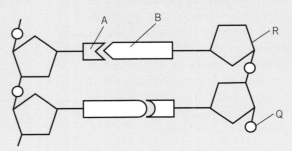

(a) Name parts **R** and **Q**.   (*2 marks*)
(b) Name the type of bonds that join **A** and **B**.   (*1 mark*)
(c) Ribonuclease is an enzyme. It is 127 amino acids long. What is the
     minimum number of DNA bases needed to code for ribonuclease?   (*1 mark*)
(d) **Figure 2** shows the sequence of DNA bases coding for seven amino acids
     in the enzyme ribonuclease.

Figure 2
GTT TAC TAC TCT TCT TCT TTA

The number of each type of amino acid coded for by this sequence of DNA bases is shown in
the table.

Amino acid	Number present
Arg	3
Met	2
Gln	1
Asn	1

Use the table and **Figure 2** to work out the sequence of amino acids in this part of the
enzyme. Write your answer in the boxes below.

Gln						

                                                                                                       (*1 mark*)
(e) Explain how a change in a sequence of DNA bases could result in a
     non-functional enzyme.   (*3 marks*)

AQA Jan 2010

3   (a) Hydrogen bonds occur between water molecules. Explain how these
         affect the properties of water as a habitat for organisms.   (*2 marks*)
    (b) Give two inorganic ions within the human body and describe one function
         of each.   (*4 marks*)
    (c) Explain the importance of the hydrolysis reaction of ATP.   (*2 marks*)

HAMPTON SCHOOL
BIOLOGY DEPARTMENT

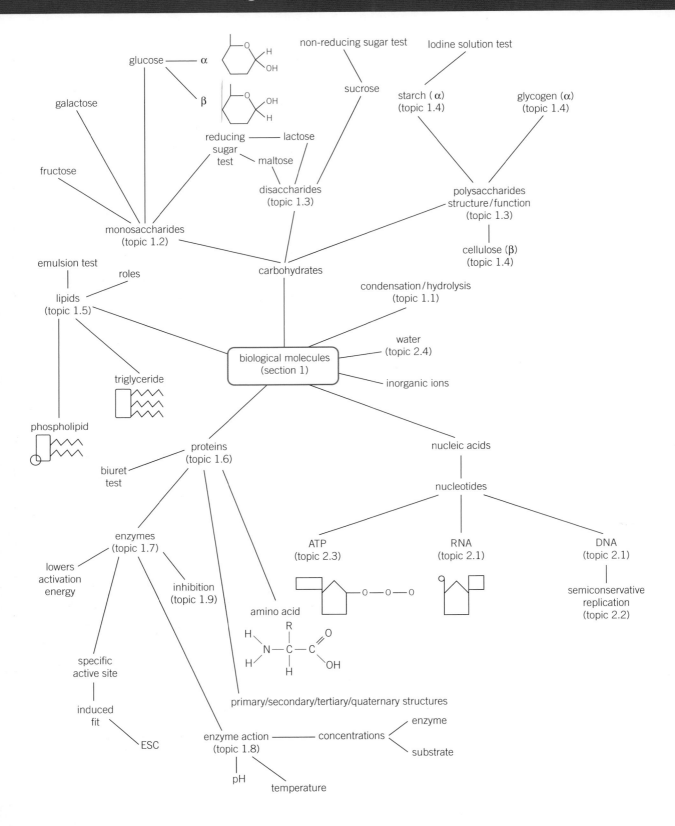

Practical skills

In this section you have met the following practical skills:

- How to carry out tests for a variety of food substances. For example:
    - Benedict's reagent to test for reducing and non-reducing sugars
    - iodine in potassium iodide solution to test for starch
    - the Biuret test for proteins
    - the emulsion test for lipids.
- The importance of keeping all other variables constant except the one being investigated, when carrying out investigations such as the effects of different factors on enzyme action.

Maths skills

In this section you have met the following maths skills:

- How to measure change in the rate of a reaction using a tangent to a curve.
- Finding the frequency of all the bases on a DNA strand when given only information on some of the bases present.
- Choosing the best way of presenting your results when investigating the rate of enzyme-controlled reactions.

Extension task

Research the following two topics using any source of information available to you, for example, textbooks, journals, the internet etc.

Proteins are very diverse in their structure and functions and so are involved in all processes of living organisms including: nutrition, respiration, transport, growth, excretion, support, movement, sensory perception, coordination and reproduction. For each of these, state the name of one different specific protein that is involved in each process and state its function.

A common riddle is: which came first, the chicken or the egg? A similar conundrum for scientists is: which came first, the enzyme needed to make the nucleic acid or the nucleic acid needed to make the enzyme?

It used to be thought that all enzymes were proteins but we now know that some reactions in cells are catalysed by non-protein molecules. Find out about these molecules and use the information to explain why these non-protein molecules provide an answer to the question, which came first – the enzyme or the nucleic acid?

1 (a) Some seeds contain lipids. Describe how you could use the emulsion test to show that a seed contains lipids. *(3 marks)*

   (b) A triglyceride is one type of lipid. The diagram shows the structure of a triglyceride molecule.

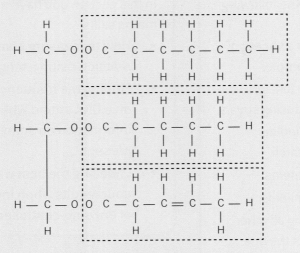

   (i)   A triglyceride molecule is formed by condensation. From how many molecules is this triglyceride formed? *(1 mark)*

   (ii) The structure of a phospholipid molecule is different from that of a triglyceride. Describe how a phospholipid is different. *(2 marks)*

   (iii) Use the diagram to explain what is meant by an unsaturated fatty acid. *(2 marks)*

   AQA Jan 2012

2 Read the following passage.

   Aspirin is a very useful drug. One of its uses is to reduce fever and inflammation. Aspirin does this by preventing cells from producing substances called prostaglandins. Prostaglandins are produced by an enzyme-controlled pathway. Aspirin works by inhibiting one of the enzymes in this pathway. Aspirin attaches permanently to a   5
   chemical group on one of the monomers that make up the active site of this enzyme.

   The enzyme that is involved in the pathway leading to the production of prostaglandins is also involved in the pathway leading to the production of thromboxane. This is a substance that promotes blood clotting. A small daily dose of aspirin may reduce the risk of   10
   myocardial infarction (heart attack).

   Use information from the passage and your own knowledge to answer the following questions

   (a) Name the monomers that make up the active site of the enzyme (lines 6 – 7). *(1 mark)*

   (b) The diagram shows the pathways by which prostaglandins and thromboxane are formed.

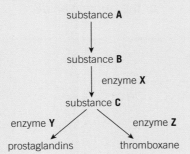

(i) Aspirin only affects one of the enzymes in this pathway. Use information in lines 5 – 7 to explain why aspirin does **not** affect the other enzymes. *(2 marks)*

(ii) Which enzyme, **X**, **Y** or **Z**, is inhibited by aspirin? Explain the evidence from the passage that supports your answer. *(2 marks)*

(c) Aspirin is an enzyme inhibitor. Explain how aspirin prevents substrate molecules being converted to product molecules. *(2 marks)*

(d) Aspirin may reduce the risk of myocardial infarction (lines 8 – 12). Explain how. *(3 marks)*

AQA Jan 2012

3 The diagram shows part of a DNA molecule

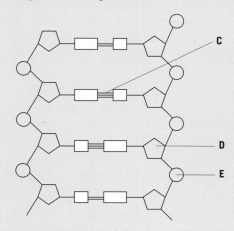

(a) (i)   DNA is a polymer. What is the evidence from the diagram that DNA is a polymer? *(1 mark)*

(ii) Name the parts of the diagram labelled **C**, **D** and **E**. *(3 marks)*

(iii) In a piece of DNA, 34% of the bases were thymine.

Complete the table to show the names and percentages of the other bases.

Name of base	Percentage
Thymine	34
	34

(2 marks)

(b) A polypeptide has 51 amino acids in its primary structure.

(i)   What is the minimum number of DNA bases required to code for the amino acids in this polypeptide? *(1 mark)*

AQA Jun 2012

Section 2
Cells

Introduction

The cell is the fundamental unit of life. All organisms, whatever their type or size, are composed of cells. All new cells are derived from existing ones by one of the following the processes of binary fission (prokaryotic cells), mitosis and meiosis (eukaryotic cells). Cells contain the genetic material of an organism and metabolic processes take place within them.

Cells all share certain basic features and yet show remarkable diversity in both structure and function. Their differences are the result of additional features that have arisen over time. This provides indirect evidence for evolution. Eukaryotic cells have a nucleus, or nuclear region, at some stage of their existence. Cells are all surrounded by a cell-surface membrane and in eukaryotic cells there are also internal membranes.

The structure of the plasma membrane is basically the same whether it forms the cell-surface membrane or any of the internal membranes within the cell. It controls the passage of substances across it by passive and active transport. The cell-surface membrane acts as the boundary between the cell and its environment. It may exclude some substances while retaining others. Some substances may pass freely across it, while others are prevented from doing so at one moment, only to pass freely across on another occasion.

The cell-surface membrane is made up almost entirely of proteins and phospholipids. Certain proteins that are embedded in the membrane are involved in communication between cells (cell signalling) while others act as antigens, allowing the cell to be recognised by the immune system as either 'self' or 'non-self' (foreign). Antigens play an important role in defence against disease and immunity. Various cells such as T lymphocytes and B lymphocytes interact to combat infections and to prevent symptoms arising when there are future infections by the same pathogen. These immune responses can be artificially induced by the process of vaccination.

Working scientifically

The study of cells gives plenty of scope to carry out practical work and develop practical skills. Required practical activities are:

- The preparation of stained squashes of cells from plant root tips and the setting up and use of an optical microscope to identify the stages of mitosis in these stained squashes. You will also be required to use your observations to calculate a mitotic index.

- The production of a dilution series of a solute to produce a calibration curve with which to identify the water potential of a plant tissue.

- An investigation into the effect of a named variable on the permeability of cell-surface membranes.

You will require a range of mathematical skills – in particular the ability to use percentages, make order of magnitude calculations, plot two variables from experimental data and draw and determine the intercept of a graph.

What you already know

While the material in this unit should be understood without much prior knowledge, there is certain information from GCSE that will prove helpful. This information includes:

- ☐ Most human and animal cells have a nucleus, cytoplasm, cell membrane, mitochondria and ribosomes.

- ☐ Plant and algal cells also have a cell wall made of cellulose, which strengthens the cell. Plant cells often have chloroplasts and a permanent vacuole filled with cell sap.

- ☐ A bacterial cell consists of cytoplasm and a membrane surrounded by a cell wall – the genes are not in a distinct nucleus.

- ☐ In body cells the chromosomes are generally found in pairs. Body cells divide by mitosis. When a body cell divides by mitosis, copies of the genetic material are made then the cell divides once to form two genetically identical body cells. Mitosis occurs during asexual reproduction or growth or to produce replacement cells.

- ☐ Diffusion is the net movement of molecules from a region where they are of a higher concentration to a region with a lower concentration.

- ☐ Osmosis is the diffusion of water from a dilute to a more concentrated solution through a selectively permeable membrane that allows the passage of water molecules.

- ☐ Substances are sometimes absorbed against a concentration gradient. This involves the use of ATP from respiration. The process is called active transport. Active transport enables cells to absorb ions from very dilute solutions.

- ☐ Microorganisms that cause infectious disease are called pathogens. White blood cells help to defend against pathogens by ingesting them, producing antibodies and antitoxins.

- ☐ The immune system of the body produces specific antibodies that lead to the death of a particular pathogen. This leads to immunity from that pathogen.

- ☐ People can be immunised against a disease by introducing small quantities of dead or inactive forms of the pathogen into the body (vaccination).

3 Cell structure
3.1 Methods of studying cells

Learning objectives

→ Explain the principles of magnification and resolution.

→ Describe what cell fractionation is.

→ Explain how ultracentrifugation works.

Specification reference: 3.2.1.3

Study tip

Make sure that you use scientific terms correctly. For example, light has a longer wavelength than a beam of electrons. It's not correct to say that optical microscopes have a longer wavelength than electron microscopes though.

Maths link $\sqrt{x}$

MS 1.8, 0.1, 0.2 and 2.2, see Chapter 11.

▼ **Table 1** *Units of length*

Unit	Symbol	Equivalent in metres
kilometre	km	10^3
metre	m	1
millimetre	mm	10^{-3}
micrometre	μm	10^{-6}
nanometre	nm	10^{-9}

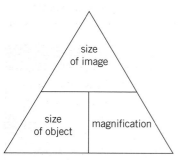

▲ **Figure 1** *The equation triangle for calculating the size of image, magnification and size of object*

The cell is the basic unit of life. However, with a few exceptions, cells are not visible to the naked eye and their structure is only apparent when seen under a microscope.

Microscopy 🔬

Microscopes are instruments that produce a magnified image of an object. A simple convex glass lens can act as a magnifying glass but such lenses work more effectively if they are used in pairs in a compound light microscope. The relatively long wavelength of light rays means that a light microscope can only distinguish between two objects if they are 0.2 μm, or further, apart. This limitation can be overcome by using beams of **electrons** rather than beams of light. With their shorter wavelengths, the beam of electrons in the electron microscope can distinguish between two objects only 0.1 nm apart.

Magnification $\sqrt{x}$

The material that is put under a microscope is referred to as the **object**. The appearance of this material when viewed under the microscope is referred to as the **image**.

The magnification of an object is how many times bigger the image is when compared to the object.

$$\text{magnification} = \frac{\text{size of image}}{\text{size of real object}}$$

In practice, it is more likely that you will need to calculate the size of an object when you know the size of the image and the magnification. In this case:

$$\text{size of real object} = \frac{\text{size of image}}{\text{magnification}}$$

The important thing to remember when calculating the magnification is to ensure that the units of length (Table 1) are the same for both the object and the image.

Worked example

An object that measures 100 nm in length appears 10 mm long in a photograph. What is the magnification of the object?

$$\frac{\text{size of image}}{\text{size of real object}} = \frac{10\,\text{mm}}{100\,\text{nm}}$$

Now convert the measurements to the same units – normally the smallest – which in this case is nanometres. There are 10 000 000 nanometres in 10 millimetres and therefore the magnification is:

$$\frac{\text{size of image}}{\text{size of real object}} = \frac{10\,000\,000\,\text{nm}}{100\,\text{nm}} = \frac{100\,000}{1} = \times 100\,000 \text{ times}$$

These figures can also be expressed in standard form as follows:

$$\frac{\text{size of image}}{\text{size of real object}} = \frac{10^7}{10^2} = \frac{10^5}{1} = \times 10^5$$

Resolution

The resolution, or resolving power, of a microscope is the minimum distance apart that two objects can be in order for them to appear as separate items. Whatever the type of microscope, the resolving power depends on the wavelength or form of radiation used. In a light microscope it is about 0.2 μm. This means that any two objects which are 0.2 μm or more apart will be seen separately, but any objects closer than 0.2 μm will appear as a single item. In other words, greater resolution means greater clarity, that is the image produced is clearer and more precise.

Increasing the magnification increases the size of an image, but does not always increase the resolution. Every microscope has a limit of resolution. Up to this point increasing the magnification will reveal more detail but beyond this point increasing the magnification will not do this. The object, while appearing larger, will just be more blurred.

Cell fractionation 🧪

In order to study the structure and function of the various organelles that make up cells, it is necessary to obtain large numbers of isolated organelles.

Cell fractionation is the process where cells are broken up and the different organelles they contain are separated out.

Before cell fractionation can begin, the tissue is placed in a cold, buffered solution of the same water potential as the tissue. The solution is:

- cold – to reduce enzyme activity that might break down the organelles
- is of the same water potential as the tissue – to prevent organelles bursting or shrinking as a result of osmotic gain or loss of water
- buffered – so that the pH does not fluctuate. Any change in pH could alter the structure of the organelles or affect the functioning of enzymes.

There are two stages to cell fractionation:

Homogenation

Cells are broken up by a homogeniser (blender).This releases the organelles from the cell. The resultant fluid, known as homogenate, is then filtered to remove any complete cells and large pieces of debris.

Ultracentrifugation

Ultracentrifugation is the process by which the fragments in the filtered homogenate are separated in a machine called a centrifuge. This spins tubes of homogenate at very high speed in order to create a centrifugal force. For animal cells, the process is as follows:

- The tube of filtrate is placed in the centrifuge and spun at a slow speed.
- The heaviest organelles, the nuclei, are forced to the bottom of the tube, where they form a thin sediment or pellet.
- The fluid at the top of the tube (supernatant) is removed, leaving just the sediment of nuclei.
- The supernatant is transferred to another tube and spun in the centrifuge at a faster speed than before.

Hint

Practise working out actual sizes from diagrams and photographs with a given scale. Practice makes it easy.

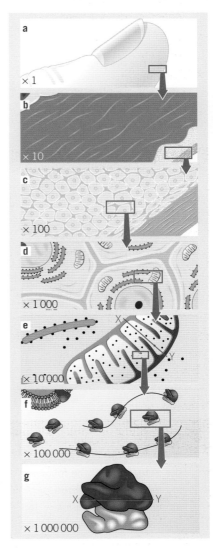

▲ **Figure 2** *The effect of progressive magnification of a portion of human skin*

Study tip

Remember that the solution used during cell fractionation prevents *organelles* bursting or shrinking as a result of osmotic gain or loss of water. Don't refer to *cells* bursting or shrinking. This is a common error!

Summary questions

1 Distinguish between magnification and resolution.

2 $\sqrt{x}$ An organelle that is 5 μm in diameter appears under a microscope to have a diameter of 1 mm. Calculate how many times the organelle has been magnified.

3 $\sqrt{x}$ A cell organelle called a ribosome is typically 25 nm in diameter. Calculate its diameter when viewed under an electron microscope that magnifies it 400 000 times.

4 $\sqrt{x}$ At a magnification of ×12 000 a structure appears to be 6 mm long. Determine its actual length.

5 ⚠ Chloroplasts have a greater mass than mitochondria but a smaller mass than nuclei. Starting with a sample of plant cells, describe briefly how you would obtain a sample rich in chloroplasts. Use Figure 3 to help you.

6 $\sqrt{x}$ Using the magnifications given in Figure 2, calculate the actual size of the following organelles as measured along the line labelled X−Y. In your answer, use the most appropriate units from Table 1.
  a The organelle in box e
  b The organelle in box g

- The next heaviest organelles, the mitochondria, are forced to the bottom of the tube.
- The process is continued in this way so that, at each increase in speed, the next heaviest organelle is sedimented and separated out.

A summary of cell fractionation is given in Figure 3.

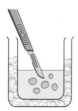

1. Tissue is cut up and kept in a cold, buffered solution.

2. Cut-up tissue further broken up in a homogeniser.

3. Homogenised tissue is spun in an ultracentrifuge at a low speed for 10 minutes.

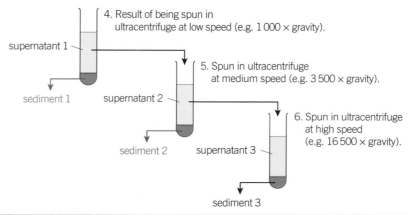

supernatant 1

4. Result of being spun in ultracentrifuge at low speed (e.g. 1 000 × gravity).

5. Spun in ultracentrifuge at medium speed (e.g. 3 500 × gravity).

sediment 1   supernatant 2

6. Spun in ultracentrifuge at high speed (e.g. 16 500 × gravity).

sediment 2   supernatant 3

sediment 3

Organelles to be separated out	Speed of centrifugation /revolutions min^{-1}
nuclei	1 000
mitochondria	3 500
lysosomes	16 500

▲ **Figure 3** *Summary of cell fractionation*

The techniques of cell fractionation and ultracentrifugation enabled considerable advances in biological knowledge. They allowed a detailed study of the structure and function of organelles, by showing what isolated components do.

◀ **Figure 4** *An ultracentrifuge used to separate the various components of cell homogenate*

Maths link $\sqrt{x}$

MS 1.8 and 2.2, see Chapter 11.

Light microscopes have poor resolution as a result of the relatively long wavelength of light. In the 1930s, however, a microscope was developed that used a beam of electrons instead of light. This is called an electron microscope and it has two main advantages:

- The electron beam has a very short wavelength and the microscope can therefore resolve objects well – it has a high resolving power.
- As electrons are negatively charged the beam can be focused using electromagnets (Figure 2).

The best modern electron microscopes can resolve objects that are just 0.1 nm apart – 2000 times better than a light microscope. Because electrons are absorbed or deflected by the molecules in air, a near-vacuum has to be created within the chamber of an electron microscope in order for it to work effectively.

Learning objectives

→ Explain how electron microscopes work.

→ Explain the differences between a transmission electron microscope and a scanning electron microscope.

→ Describe the limitations of the transmission and the scanning electron microscopes.

Specification reference: 3.2.1.3

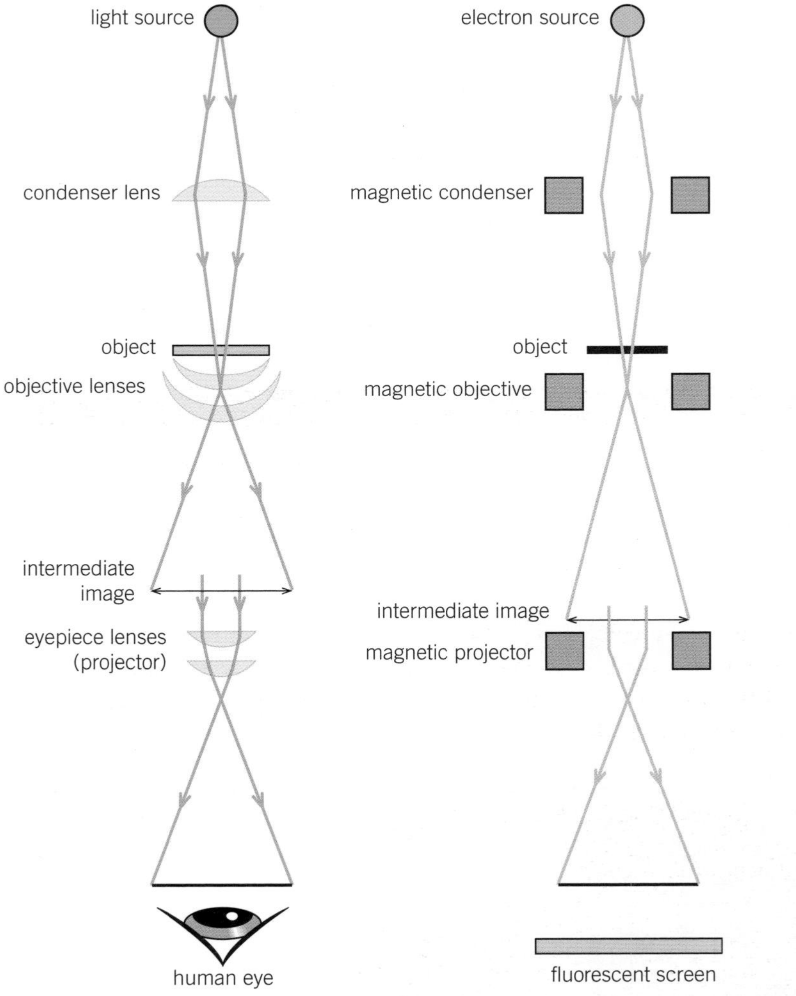

light source

condenser lens

object
objective lenses

intermediate
image

eyepiece lenses
(projector)

human eye

electron source

magnetic condenser

object
magnetic objective

intermediate image

magnetic projector

fluorescent screen

▲ **Figure 2** *Comparison of radiation pathways in light and electron microscopes*

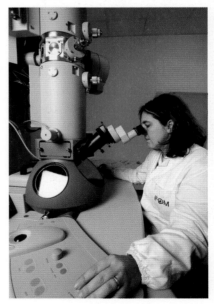

▲ **Figure 1** *Scientist looking at a sample using a transmission electron microscope (TEM)*

There are two types of electron microscope:

- the transmission electron microscope (TEM)
- the scanning electron microscope (SEM).

The transmission electron microscope

The TEM consists of an electron gun that produces a beam of electrons that is focused onto the specimen by a condenser electromagnet. In a TEM, the beam passes through a thin section of the specimen. Parts of this specimen absorb electrons and therefore appear dark. Other parts of the specimen allow the electrons to pass through and so appear bright. An image is produced on a screen and this can be photographed to give a **photomicrograph**. The resolving power of the TEM is 0.1 nm although this cannot always be achieved in practice because:

- difficulties preparing the specimen limit the resolution that can be achieved
- a higher energy electron beam is required and this may destroy the specimen.

The main limitations of the TEM are:

- The whole system must be in a vacuum and therefore living specimens cannot be observed.
- A complex 'staining' process is required and even then the image is not in colour.
- The specimen must be extremely thin.
- The image may contain artefacts. Artefacts are things that result from the way the specimen is prepared. Artefacts may appear on the finished photomicrograph but are not part of the natural specimen. It is therefore not always easy to be sure that what we see on a photomicrograph really exists in that form.

In the TEM the specimens must be extremely thin to allow electrons to penetrate. The result is therefore a flat, 2-D image. We can partly get over this by taking a series of sections through a specimen. We can then build up a 3-D image of the specimen by looking at the series of

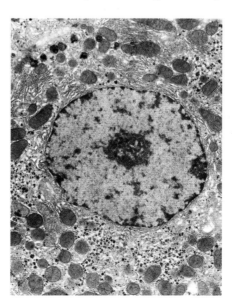

◀ **Figure 3** *Part of an animal cell seen under a TEM*

photomicrographs produced. However, this is a slow and complicated process. One way in which this problem has been overcome is the development of the SEM.

The scanning electron microscope

All the limitations of the TEM also apply to the SEM, except that specimens need not be extremely thin as electrons do not penetrate. Basically similar to a TEM, the SEM directs a beam of electrons on to the surface of the specimen from above, rather than penetrating it from below. The beam is then passed back and forth across a portion of the specimen in a regular pattern. The electrons are scattered by the specimen and the pattern of this scattering depends on the contours of the specimen surface. We can build up a 3-D image by computer analysis of the pattern of scattered electrons and secondary electrons produced. The basic SEM has a lower resolving power than a TEM, around 20 nm, but is still ten times better than a light microscope.

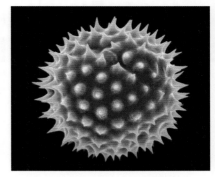
▲ **Figure 4** *False-colour (SEM) of a pollen grain from a marigold plant*

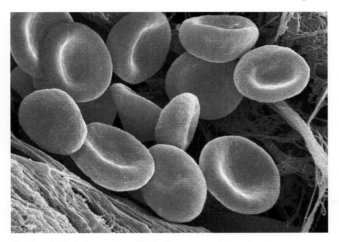
▲ **Figure 5** *False-colour SEM of human red blood cells*

Summary questions

1 Explain how the electron microscope is able to resolve objects better than the light microscope.

2 Explain why specimens have to be kept in a near-vacuum in order to be viewed effectively using an electron microscope.

3 State which of the biological structures in the following list can be resolved using each of the microscopes below:
plant cell (100 μm)   DNA molecule (2 nm)   virus (100 nm)
actin molecule (3.5 nm)   a bacterium (1 μm)

   a a light microscope

   b a transmission electron microscope

   c a scanning electron microscope.

4 In practice, the theoretical resolving power of an electron microscope cannot always be achieved. Explain why not.

5 ⓥ In a photomicrograph, an organelle measures 25 mm when its actual size is 5 μm. Calculate the magnification of this photomicrograph.

Maths link √x̄

MS 1.8, see Chapter 11.

Maths link √x̄

MS 2.2 and 1.8, see Chapter 11.

Learning outcomes

→ Explain how to calibrate an eyepiece graticule.

→ Explain how to measure cell size using an eyepiece graticule.

→ Learn how to calculate the size of a specimen and/or magnifications from drawings and photographs.

Specification reference: 3.2.1.3

Measuring cells

When using a light microscope, we can measure the size of objects using an **eyepiece graticule**. The graticule is a glass disc that is placed in the eyepiece of a microscope. A scale is etched on the glass disc. This scale is typically 10 mm long and is divided into 100 sub-divisions as shown in Figure 1. The scale is visible when looking down the eyepiece of the microscope.

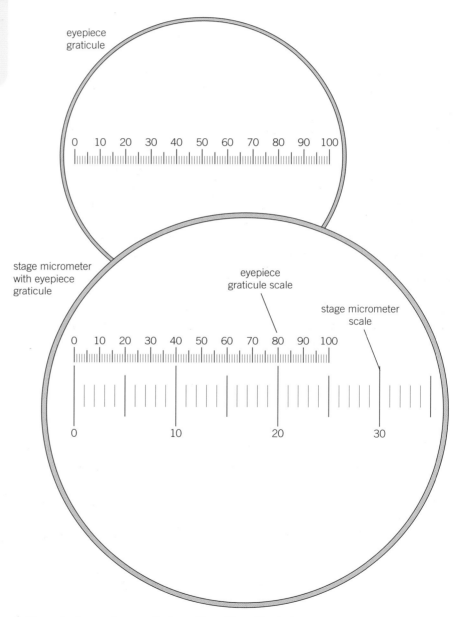

▲ **Figure 1** *An eyepiece graticule and how it is calibrated*

The scale on the eyepiece graticule cannot be used directly to measure the size of objects under a microscope's objective lens because each objective lens will magnify to a different degree. The graticule must first be calibrated for a particular objective lens. Once calibrated in this way, the graticule can remain in position for future use, provided the same objective lens is used.

It is therefore sensible to record the results of the calibration for a particular objective lens and to leave this attached to the microscope. This will save you having to recalibrate each time you want to measure the size of the object being viewed under the microscope.

Calibrating the eyepiece graticule

To calibrate an eyepiece graticule you need to use a special microscope slide called a **stage micrometer**. This slide also has a scale etched onto it. Usually the scale is 2 mm long and its smallest sub-divisions are 0.01 mm (10 μm).

When the eyepiece graticule scale and the stage micrometer scales are lined-up as shown in Figure 1, it is possible to calculate the length of the divisions on the eyepiece graticule. For example, you can see in Figure 1, that:

- 10 units on the micrometer scale are equivalent to 40 units on the graticule scale

- therefore 1 unit on the micrometer scale equals 4 units on the graticule scale

- as each unit on the micrometer scale equals 10 μm, each unit on the graticule equals 10 ÷ 4 = 2.5 μm.

It is easy to calculate the scale for different objective lenses by dividing the differences in magnification. For example, if an objective lens magnifying ×40 gives a calibration of 25 μm per graticule unit, then an objective lens magnifying ×400 (10 times greater) will mean a graticule unit is equivalent to 25 μm ÷ 10 = 2.5 μm.

Calculating linear magnifications of drawings and photographs

You may need to calculate the magnification of a drawing or photograph of an object under a microscope. You have looked at how the calculation is made. Now try an example using an actual photograph.

Look at Figure 2 of part of an animal cell seen under a transmission electron microscope. On this photograph a red line X—Y is marked. This line represents a length of 5 μm of the actual cell. Using this information, calculate the magnification of the photograph as follows:

- If you measure the length of X—Y as drawn on the photograph you find it is 23 mm long.

- As the line represents 5 μm in the cell, you also need to convert your measurement in the photograph to microns (μm). 23 μm = 23 000 μm.

- If 23 000 μm on the photograph is equivalent to 5 μm in the cell, then the magnification must be 23 000 ÷ 5 = 4600 times.

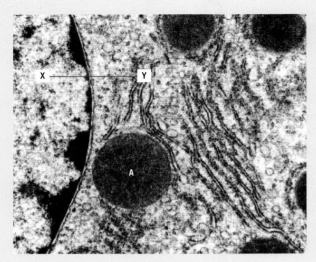

▲ Figure 2 *Part of an animal cell seen under a TEM.*

Calculating actual sizes of specimens from drawings and photographs

You have seen how the size of an object can be calculated when you know the magnification and the image size. You can see how this is done again using a photograph. In Figure 2, the magnification is ×4600. Calculate the actual size of a mitochondrion shown in the photo.

This mitochondrion is labelled A. To calculate its actual size:

* Measure the diameter of mitochondrion A. As it is not truly spherical you need to calculate the mean of a number of different diameters. The mean is 20 mm.

* The size of the image is 20 mm (20 000 μm).

* The actual size of the mitochondrion equals the size of the image ÷ magnification or 20 000 μm ÷ 4 600 = 4.3 μm.

Maths link $\sqrt{x}$

MS 1.2 and 1.8, see Chapter 11.

Maths link $\sqrt{x}$

MS 1.8, see Chapter 11.

Summary questions

In the following passage, state the missing word indicated by each letter **a–h**.

To measure the size of an object under a **a** microscope you can use an **b** graticule and a **c** micrometer. Before you can use the graticule to measure the size of objects it must first be **d**. To do this you line up the scale on the eyepiece with that on the micrometer using an objective lens that magnifies 400 times. Suppose this shows that 50 graticule units are equivalent to 10 micrometer units. If each micrometer unit is 10 μm, then each graticule unit equals **e** μm. If an objective lens magnifying 100 times is used, each graticule unit would be equivalent to **f** μm. A photograph of a cell under an electron microscope is magnified 5000 times. On the photograph the nucleus measures 100 mm in diameter. The actual size of the nucleus is therefore **g** μm. A chloroplast that is 5 μm in diameter measures 15 mm in a drawing made of a plant cell as seen under a microscope. The magnification of this drawing is therefore **h** times.

Each cell can be regarded as a metabolic compartment, a separate place where the chemical processes of that cell occur. Cells are often adapted to perform a particular function. Depending on that function, each cell type has an internal structure that suits it for its job. This is known as the **ultrastructure** of the cell. **Eukaryotic** cells have a distinct nucleus and possess membrane-bounded organelles. They differ from prokaryotic cells, such as bacteria. More details of these differences are given in Topic 3.6. Using an electron microscope, we can see the structure of organelles within cells, details of which are described below. The most important of these organelles are described below, with the exception of the cell-surface membrane.

The nucleus

The nucleus (Figure 1) is the most prominent feature of a eukaryotic cell, such as an epithelial cell. The nucleus contains the organism's hereditary material and controls the cell's activities. Usually spherical and between 10 and 20 µm in diameter, the nucleus has a number of parts.

- The **nuclear envelope** is a double membrane that surrounds the nucleus. Its outer membrane is continuous with the endoplasmic reticulum of the cell and often has ribosomes on its surface. It controls the entry and exit of materials in and out of the nucleus and contains the reactions taking place within it.
- **Nuclear pores** allow the passage of large molecules, such as messenger RNA, out of the nucleus. There are typically around 3000 pores in each nucleus, each 40–100 nm in diameter.
- **Nucleoplasm** is the granular, jelly-like material that makes up the bulk of the nucleus.
- **Chromosomes** consist of protein-bound, linear DNA.
- The **nucleolus** is a small spherical region within the nucleoplasm. It manufactures ribosomal RNA and assembles the ribosomes. There may be more than one nucleolus in a nucleus.

The functions of the nucleus are to:

- act as the control centre of the cell through the production of mRNA and tRNA and hence protein synthesis (see Topic 8.4)
- retain the genetic material of the cell in the form of DNA and chromosomes
- manufacture ribosomal RNA and ribosomes.

The mitochondrion

Mitochondria (Figures 2 and 3) are usually rod-shaped and 1–10 µm in length. They are made up of the following structures:

- Around the organelle is a **double membrane** that controls the entry and exit of material. The inner of the two membranes is folded to form extensions known as cristae.

Learning objectives

→ Describe the structure and functions of the nucleus, mitochondria, chloroplasts, rough and smooth endoplasmic reticulum, Golgi apparatus, Golgi vesicles and lysosomes.

→ Describe the structure and function of the cell wall in plants, algae and fungi.

→ Describe the structure and function of the cell vacuole in plants.

Specification reference: 3.2.1.1

Hint

When you look at a group of animal cells, such as epithelial cells, under a light microscope you cannot see the cell-surface membrane because it is too thin to be observed. What you actually see is the boundary between cells.

Synoptic link

The cell-surface membrane is covered in Topic 4.1, and DNA is covered in Topics 2.1, 2.2 and 8.2.

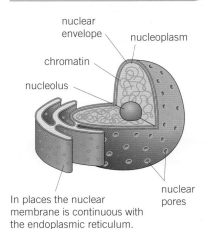

In places the nuclear membrane is continuous with the endoplasmic reticulum.

▲ **Figure 1** *The nucleus*

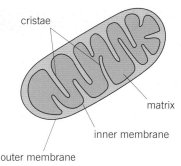

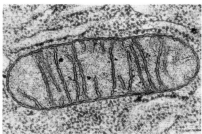

▲ **Figure 2** *The basic structure of a mitochondrion (top); false-colour TEM of a mitochondrion (bottom)*

- **Cristae** are extensions of the inner membrane, which in some species extend across the whole width of the mitochondrion. These provide a large surface area for the attachment of enzymes and other proteins involved in respiration.
- The **matrix** makes up the remainder of the mitochondrion. It contains protein, lipids, ribosomes and DNA that allows the mitochondria to control the production of some their own proteins. Many enzymes involved in respiration are found in the matrix.

Mitochondria are the sites of the aerobic stages of respiration (the Krebs cycle and the oxidative phosphorylation pathway. They are therefore responsible for the production of the energy-carrier molecule, ATP, from respiratory substrates such as glucose. Because of this, the number and size of the mitochondria, and the number of their cristae, are high in cells that have a high level of metabolic activity and therefore require a plentiful supply of ATP. Examples of metabolically active cells include muscle and epithelial cells. Epithelial cells in the intestines require a lot of ATP in the process of absorbing substances from the intestines by active transport.

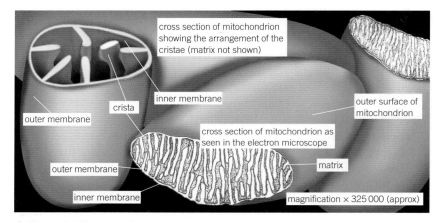

▲ **Figure 3** *Mitochondria*

Chloroplasts

Chloroplasts (Figure 4) are the organelles that carry out photosynthesis (see Topic 11.2). They vary in shape and size but are typically disc-shaped, 2–10 μm long and 1 μm in diameter. The following are their main features:

- **The chloroplast envelope** is a double plasma membrane that surrounds the organelle. It is highly selective in what it allows to enter and leave the chloroplast.
- **The grana** are stacks of up to 100 disc-like structures called **thylakoids**. Within the thylakoids is the photosynthetic pigment called **chlorophyll**. Some thylakoids have tubular extensions that join up with thylakoids in adjacent grana. The grana are where the first stage of photosynthesis (light absorption) takes place.
- **The stroma** is a fluid-filled matrix where the second stage of photosynthesis (synthesis of sugars) takes place. Within the stroma are a number of other structures, such as starch grains.

Chloroplasts are adapted to their function of harvesting sunlight and carrying out photosynthesis in the following ways:

Hint

Chloroplasts have DNA and may have evolved from free-living prokaryotic cells, but they are organelles, not cells.

- The granal membranes provide a large surface area for the attachment of chlorophyll, electron carriers and enzymes that carry out the first stage of photosynthesis. These chemicals are attached to the membrane in a highly ordered fashion.
- The fluid of the stroma possesses all the enzymes needed to make sugars in the second stage of photosynthesis.
- Chloroplasts contain both DNA and ribosomes so they can quickly and easily manufacture some of the proteins needed for photosynthesis.

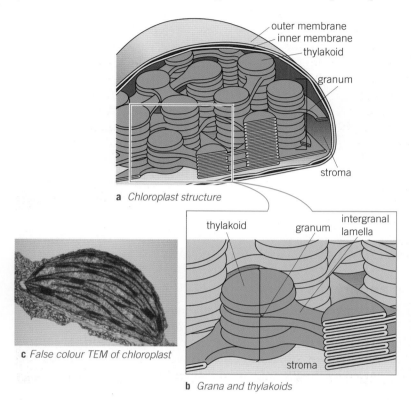

a Chloroplast structure

c False colour TEM of chloroplast

b Grana and thylakoids

▲ **Figure 4** Chloroplast structure

Endoplasmic reticulum

The endoplasmic reticulum (ER) is an elaborate, three-dimensional system of sheet-like membranes, spreading through the cytoplasm of the cells. It is continuous with the outer nuclear membrane. The membranes enclose a network of tubules and flattened sacs called cisternae (see Figure 5). There are two types of ER:

- **Rough endoplasmic reticulum (RER)** has ribosomes present on the outer surfaces of the membranes. Its functions are to:
  - **a** provide a large surface area for the synthesis of proteins and glycoproteins
  - **b** provide a pathway for the transport of materials, especially proteins, throughout the cell.

- **Smooth endoplasmic reticulum (SER)** lacks ribosomes on its surface and is often more tubular in appearance. Its functions are to:
  - **a** synthesise, store and transport lipids
  - **b** synthesise, store and transport carbohydrates.

protein-containing vesicle from rough endoplasmic reticulum transferring substances to the Golgi apparatus

primary lysosomes formed by Golgi apparatus

hydrolytic enzymes

particle to be broken down

vesicle (phagosome) formed from phagocytosis of bacterium or membrane formed around worn-out organelle

enzymes hydrolyse the particle

soluble products are absorbed into cytoplasm

cell-surface membrane of cell

insoluble debris is egested

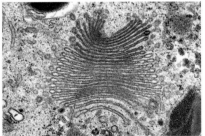

▲ **Figure 6** *The Golgi apparatus and the formation and functioning of a lysosome (top); false-colour TEM of a Golgi apparatus (orange) (bottom)*

It follows that cells that manufacture and store large quantities of carbohydrates, proteins and lipids have a very extensive ER. Such cells include liver and secretory cells, for example the epithelial cells that line the intestines.

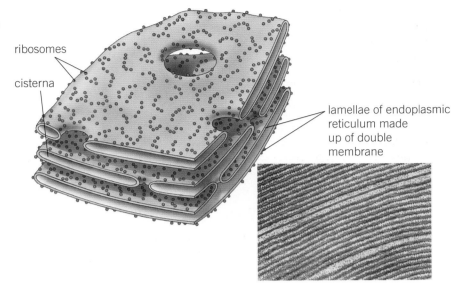

ribosomes

cisterna

lamellae of endoplasmic reticulum made up of double membrane

▲ **Figure 5** *Structure of RER (above); false-colour TEM of a section through RER (RER; red) (right)*

Golgi apparatus

The Golgi apparatus occurs in almost all eukaryotic cells and is similar to the SER in structure except that it is more compact. It consists of a stack of membranes that make up flattened sacs, or **cisternae**, with small rounded hollow structures called vesicles. The proteins and lipids produced by the ER are passed through the Golgi apparatus in strict sequence. The Golgi modifies these proteins often adding non-protein components, such as carbohydrate, to them. It also 'labels' them, allowing them to be accurately sorted and sent to their correct destinations. Once sorted, the modified proteins and lipids are transported in Golgi vesicles which are regularly pinched off from the ends of the Golgi cisternae (Figure 6). These vesicles may move to the cell surface, where they fuse with the membrane and release their contents to the outside.

The functions of the Golgi apparatus are to:

- add carbohydrate to proteins to form glycoproteins
- produce secretory enzymes, such as those secreted by the pancreas
- secrete carbohydrates, such as those used in making cell walls in plants
- transport, modify and store lipids
- form lysosomes.

The Golgi apparatus is especially well developed in secretory cells, such as the epithelial cells that line the intestines.

Lysosomes

Lysosomes are formed when the vesicles produced by the Golgi apparatus contain enzymes such as proteases and lipases. They also contain lysozymes, enzymes that hydrolyse the cell walls of certain bacteria. As many as 50 such enzymes may be contained in a single lysosome. Up to 1.0 μm in diameter, lysosomes isolate these enzymes from the rest of the cell before releasing them, either to the outside or into a **phagocytic** vesicle within the cell (Figure 6).

The functions of lysosomes are to:

- hydrolyse material ingested by phagocytic cells, such as white blood cells and bacteria
- release enzymes to the outside of the cell (exocytosis) in order to destroy material around the cell
- digest worn out organelles so that the useful chemicals they are made of can be re-used
- completely break down cells after they have died (autolysis).

Given the roles that lysosomes perform, it is not surprising that they are especially abundant in secretory cells, such as epithelial cells, and in **phagocytic** cells.

Ribosomes

Ribosomes are small cytoplasmic granules found in all cells. They may occur in the cytoplasm or be associated with the RER. There are two types, depending on the cells in which they are found:

- **80S** – found in eukaryotic cells, is around 25 nm in diameter.
- **70S** – found in prokaryotic cells, mitochondria and chloroplasts, is slightly smaller.

Ribosomes have two subunits – one large and one small (Figure 7) – each of which contains ribosomal RNA and protein. Despite their small size, they occur in such vast numbers that they can account for up to 25 % of the dry mass of a cell. Ribosomes are the site of in protein synthesis.

Cell wall

Characteristic of all plant cells, the cell wall consists of microfibrils of the polysaccharide cellulose, embedded in a matrix. Cellulose microfibrils have considerable strength and so contribute to the overall strength of the cell wall. Cell walls have the following features:

- They consist of a number of polysaccharides, such as cellulose.
- There is a thin layer, called the **middle lamella**, which marks the boundary between adjacent cell walls and cements adjacent cells together.

The functions of the cellulose cell wall are:

- to provide mechanical strength in order to prevent the cell bursting under the pressure created by the osmotic entry of water

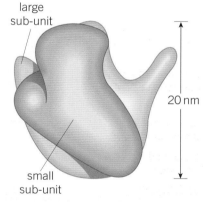

large sub-unit

20 nm

small sub-unit

▲ **Figure 7** *Structure of a ribosome*

> **Hint**
>
> To help you understand the functions of the Golgi apparatus, think of it as the cell's post office, but receiving, sorting and delivering proteins and lipids, rather than letters.

> **Hint**
>
> Lysosomes can be thought of as refuse disposal operatives. They remove useless and potentially dangerous material (e.g., bacteria) and reuse the useful parts, disposing of only that which cannot be recycled.

> **Synoptic link**
>
> Look back to Topic 1.4, to refresh your knowledge of cellulose. Osmosis will be covered in Topic 4.3.

> **Study tip**
>
> Plant cells have a cell-surface membrane *and* a cell wall, not just a cell wall.

- to give mechanical strength to the plant as a whole
- to allow water to pass along it and so contribute to the movement of water through the plant.

The cell walls of algae are made up of either cellulose or glycoproteins, or a mixture of both.

The cell walls of fungi do not contain cellulose but comprise a mixture of a nitrogen-containing polysaccharide called **chitin**, a polysaccharide called glycan and glycoproteins.

Vacuoles

A fluid-filled sac bounded by a single membrane may be termed a vacuole. Within mature plant cells there is usually one large central vacuole. The single membrane around it is called the **tonoplast**. A plant vacuole contains a solution of mineral salts, sugars, amino acids, wastes and sometimes pigments such as anthocyanins.

Plant vacuoles serve a variety of functions:

- They support herbaceous plants, and herbaceous parts of woody plants, by making cells turgid.
- The sugars and amino acids may act as a temporary food store.
- The pigments may colour petals to attract pollinating insects.

Relating cell ultrastructure to function

As each organelle has its own function, it is possible to deduce, with reasonable accuracy, the role of a cell by looking at the number and size of the organelles it contains. For example, as mitochondria produce ATP that is used as a temporary energy store, it follows that cells with many mitochondria are likely to require a lot of ATP and therefore have a high rate of metabolism. Even within each mitochondrion, the more dense and numerous the cristae, the greater the metabolic rate of the cell possessing these mitochondria.

Summary questions

1 State in which process ribosomes are important.

2 List **three** carbohydrates that are absorbed by an epithelial cell of the small intestine.

3 State the organelle that is being referred to in each of the following descriptions:

   a It possesses structures called cristae.

   b It contains chromatin.

   c It synthesises glycoproteins.

   d It digests worn out organelles.

4 The following list gives a type of cell and a brief description of its role. Suggest *two* organelles that might be numerous and/or well developed in each of the cells.

   a A sperm cell swims a considerable distance carrying the male chromosomes.

   b One type of white blood cell engulfs and digests foreign material.

   c Liver cells manufacture proteins and lipids at a rapid rate.

In multicellular organisms, cells are specialised to perform specific functions. Similar cells are then grouped together into tissues, tissues into organs and organs into systems for increased efficiency.

Cell specialisation

To stay alive, all cells of a multicellular organism perform certain basic functions. However, no one cell can provide the best conditions for all functions. Therefore the cells of multicellular organisms are each specialised in different ways to perform a particular role. Each specialised cell has evolved more or fewer of certain organelles and structures to suit the role it carries out.

The first group of cells in an embryo are all initially identical. As it matures, each cell takes on its own individual characteristics that suit it to the function that it will perform when it is mature. In other words, each cell becomes specialised in structure to suit the role that it will carry out.

All the cells in an organism, such as a human, are produced by mitotic divisions from the fertilised egg. It follows that they all contain exactly the same **genes**. How then does the cell become specialised? Every cell contains the genes needed for it to develop into any one of the many different cells in an organism. But only some of these genes are switched on (expressed) in any one cell, at any one time. Different genes are switched on in each type of specialised cell. The rest of the genes are switched off.

It is not just the shape of different cells that varies, but also the numbers of each of their organelles. For example, a muscle or sperm cell will have many mitochondria, while a bone cell has very few. White blood cells have many lysosomes while a muscle cell has very few.

The cells of a multicellular organism have therefore evolved to become more and more suited to one specialised function. These cells are adapted to their own particular function and perform it more effectively. As a result, the whole organism functions efficiently.

Tissues

For working efficiency, cells are normally aggregated together. Such a collection of similar cells that perform a specific function is known as a **tissue**. Examples of tissues include:

- **epithelial tissues** (see Topic 4.5), which are found in animals and consist of sheets of cells. They line the surfaces of organs and often have a protective or secretory function. There are many similar types, including those made up of thin, flat cells that line organs where diffusion takes place, for example the alveoli of the lungs (see Topic 6.8), and ciliated epithelium that lines a duct such as the trachea (see Topic 6.6). The cilia are used to move mucus over the epithelial surface.

Learning objectives

→ Discuss the advantages of cellular differentiation.

→ Describe how cells are arranged into tissues.

→ Describe how tissues are arranged into organs.

→ Describe how organs are arranged into organ systems.

Specification reference: 3.2.1.1

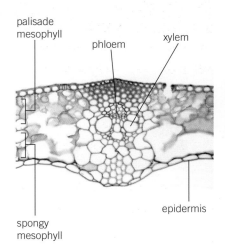

palisade mesophyll

phloem

xylem

epidermis

spongy mesophyll

▲ **Figure 1** *Some of the various tissues that make up the organ called a leaf*

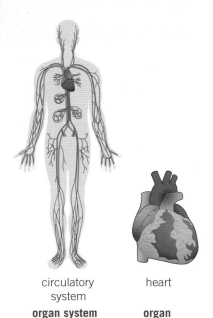

circulatory
system

organ system

heart

organ

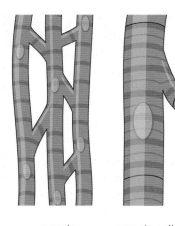

muscle

tissue

muscle cell

cell

▲ **Figure 2** *The circulatory system as an example of an organ system*

- **xylem** (see Topic 7.8), which occurs in plants and is made up of a number of similar cell types. It is used to transport water and mineral ions throughout the plant and also gives mechanical support.

Organs

Just as cells are aggregated into tissues, so tissues are aggregated into organs. An organ is a combination of tissues that are coordinated to perform a variety of functions, although they often have one predominant major function. In animals, for example, the stomach is an organ that is involved in the digestion of certain types of food. It is made up of tissues such as:

- muscle to churn and mix the stomach contents
- epithelium to protect the stomach wall and produce secretions
- connective tissue to hold together the other tissues.

In plants, a leaf (Figure 1) is an organ made up of the following tissues:

- palisade mesophyll made up of leaf palisade cells that carry out photosynthesis
- spongy mesophyll adapted for gaseous diffusion
- epidermis to protect the leaf and allow gaseous diffusion
- phloem to transport organic materials away from the leaf
- xylem to transport water and ions into the leaf.

It is not always easy to determine which structures are organs. Blood capillaries, for example, are not organs whereas arteries and veins are both organs. All three structures have the same major function, namely the transport of blood. However, capillaries are made up of just one tissue – epithelium – whereas arteries and veins are made up of many tissues, for example, epithelial, muscle and other tissues.

Organ systems

Organs work together as a single unit known as an organ system. These systems may be grouped together to perform particular functions more efficiently. There are a number of organ systems in humans.

- The **digestive system** digests and processes food. It is made up of organs that include the salivary glands, oesophagus, stomach, duodenum, ileum, pancreas and liver.
- The **respiratory system** is used for breathing and gas exchange. It is made up of organs that include the trachea, bronchi and lungs.
- The **circulatory system** (Figure 2) pumps and circulates blood. It is made up of organs that include the heart, arteries and veins.

Summary questions

1   Explain what is meant by a tissue.

2   Explain why an artery is described as an organ whereas a blood capillary is not.

3   State whether each of the following is a tissue or an organ.

    a heart   **b** xylem   **c** lungs   **d** epithelium.

Although cells come in a diverse variety of size, shape and function, they are of two main types:

- **Eukaryotic cells** are larger and have a nucleus bounded by nuclear membranes (nuclear envelope).
- **Prokaryotic** cells are smaller and have no nucleus or nuclear envelope.

The structure of a generalised prokaryotic cell is shown in Figure 1. The differences between prokaryotic and eukaryotic cells are listed in Table 1.

Structure of a bacterial cell

Bacteria occur in every habitat in the world – they are versatile, adaptable and very successful. Much of their success is a result of their small size, normally ranging from 0.1 to10 μm in length. Their cellular structure is relatively simple (Figure 1). All bacteria possess a **cell wall**, which is made up of murein. This is a polymer of polysaccharides and peptides. Many bacteria further protect themselves by secreting a **capsule** of mucilaginous slime around this wall.

Inside the cell wall is the **cell-surface membrane**, within which is the cytoplasm that contains 70S ribosomes. These ribosomes are smaller than those in the cytoplasm of eukaryotic cells (80S), but nevertheless still synthesise proteins. Bacteria store food reserves as glycogen granules and oil droplets. The genetic material in bacteria is in the form of a **circular strand of DNA**. Separate from this are smaller circular pieces of DNA, called **plasmids**. These can reproduce themselves independently and may give the bacterium resistance to harmful chemicals, such as antibiotics. Plasmids are used extensively as vectors (carriers of genetic information) in genetic engineering. The roles of the main structures in a bacterial cell are summarised in Table 2.

Learning objectives

→ Describe the structure of prokaryotic cells.

→ Distinguish prokaryotic cells from eukaryotic ones.

Specification reference: 3.2.1.2

▼ **Table 2** *Roles of structures found in a bacterial cell*

Cell structure	Role
cell wall	physical barrier that excludes certain substances and protects against mechanical damage and osmotic lysis
capsule	protects bacterium from other cells and helps groups of bacteria to stick together for further protection
cell-surface membrane	acts as a differentially permeable layer, which controls the entry and exit of chemicals
circular DNA	possesses the genetic information for the replication of bacterial cells
plasmid	possesses genes that may aid the survival of bacteria in adverse conditions, e.g. produces enzymes that break down antibiotics

▼ **Table 1** *Comparison of prokaryotic and eukaryotic cells*

Prokaryotic cells	Eukaryotic cells
no true nucleus, only an area where DNA is found	distinct nucleus, with a nuclear envelope
(Pro) DNA is not associated with proteins	DNA is associated with proteins called histones.
some DNA may be in the form of circular strands called plasmids	There are no plasmids and DNA is linear.
no membrane-bounded organelles	membrane-bounded organelles, such as mitochondria, are present
no chloroplasts, only bacterial chlorophyll associated with the cell-surface membrane in some bacteria	chloroplasts present in plants and algae
ribosomes are smaller (70S)	ribosomes are larger (80S)
cell wall made of **murein** (peptidoglycan)	where present, cell wall is made mostly of cellulose (or chitin in fungi)
may have an outer mucilaginous layer called a capsule	no capsule

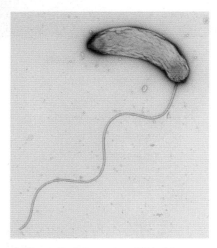

▲ **Figure 2** *False-colour TEM of the cholera bacterium, Vibrio cholerae*

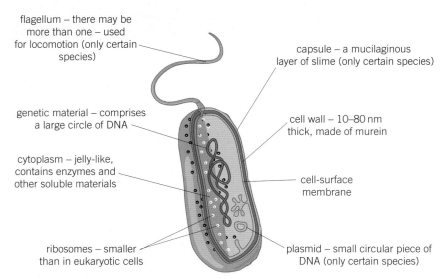

flagellum – there may be more than one – used for locomotion (only certain species)

capsule – a mucilaginous layer of slime (only certain species)

genetic material – comprises a large circle of DNA

cell wall – 10–80 nm thick, made of murein

cytoplasm – jelly-like, contains enzymes and other soluble materials

cell-surface membrane

ribosomes – smaller than in eukaryotic cells

plasmid – small circular piece of DNA (only certain species)

▲ **Figure 1** *Structure of a generalised bacterial cell*

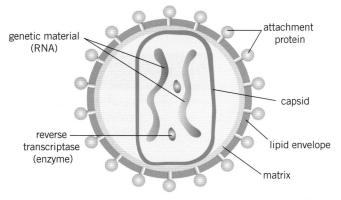

genetic material (RNA)

attachment protein

capsid

reverse transcriptase (enzyme)

lipid envelope

matrix

▲ **Figure 3** *Structure of the human immunodeficiency virus (HIV)*

Viruses

Viruses are acellular, non-living particles. They are smaller than bacteria, ranging in size from 20–300 nm. They contain **nucleic acids** such as DNA or RNA as genetic material but can only multiply inside living host cells. The nucleic acid is enclosed within a protein coat called the **capsid**. Some viruses, like the human immunodeficiency virus, are further surrounded by a lipid envelope. The lipid envelope, or if this is not present, the capsid, have **attachment proteins** which are essential to allow the virus to identify and attach to a host cell.

Summary questions

1 Table 3 lists some of the features of cells. For the letter in each box, write down **one** of the following:
'present' if the feature always occurs
'absent' if it never occurs
'sometimes' if it occurs in some cells but not others.

▼ **Table 3** *Features of prokaryotic and eukaryotic cells*

Feature	Prokaryotic cell	Eukaryotic cell
nuclear envelope	A	B
cell wall	C	D
flagellum	E	F
ribosomes	G	H
plasmid	I	J
cell-surface membrane	K	L
mitochondria	M	N

2  If a bacterium is 6 μm long and a virus is 150 nm long, calculate how many times larger the bacterium is than the virus.

Maths link √x̄

MS 0.1, see Chapter 11.

Cell division can take place by either mitosis or meiosis:

- **Mitosis** produces two daughter cells that have the same number of chromosomes as the parent cell and each other.
- **Meiosis** produces four daughter cells, each with half the number of chromosomes of the parent cell. We shall consider meiosis in Topic 9.2.

The structure of a chromosome is shown in Figure 1.

Mitosis

Mitosis is division of a cell that results in each of the daughter cells having an exact copy of the DNA of the parent cell. Except in the rare event of a mutation, the genetic make-up of the two daughter nuclei is also identical to that of the parent nucleus. Mitosis is always preceded by a period during which the cell is not dividing. This period is called **interphase**. It is a period of considerable cellular activity that includes a very important event, the replication of DNA. The two copies of DNA after replication remain joined at a place called the centromere. Although mitosis is a continuous process, it can be divided into four stages for convenience:

Prophase

In prophase, the chromosomes first become visible, initially as long thin threads, which later shorten and thicken. Animal cells contain two cylindrical organelles called centrioles, each of which moves to opposite ends (called poles) of the cell. From each of the centrioles, **spindle fibres** develop, which span the cell from pole to pole. Collectively, these spindle fibres are called the **spindle apparatus**. As plant cells lack centrioles but do develop a spindle apparatus, centrioles are clearly not essential to spindle fibre formation. The nucleolus disappears and the nuclear envelope breaks down, leaving the chromosomes free in the cytoplasm of the cell. These chromosomes are drawn towards the equator of the cell by the spindle fibres attached to the centromere.

Metaphase

By metaphase the chromosomes are seen to be made up of two chromatids. Each chromatid is an identical copy of DNA from the parent cell. The chromatids are joined by the centromere (Topic 8.2). It is to this centromere that some microtubules from the poles are attached, and the chromosomes are pulled along the spindle apparatus and arrange themselves across the equator of the cell.

Anaphase

In anaphase, the centromeres divide into two and the spindle fibres pull the individual chromatids making up the chromosome apart. The chromatids move rapidly to their respective, opposite poles of the cell and we now refer to them as chromosomes. The energy for the process is provided by mitochondria, which gather around the spindle fibres. If cells are treated with chemicals that destroy the spindle, the chromosomes remain at the equator, unable to reach the poles.

Learning objectives

→ Describe what mitosis is.
→ State when DNA replication takes place.
→ Explain the importance of mitosis.

Specification reference: 3.2.2

Study tip

It is important to remember that the replication of DNA takes place during interphase before the nucleus and the cell divide.

Synoptic link

The replication of DNA was covered in Topic 2.2, DNA replication.

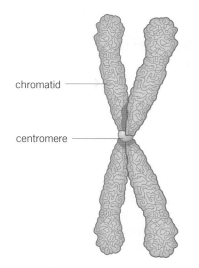

chromatid

centromere

▲ **Figure 1** *Structure of a chromosome*

Synoptic link

We will learn about the centromere in Topic 8.2, DNA and chromosomes.

Telophase and cytokinesis

In this stage, the chromosomes reach their respective poles and become longer and thinner, finally disappearing altogether, leaving only widely spread chromatin. The spindle fibres disintegrate and the nuclear envelope and nucleolus re-form. Finally the cytoplasm divides in a process called **cytokinesis**. The process is illustrated and explained in Figure 1.

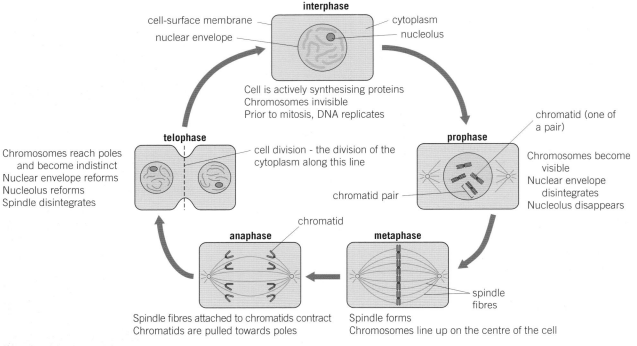

▲ **Figure 1** *The stages of mitosis in an animal cell*

Cell division in prokaryotic cells

Cell division in prokaryotic cells takes place by a process called **binary fission** as follows:

- The circular DNA molecule replicates and both copies attach to the cell membrane.
- The plasmids also replicate.
- The cell membrane begins to grow between the two DNA molecules and begins to pinch inward, dividing the cytoplasm into two.
- A new cell wall forms between the two molecules of DNA, dividing the original cell into two identical daughter cells, each with a single copy of the circular DNA and a variable number of copies of the plasmids.

Replication of viruses

As viruses are non-living, they cannot undergo cell division. Instead they replicate by attaching to their host cell with the attachment proteins on their surface. They then inject their nucleic acid into the host cell. The genetic information on the injected viral nucleic acid then provides the 'instructions' for the host cell's metabolic processes to start producing the viral components, nucleic acid, enzymes and structural proteins, which are then assembled into new viruses.

> ### Practical link ⚗
>
> Required practical 2. Preparation of stained squashes of cells from plant root tips; set-up and use of an optical microscope to identify the stages of mitosis in these stained squashes and calculation of a mitotic index. Also measuring the apparent size of cells in the root tip and calculating their actual size.

The importance of mitosis

Mitosis is important in organisms as it produces daughter cells that are genetically identical to the parent cells. Why is it essential to make exact copies of existing cells? There are three reasons:

- **growth**. When two **haploid** cells (e.g., a sperm and an ovum) fuse together to form a **diploid** cell, it has all the genetic information needed to form the new organism. If the new organism is to resemble its parents, all the cells that grow from this original cell must be genetically identical. Mitosis ensures that this happens.

- **repair**. If cells are damaged or die it is important that the new cells produced have an identical structure and function to the ones that have been lost.

- **reproduction** single-celled organisms divide by mitosis to give two new organisms. Each new organism is genetically identical to the parent organism.

> 1 Suggest an advantage and a disadvantage of having offspring that are genetically identical to their parents.

Summary question

1 In the following passage about mitosis, state the most appropriate word that is represented by each of the letters.

The period when a cell is not dividing is called **a**. The stage of mitosis when the chromosomes are first visible as distinct structures is called **b**. During this stage thin threads develop that span the cell from end to end and together form a structure called the **c**. Towards the end of this stage, the **d** breaks down and the **e** disappears. The stage when the chromosomes arrange themselves across the centre of the cell is called **f**. During the stage called **g** the chromatids move to opposite ends of the cell.

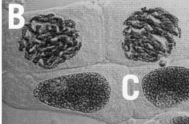

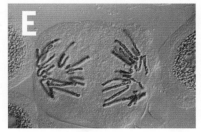

▲ **Figure 2** *Stages of mitosis*

Recognising the stages of mitosis $\sqrt{x}$

The photographs in Figure 2 show cells at various stages of mitosis.

Mitosis is a continuous process. When mitosis is viewed under a microscope, the observer only gets a snapshot of the process at one moment in time. The ratio of the number of cells undergoing mitosis to the total number of cells is called the mitotic index. The number of cells at each stage of mitosis is proportional to the time each cell spends undergoing that stage. Table 1 shows the number of cells at each stage of mitosis during one observation.

▼ **Table 1**

Stage	Number of cells
interphase	890
prophase	73
metaphase	20
anaphase	9
telophase	8

> **Maths link** $\sqrt{x}$
>
> MS 0.3, see Chapter 11.

1 State the names of the five different stages represented by the letters A–E in Figure 2. In each case give a reason for choosing your answer.
2 $\sqrt{x}$ From Table 1, if one complete cycle takes 20 hours, calculate how many minutes were spent in metaphase. Show your working.
3 $\sqrt{x}$ Calculate in what percentage of the cells the chromosomes would have been visible. Show your working.

Learning objectives

→ Describe the three stages of the cell cycle.

→ Describe what happens during interphase.

→ Explain how mitosis is controlled.

→ Describe how cancer and its treatment relate to the cell cycle.

Specification reference: 3.2.2

Hint

Interphase is sometimes known as the resting phase because no division takes place. In one sense, this description could hardly be further from the truth because interphase is a period of intense chemical activity.

Only some cells in multicellular organisms retain the ability to divide. Those that do not divide continuously, but undergo a regular cycle of division separated by periods of cell growth. This is known as the **cell cycle** and has three stages:

1   **interphase**, which occupies most of the cell cycle, and is sometimes known as the resting phase because no division takes place

2   **nuclear division**, when the nucleus divides either into two (mitosis) or four (meiosis)

3   **division of the cytoplasm (cytokinesis)**, which follows nuclear division and is the process by which the cytoplasm divides to produce two new cells (mitosis) or four new cells (meiosis) (Topic 9.2).

The length of a complete cell cycle varies greatly amongst organisms. Typically, a mammalian cell takes about 24 hours to complete a cell cycle, of which about 90% is interphase.

The various stages of the cell cycle are shown in Figure 1.

Figure 2 shows the variations in mass of a diploid cell and the DNA within it during the cell cycle.

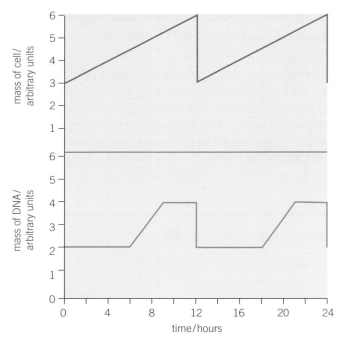

▲ **Figure 2** *Variation in the mass of a diploid cell and the DNA within it during the cell cycle*

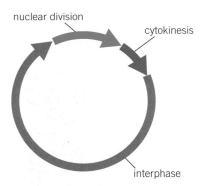

▲ **Figure 1** *The cell cycle*

Cancer and the control of mitosis

Cancer is a group of diseases (around 200 in total) caused by a growth disorder of cells. It is the result of damage to the genes that regulate mitosis and the cell cycle. This leads to uncontrolled growth and division of cells. As a consequence, a group of abnormal cells, called a tumour, develops and constantly expands in size. Tumours can develop in any organ of the body, but are most commonly found in the lungs, prostate gland (male), breast and ovaries (female), large intestine, stomach, oesophagus and pancreas. A tumour becomes cancerous if it changes from benign to malignant.

Most cells divide by mitosis, either to increase the size of a tissue during development (growth) or to replace dead and worn out cells (repair). The rate of mitosis can be affected by the environment of the cell and by growth factors. It is also controlled by two types of gene. A mutation to one of these genes results in uncontrolled mitosis. The mutant cells so formed are usually structurally and functionally different from normal cells. Most mutated cells die. However, any that survive are capable of dividing to form clones of themselves and forming tumours. Malignant tumours grow rapidly, are less compact and are more likely to be life-threatening, while benign ones grow more slowly, are more compact and are less likely to be life-threatening.

Treatment of cancer

The treatment of cancer often involves killing dividing cells by blocking a part of the cell cycle. In this way the cell cycle is disrupted and cell division, and hence cancer growth, ceases. Drugs used to treat cancer (chemotherapy) usually disrupt the cell cycle by:

- preventing DNA from replicating
- inhibiting the metaphase stage of mitosis by interfering with spindle formation.

The problem with such drugs is that they also disrupt the cell cycle of normal cells. However, the drugs are more effective against rapidly dividing cells. As cancer cells have a particularly fast rate of division, they are damaged to a greater degree than normal cells. Those normal body cells, such as hair-producing cells, that divide rapidly are also vulnerable to damage. This explains the hair loss frequently seen in patients undergoing cancer treatment.

Summary questions

1   List the three main stages of the cell cycle.

2   $\sqrt{x}$ Using Figure 2, state at what time(s), or during which period, each of the following occur:

   a   cell division

   b   replication of DNA.

Treating cancer $\sqrt{x}$

The graph in Figure 3 shows the effect of a chemotherapy drug that kills dividing cells. It was given to a cancer patient once every three weeks starting at time zero. The graph plots the changes in the number of healthy cells and cancer cells in a tissue over the treatment period of 12 weeks.

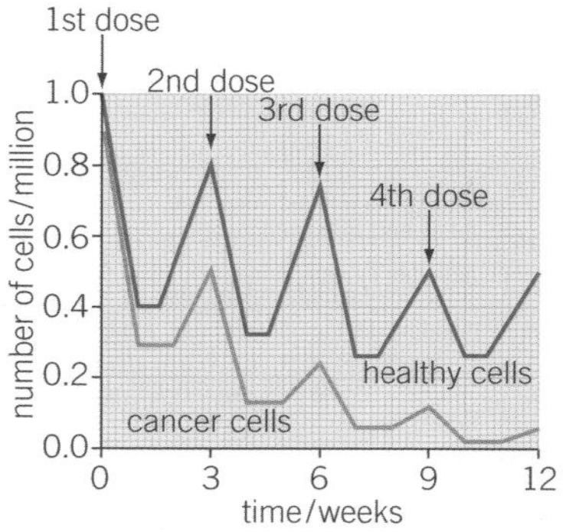

5   It would be possible to kill more cancer cells if the same dose of the drug was given more frequently or the frequency was kept the same but a larger dose of the drug was used each time. Suggest why

   a   the drug was not given more frequently

   b   the dose of the drug was not increased.

▲ **Figure 3** *Changes in the number of healthy cells and cancer cells in a tissue during a chemotherapy treatment of 12 weeks*

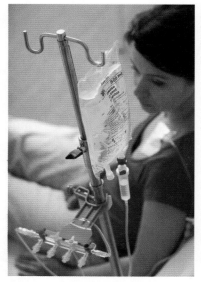

▲ **Figure 4** *Patient undergoing chemotherapy treatment*

1   An amoeba is a single-celled, eukaryotic organism. Scientists used a transmission electron microscope to study an amoeba. The diagram shows its structure.

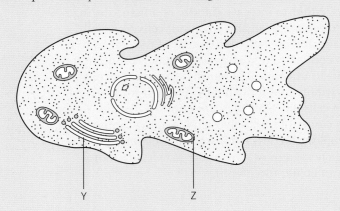

(a) (i)   Name organelle **Y**.   (*1 mark*)
     (ii) Name **two** other structures in the diagram which show that the amoeba
           is a eukaryotic cell.   (*2 marks*)
(b) What is the function of organelle **Z**?   (*1 mark*)
(c) The scientists used a transmission electron microscope to study the structure
     of the amoeba. Explain why.   (*2 marks*)

AQA June 2012

2   The photograph shows part of the cytoplasm of a cell.

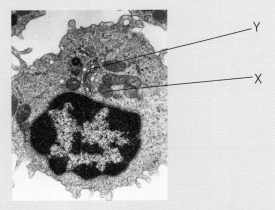

2   (a) (i)   Organelle X is a mitochondrion. What is the function of this organelle?   (*1 mark*)
         (ii) Name organelle Y.   (*1 mark*)
    (b) This photograph was taken using a transmission electron microscope. The
         structure of the organelles visible in the photograph could not have been
         seen using an optical (light) microscope. Explain why.   (*2 marks*)

AQA Jan 2013

3   The diagram shows a chloroplast as seen with an electron microscope.

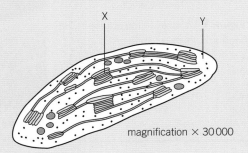

magnification × 30 000

(a) Name **X** and **Y**. (*2 marks*)
(b) Describe the function of a chloroplast. (*2 marks*)
(c) Calculate the maximum length of this chloroplast in micrometres (μm).
Show your working. (*2 marks*)

AQA Jan 2012

4 (a) The table shows some features of cells. Complete the table by putting a tick in the box
if the feature is present in the cell.

Feature	Cell		
	Cholera bacterium	Epithelial cell from intestine	Epithelial cell from alveolus of lung
Cell-surface membrane			
Flagellum			
Nucleus			

(*3 marks*)

(b) The diagram shows part of an epithelial cell from an insect's gut.

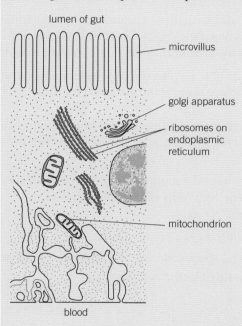

This cell is adapted for the three functions listed below. Use the diagram to explain
how this cell is adapted for each of these functions.

Use a **different** feature in the diagram for each of your answers.
(i) the active transport of substances from the cell into the blood (*2 marks*)
(ii) the synthesis of enzymes (*2 marks*)
(iii) rapid diffusion of substances from the lumen of the gut into the
cytoplasm (*1 mark*)

AQA Jan 2012

Learning objectives

→ Describe the structure of the cell-surface membrane.

→ Describe the functions of the various components of the cell-surface membrane.

→ Explain the fluid-mosaic model of cell membrane structure.

Specification reference: 3.2.3

All membranes around and within all cells (including those around and within cell organelles) have the same basic structure and are known as **plasma membranes**.

The cell-surface membrane is the name specifically given to the plasma membrane that surrounds cells and forms the boundary between the cell cytoplasm and the environment. It allows different conditions to be established inside and outside a cell. It controls the movement of substances in and out of the cell. Before we look at how the cell-surface membrane achieves this, we need first to look in more detail at the molecules that form its structure.

Phospholipids

We looked at the molecular structure of a phospholipid in Topic 1.5. Phospholipids form a bilayer (see Figure 1). They are important components of cell-surface membranes for the following reasons:

- The hydrophilic heads of both phospholipid layers point to the outside of the cell-surface membrane attracted by water on both sides.

- The hydrophobic tails of both phospholipid layers point into the centre of the cell membrane, repelled by the water on both sides.

Lipid-soluble material moves through the membrane via the phospholipid portion. The functions of phospholipids in the membrane are to:

- allow lipid-soluble substances to enter and leave the cell

- prevent water-soluble substances entering and leaving the cell

- make the membrane flexible and self-sealing.

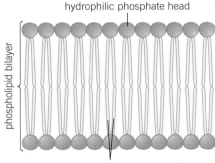

hydrophilic phosphate head

phospholipid bilayer

hydrophobic hydrocarbon tails

▲ **Figure 1** *A simplified diagram of a phospholipid bilayer.*

Proteins

Proteins are interspersed throughout the cell surface membrane. They are embedded in the phospholipid bilayer in two main ways:

- Some proteins occur in the surface of the bilayer and never extend completely across it. They act either to give mechanical support to the membrane or, in conjunction with glycolipids, as cell receptors for molecules such as hormones.

- Other proteins completely span the phospholipid bilayer from one side to the other. Some are **protein channels**, which form water-filled tubes to allow water-soluble ions to diffuse across the membrane. Others are **carrier proteins** that bind to ions or molecules like glucose and amino acids, then change shape in order to move these molecules across the membrane.

Hint

Organelles such as mitochondria and chloroplasts are surrounded by two plasma membranes. The term *cell-surface membrane* is reserved only for the plasma membrane around the cell.

The functions of the proteins in the membrane are to:

- provide structural support
- act as channels transporting water-soluble substances across the membrane
- allow active transport across the membrane through carrier proteins
- form cell-surface receptors for identifying cells
- help cells adhere together
- act as receptors, for example for hormones.

Cholesterol

Cholesterol molecules occur within the phospholipid bilayer of the cell-surface membrane. They add strength to the membranes. Cholesterol molecules are very hydrophobic and therefore play an important role in preventing loss of water and dissolved ions from the cell. They also pull together the fatty acid tails of the phospholipid molecules, limiting their movement and that of other molecules but without making the membrane as a whole too rigid.

The functions of cholesterol in the membrane are to:

- reduce lateral movement of other molecules including phospholipids
- make the membrane less fluid at high temperatures
- prevent leakage of water and dissolved ions from the cell.

Glycolipids

Glycolipids are made up of a carbohydrate covalently bonded with a lipid. The carbohydrate portion extends from the phospholipid bilayer into the watery environment outside the cell where it acts as a cell-surface receptor for specific chemicals, for example the human ABO blood system operates as a result of glycolipids on the cell-surface membrane.

The functions of glycolipids in the membrane are to:

- act as recognition sites
- help maintain the stability of the membrane
- help cells to attach to one another and so form tissues.

Glycoproteins

Carbohydrate chains are attached to many extrinsic proteins on the outer surface of the cell membrane. These glycoproteins also act as cell-surface receptors, more specifically for hormones and neurotransmitters.

The functions of glycoproteins in the membrane are to:

- act as recognition sites
- help cells to attach to one another and so form tissues
- allows cells to recognise one another, for example **lymphocytes** can recognise an organism's own cells.

> **Study tip**
>
> When representing a phospholipid it is important to be accurate. It has a *single* phosphate head and *two* fatty acid tails. All too often students show too many heads and/or too many tails.

> **Hint**
>
> All plasma membranes found around and inside cells have the same phospholipid bilayer structure. What gives plasma membranes their different properties are the different substances they contain — especially proteins.

Functions of membranes within cells
control the entry and exit of materials in discrete organelles such as mitochondria and chloroplasts
separate organelles from cytoplasm so that specific metabolic reactions can take place within them
provide an internal transport system, e.g., endoplasmic reticulum
isolate enzymes that might damage the cell, e.g., lysosomes
provide surfaces on which reactions can occur, e.g., protein synthesis using ribosomes on rough endoplasmic reticulum

Practical link

Required practical 4. Investigation into the effect of a named variable on the permeability of cell-surface membranes.

Summary questions

1 State the overall function of the cell-surface membrane.

2 State which end of the phospholipid molecule lies towards the inside of the cell-surface membrane.

3 State through which molecule in the cell-surface membrane each of the following are likely to pass in order to get in or out of a cell.

   a a molecule that is soluble in lipids

   b a mineral ion

4 From your knowledge of the cell-surface membrane, suggest *two* properties that a drug should possess if it is to enter a cell rapidly.

Permeability of the cell-surface membrane

The cell-surface membrane controls the movement of substances into and out of the cell. In general most molecules do not freely diffuse across it because many are:

- not soluble in lipids and therefore cannot pass through the phospholipid layer
- too large to pass through the channels in the membrane
- of the same charge as the charge on the protein channels and so, even if they are small enough to pass through, they are repelled
- electrically charged (in other words are polar) and therefore have difficulty passing through the non-polar hydrophobic tails in the phospholipid bilayer.

Fluid-mosaic model of the cell-surface membrane

The way in which all the various molecules are combined into the structure of the cell-surface membrane is shown in Figure 2. This arrangement is known as the **fluid-mosaic model** for the following reasons:

- **fluid** because the individual phospholipid molecules can move relative to one another. This gives the membrane a flexible structure that is constantly changing in shape
- **mosaic** because the proteins that are embedded in the phospholipid bilayer vary in shape, size and pattern in the same way as the stones or tiles of a mosaic.

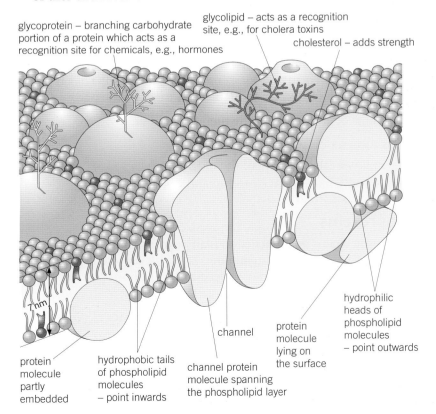

▲ **Figure 2** *The fluid-mosaic model of the cell-surface membrane*

The exchange of substances between cells and the environment occurs in ways that require metabolic energy (active transport) and in ways that do not (passive transport). Diffusion is an example of passive transport.

Explanation of simple diffusion

As all movement involves energy, it is possibly confusing to describe diffusion as passive transport. In this sense, 'passive' means that the energy comes from the natural, inbuilt motion of particles, rather than from some external source such as ATP. To help understand diffusion and other passive forms of transport it is necessary to understand that:

- all particles are constantly in motion due to the kinetic energy that they possess
- this motion is random, with no set pattern to the way the particles move around
- particles are constantly bouncing off one another as well as off other objects, for example, the sides of the vessel in which they are contained.

Given these facts, particles that are concentrated together in part of a closed vessel will, of their own accord, distribute themselves evenly throughout the vessel as a result of diffusion.

Diffusion is therefore defined as:

the net movement of molecules or ions from a region where they are more highly concentrated to one where their concentration is lower until evenly distributed.

We saw in Topic 4.1 that most molecules do not easily pass across the cell-surface membrane. Amongst the few molecules that can diffuse across membranes are small, non-polar molecules such as oxygen and carbon dioxide.

Facilitated diffusion

We saw in Topic 4.1 that plasma membranes are not readily permeable to molecules. Only small, non-polar molecules like oxygen can diffuse across them easily. Charged ions and polar molecules do not diffuse easily because of the hydrophobic nature of the fatty-acid tails of the phospholipids in the membrane. The movement of these molecules is made easier (facilitated) by transmembrane channels and carriers that span the membrane. The process is therefore called facilitated diffusion.

Facilitated diffusion is a passive process. It relies only on the inbuilt motion (kinetic energy) of the diffusing molecules. There is no external input of ATP from respiration. Like diffusion, it occurs down a concentration gradient, but it differs in that it occurs at specific points on the plasma membrane where there are special protein molecules. Two types of protein are involved – **protein channels** and **carrier proteins**. Each has a different mechanism.

Learning objectives

→ Explain what diffusion is and how it occurs.

→ Explain what affects the rate of diffusion.

→ Distinguish between facilitated diffusion and diffusion.

Specification reference: 3.2.3

Hint

Remember that diffusion is the **net** movement of particles. All particles move at random in diffusion; it is just that more move in one direction than in the other. This is due to concentration differences.

Hint

Diffusion only occurs between different concentrations of the *same* substance. For example, it may occur between different concentrations of oxygen or between different concentrations of carbon dioxide. It *never* occurs between different concentrations of oxygen and carbon dioxide.

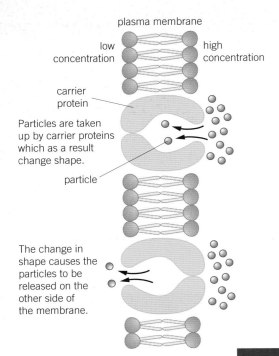

plasma membrane

low concentration

high concentration

carrier protein

Particles are taken up by carrier proteins which as a result change shape.

particle

The change in shape causes the particles to be released on the other side of the membrane.

▲ **Figure 2** *Facilitated diffusion involving carrier proteins*

Protein channels

These proteins form water-filled hydrophilic channels across the membrane. They allow specific water-soluble ions to pass through. The channels are selective, each opening in the presence of a specific ion. If the particular ion is not present, the channel remains closed. In this way, there is control over the entry and exit of ions. The ions bind with the protein causing it to change shape in a way that closes it to one side of the membrane and opens it to the other side.

Carrier proteins

An alternative form of facilitated diffusion involves carrier proteins that span the plasma membrane. When a molecule such as glucose that is specific to the protein is present, it binds with the protein. This causes it to change shape in such a way that the molecule is released to the inside of the membrane (Figure 2). No external energy is needed for this. The molecules move from a region where they are highly concentrated to one of lower concentration, using only the kinetic energy of the molecules themselves.

Study tip

Diffusion is proportional to the *difference* in concentration between two regions (the concentration gradient). It is incorrect to state that diffusion is proportional to concentration.

Hint

Remember that protein channels and carrier proteins have binding sites, but these are different to active sites.

Summary questions

1 State *three* factors that affect the rate of diffusion.

2 Contrast facilitated diffusion and diffusion.

3 Explain why facilitated diffusion is a passive process.

4 Glucose molecules are transported into cells through the pores in the proteins that span the phospholipid bilayer. Explain why they do not pass easily through the phospholipid bilayer.

5 List *two* changes to the structure of cell-surface membranes that would increase the rate at which glucose is transported into a cell.

6 √x̄ Oxygen is required by cells for respiration. This diffuses into the blood through the epithelial layers of the alveoli and blood capillaries. Calculate by how much each of the following changes would increase or decrease the rate of diffusion of oxygen.

   a The surface area of the alveoli is doubled.

   b The surface area of the alveoli is halved and the oxygen concentration gradient is doubled.

   c The oxygen concentration gradient is halved and the total thickness of the epithelial layers is doubled.

   d The oxygen concentration of the blood is halved and the carbon dioxide concentration of the alveoli is doubled.

4.3 Osmosis

In the last topic we learned about diffusion. We now turn our attention to a special case of diffusion, known as osmosis. Osmosis only involves the movement of water molecules.

What is osmosis?

Osmosis is defined as:

> **the passage of water from a region where it has a higher water potential to a region where it has a lower water potential through a selectively permeable membrane.**

Cell-surface membranes and other plasma membranes such as those around organelles are selectively permeable, that is, they are permeable to water molecules and a few other small molecules, but not to larger molecules.

Solutions and water potential

A solute is any substance that is dissolved in a solvent, for example, water. The solute and the solvent together form a solution.

Water potential is represented by the Greek letter psi (Ψ), and is measured in units of pressure, usually kiloPascals (kPa). Water potential is the pressure created by water molecules. Under standard conditions of temperature and pressure (25 °C and 100 kPa), pure water is said to have a water potential of zero.

It follows that:

- the addition of a solute to pure water will lower its water potential
- the water potential of a solution (water + solute) must always be less than zero, that is, a negative value
- the more solute that is added (i.e., the more concentrated a solution), the lower (more negative) its water potential
- water will move by osmosis from a region of higher (less negative) water potential (e.g., −20 kPa) to one of lower (more negative) water potential (e.g., −30 kPa).

One way of finding the water potential of cells or tissues is to place them in a series of solutions of different water potentials. Where there is no net gain or loss of water from the cells or tissues, the water potential inside the cells or tissues must be the same as that of the external solution.

Explanation of osmosis

Consider the hypothetical situation in Figure 1 overleaf, in which a selectively permeable plasma membrane separates two solutions.

- The solution on the left has a low concentration of solute molecules while the solution on the right has a high concentration of solute molecules.
- Both the solute and water molecules are in random motion due to their **kinetic energy**.

Study tip

Remember that *all* water potential values are negative. The highest water potential is *zero*. Therefore the lower the water potential, the more negative it becomes.

- The selectively permeable plasma membrane, however, only allows water molecules across it and not solute molecules.
- The water molecules diffuse from the left-hand side, which has the higher water potential, to the right-hand side, which has the lower water potential, that is, down a water potential gradient (Figure 2).
- At the point where the water potentials on either side of the plasma membrane are equal, a dynamic equilibrium is established and there is no net movement of water.

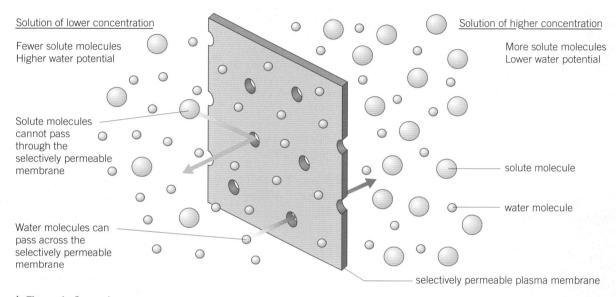

Solution of lower concentration

Fewer solute molecules
Higher water potential

Solute molecules cannot pass through the selectively permeable membrane

Water molecules can pass across the selectively permeable membrane

Solution of higher concentration

More solute molecules
Lower water potential

solute molecule

water molecule

selectively permeable plasma membrane

▲ **Figure 1** *Osmosis*

Understanding water potential

The highest value of water potential, that of pure water, is zero, and so all other values are negative. The more negative the value, the lower the water potential.

Osmosis and animal cells

Animal cells, such as red blood cells, contain a variety of solutes dissolved in their watery cytoplasm. If a red blood cell is placed in pure water it will absorb water by osmosis because it has a lower water potential. Cell surface membranes are very thin (7 nm) and, although they are flexible, they cannot stretch to any great extent. The cell-surface membrane will therefore break, bursting the cell and releasing its contents (in red blood cells this is called haemolysis). To prevent this happening, animal cells normally live in a liquid which has the same water potential as the cells. In our example, the liquid is the blood plasma. This and red blood cells have the same water potential. If a red blood cell is placed in a solution with a water potential lower than its own, water leaves by osmosis and the cell shrinks and becomes shrivelled (see Table 1).

key

x kPa   water potential of cell
⟶   direction of water movement

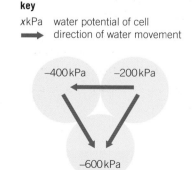

water moves from higher water potential to lower water potential. The highest water potential is zero

▲ **Figure 2** *Movement of water between cells along a water potential gradient*

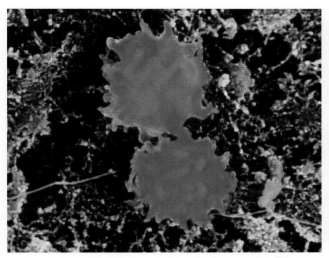

▲ **Figure 3** *SEM of red blood cells that have been placed in a solution of lower water potential. Water has left by osmosis and the cells have become shrunken and shrivelled*

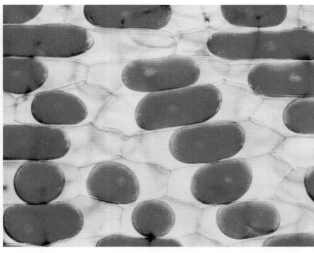

▲ **Figure 4** *Onion epidermal cells showing plasmolysis. The protoplasts, with their vacuoles containing red liquid, have shrunk and pulled away from the cell walls*

▼ **Table 1** *Summary of osmosis in an animal cell, for example, a red blood cell*

Water potential (ψ) of external solution compared to cell solution	higher (less negative)	equal	lower (more negative)
Net movement of water	enters cell	neither enters nor leaves	leaves cell
State of cell	swells and bursts	no change	shrinks
	contents, including haemoglobin, are released / remains of cell-surface membrane	normal red blood cell	haemoglobin is more concentrated, giving cell a darker appearance / cell shrunken and shrivelled

Summary questions

1   Explain what is meant by a selectively permeable membrane.

2   Under standard conditions of pressure and temperature, what is the water potential of pure water?

3   🆅 Four cells have the following water potentials:

   Cell A = −200 kPa
   Cell B = −250 kPa
   Cell C = −100 kPa
   Cell D = −150 kPa.

   Determine the order in which the cells have to be placed for water to pass from one cell to the next if they are arranged in a line.

Osmosis and plant cells

▼ **Table 2** *Summary of osmosis in a plant cell*

Water potential (ψ) of external solution compared to cell solution	higher (less negative)	equal	lower (more negative)
Net movement of water	enters cell	neither enters nor leaves	leaves cell
Protoplast	swells	no change	shrinks
Condition of cell	turgid	incipient plasmolysis	plasmolysed
	protoplast pushed against cell wall / nucleus / cellulose cell wall / protoplast	protoplast beginning to pull away from the cell wall	protoplast completely pulled away from the cell wall

For the purposes of the following explanations, the plant cell can be divided into three parts:

- the **central vacuole**, which contains a solution of salts, sugars and organic acids in water
- the **protoplast**, consisting of the outer cell-surface membrane, nucleus, cytoplasm and the inner vacuole membrane
- the **cellulose cell wall**, a tough, inelastic covering that is permeable to even large molecules.

Like animal cells, plant cells also contain a variety of solutes, mainly dissolved in the water of the large cell vacuole that each possesses. When placed in pure water they also absorb water by osmosis because of their lower (more negative) water potential. Unlike animal cells, however, they are unable to control the composition of the fluid around their cells. Indeed, plant cells are normally permanently bathed in almost pure water, which is constantly absorbed from the plant's roots. Water entering a plant cell by osmosis causes the protoplast to swell and press on the cell wall. Because the cell wall is capable of only very limited expansion, a pressure builds up on it that resists the entry of further water. In this situation, the protoplast of the cell is kept pushed against the cell wall and the cell is said to be **turgid**.

If the same plant cell is placed in a solution with a lower water potential than its own, water leaves by osmosis. The volume of the cell decreases. A stage is reached where the protoplast no longer presses on the cellulose cell wall. At this point the cell is said to be at **incipient plasmolysis**. Further loss of water will cause the cell contents to shrink further and the protoplast to pull away from the cell wall. In this condition the cell is said to be **plasmolysed**. These events are summarised in Table 2.

1 Explain why an animal cell placed in pure water bursts while a plant cell placed in pure water does not.

2 √x̄ Plant cells that have a water potential of −600 kPa are placed in solutions of different water potentials. Determine in each of the following cases whether, after 10 minutes, the cells would be turgid, plasmolysed or at incipient plasmolysis.
   a Solution A = −400 kPa
   b Solution B = −600 kPa
   c Solution C = −900 kPa
   d Solution D = pure water

3 √x̄ An animal cell with a water potential of −700 kPa was placed in each of the solutions. Deduce in which solutions the cell is likely to burst.

We have looked at diffusion and osmosis, both of which are passive processes, that is they occur without the use of metabolic energy. The transport of some molecules in and out of cells involves a process that uses metabolic energy. This process is active transport.

What is active transport?

Active transport is:

the movement of molecules or ions into or out of a cell from a region of lower concentration to a region of higher concentration using ATP and carrier proteins.

In active transport ATP is used to:

*   directly move molecules
*   individually move molecules using a concentration gradient which has already been set up by (direct) active transport. This is known as **co-transport** and is further explained in Topic 4.5.

It differs from passive forms of transport in the following ways:

*   Metabolic energy in the form of **ATP** is needed.
*   Substances are moved against a concentration gradient, that is from a lower to a higher concentration.
*   Carrier protein molecules which act as 'pumps' are involved.
*   The process is very selective, with specific substances being transported.

Direct active transport of a single molecule or ion is described below.

*   The carrier proteins span the plasma membrane and bind to the molecule or ion to be transported on one side of it.
*   The molecule or ion binds to receptor sites on the carrier protein.
*   On the inside of the cell/organelle, ATP binds to the protein, causing it to split into ADP and a phosphate molecule. As a result, the protein molecule changes shape and opens to the opposite side of the membrane.
*   The molecule or ion is then released to the other side of the membrane.
*   The phosphate molecule is released from the protein which causes the protein to revert to its original shape, ready for the process to be repeated. The phosphate molecule then recombines with the ADP to form ATP during respiration.

These events are illustrated in Figure 1. It is important to distinguish between active transport and facilitated diffusion. Both use carrier proteins but facilitated diffusion occurs *down* a concentration gradient, while active transport occurs *against* a concentration gradient. This means that facilitated diffusion does not require metabolic energy, while active transport does. The metabolic energy is provided in the form of ATP.

Learning objectives

→   Explain the process of active transport.

→   Describe the conditions required for active transport.

Specification reference: 3.2.3

Study tip

Carrier proteins have a specific tertiary structure and will only transport particular substances across a membrane. They have binding sites – these are different to the active sites of enzymes.

Sometimes more than one molecule or ion may be moved in the same direction at the same time by active transport. Occasionally, the molecule or ion is moved into a cell/organelle at the same time as a different one is being removed from it. One example of this is the **sodium–potassium pump**.

In the sodium–potassium pump, sodium ions are actively removed from the cell/organelle while potassium ions are actively taken in from the surroundings. This process is essential to a number of important processes in the organism, including the creation of a nerve impulse.

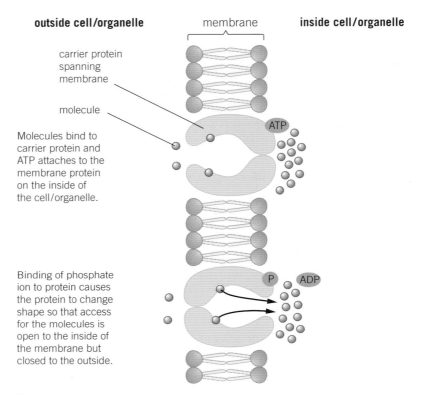

▲ **Figure 1** *Active transport*

Summary questions

1 State *one* similarity and *one* difference between active transport and facilitated diffusion.

2 The presence of many mitochondria is typical of cells that carry out active transport. Explain why this is so.

3 In the production of urine, glucose is initially lost from the blood but is then reabsorbed into the blood by cells in the kidneys. Explain why it is important that this reabsorption occurs by active transport rather than by diffusion.

4.5 Co-transport and absorption of glucose in the ileum

To illustrate how the various forms of movement across membranes occur in a particular situation, you can look at how products of digestion like glucose and amino acids are absorbed in the ileum (small intestine). Firstly let us look at how the rate of transport across membranes and into cells may be increased.

Increasing the rate of movement across membranes

The epithelial cells lining the ileum possess **microvilli** (see Figure 1). These are finger-like projections of the cell-surface membrane about 0.6 μm in length. They are collectively termed a 'brush border' because, when viewing them under a light microscope, they look like the bristles on a brush. The microvilli provide more surface area for the insertion of carrier proteins through which diffusion, facilitated diffusion and active transport can take place. Another mechanism to increase transport across membranes is to increase the number of protein channels and carrier proteins in any given area of membrane (i.e., increase their density).

The role of diffusion in absorption

Diffusion (Topic 4.2) is the net movement of molecules or ions from a region where they are highly concentrated to a region where their concentration is lower.

As carbohydrates and proteins are being digested continuously, there is normally a greater concentration of glucose and amino acids within the ileum than in the blood. There is therefore a concentration gradient down which glucose moves by facilitated diffusion from inside the ileum into the blood. Given that the blood is constantly being circulated by the heart, the glucose absorbed into it is continuously being removed by the cells as they use it up during respiration. This helps to maintain the concentration gradient between the inside of the ileum and the blood (Figure 2). This means the rate of movement by facilitated diffusion across epithelial cell-surface membranes is increased.

Role of active transport in absorption

At best, diffusion only results in the concentrations either side of the intestinal epithelium becoming equal. This means that not all the available glucose and amino acids can be absorbed in this way and some may pass out of the body. The reason why this does not happen is because glucose and amino acids are also being absorbed by active transport (see Topic 4.4). This means that all the glucose and amino acids should be absorbed into the blood.

The actual mechanism by which they are absorbed from the small intestine is an example of **co-transport**. This term is used because either glucose or amino acids are drawn into the cells along with sodium ions that have been actively transported out by the sodium–potassium pump (see Topic 4.4). It takes place in the following manner (see Figure 3):

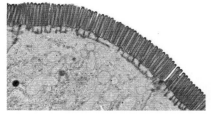
▲ **Figure 1** *Microvilli on an epithelial cell from the small intestine*

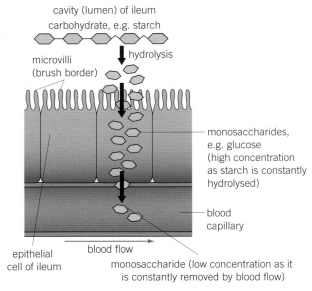

▲ **Figure 2** *Absorption of monosaccharides (e.g., glucose) by diffusion in the ileum*

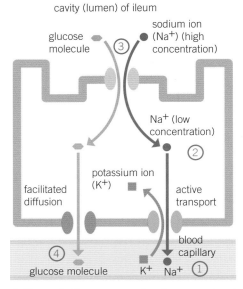

▲ **Figure 3** *Co-transport of a glucose molecule*

1 Sodium ions are actively transported out of epithelial cells, by the sodium–potassium pump, into the blood. This takes place in one type of protein-**carrier molecule** found in the cell-surface membrane of the epithelial cells.

2 This maintains a much higher concentration of sodium ions in the lumen of the intestine than inside the epithelial cells.

3 Sodium ions diffuse into the epithelial cells down this concentration gradient through a different type of protein carrier (co-transport protein) in the cell-surface membrane. As the sodium ions diffuse in through this second carrier protein, they carry either amino acid molecules or glucose molecules into the cell with them.

4 The glucose/amino acids pass into the blood plasma by facilitated diffusion using another type of carrier.

Both sodium ions and glucose/amino acid molecules move into the cell, but while the sodium ions move *down* their concentration gradient, the glucose molecules move *against* their concentration gradient. It is the sodium ion concentration gradient, rather than **ATP** directly, that powers the movement of glucose and amino acids into the cells. This makes it an indirect rather than a direct form of active transport.

Summary questions

1 State **three** ways in which the rate of movement across membranes can be increased.

2 Explain why the term 'co-transport' is used to describe the transport of glucose into cells.

3 In each of the following events in the glucose co-transport system, state whether the movements are active or passive.

   a Sodium ions move out of the epithelial cell.

   b Sodium ions move into the epithelial cell.

   c Glucose molecules move into the epithelial cell.

Synoptic link

The absorption of amino acids and lipids in the ileum will be covered in Topic 6.10, the structure of glucose was described in Topic 1.2 and the structure of amino acids in Topic 1.6.

Oral rehydration therapy

There are a number of diarrhoeal diseases that infect the intestines. Diarrhoea kills many people, especially the very young, and yet a treatment to prevent death is relatively simple. This treatment is **oral rehydration therapy**.

Diarrhoea is an intestinal disorder where watery faeces are produced frequently. The causes include:

- damage to the epithelial cells lining the intestine
- loss of microvilli due to toxins
- excessive secretion of water due to toxins, for example cholera toxin

As a result of diarrhoea, insufficient fluid is taken into, and/or excessive fluid is lost from, the body. Either way, dehydration results and may be fatal.

To treat diarrhoeal diseases it is vital to rehydrate the patient. Just drinking water is ineffective for two reasons:

- Water is not being absorbed from the intestine – indeed, as in the case of cholera, water is actually being lost from cells.
- Drinking water does not replace the electrolytes (ions) that are being lost from the intestinal cells.

It is possible to replace the water and electrolytes intravenously by a drip, but this requires trained personnel and means the patient is confined to bed for much of the time. What is required is a suitable mixture of substances that can safely be taken by mouth and which will be absorbed by the intestine.

But how can the patient be rehydrated if the intestine is not absorbing water? As it happens, there is more than one type of carrier protein in the plasma membranes of epithelial cells that absorbs sodium ions. The trick is to develop a rehydration solution that uses these alternative pathways. As sodium ions are absorbed, so the water potential of the cells falls and water enters the cells by osmosis. Therefore, a rehydration solution needs to contain:

- **water** – to rehydrate the tissues
- **sodium** ions – to replace the sodium ions lost from the epithelium of the intestine and to optimise use of the alternative sodium-glucose carrier proteins
- **glucose** – to stimulate the uptake of sodium ions from the intestine and to provide energy
- **potassium** ions – to replace lost potassium ions and to stimulate appetite
- **other electrolyes** – such as chloride ions and citrate ions to help prevent electrolyte imbalance and a condition called metabolic acidosis.

The ingredients can be mixed and packaged as a powder and then the solution made up with boiled water as needed. It can then be administered by people with minimal training. The solution must be given regularly, and in large amounts, throughout the illness.

Oral rehydration solutions do not prevent or cure diarrhoea. They simply rehydrate and nourish the patient until the diarrhoea is cured by some other means.

When commercial products are not available it is possible to use an inexpensive, home-made rehydration solution. This can be made up of eight level teaspoons of sugar + 1 level teaspoon of table salt dissolved in 1 litre of boiled water.

1 List *two* reasons why glucose is included in the mixture.
2 Table salt is sodium chloride. List *three* reasons why it is included in the mixture.
3 Explain why is it essential that the water is boiled.
4 Bananas are rich in potassium. It is sometimes recommended that mashed banana is added to the mixture. List *two* reasons why this might help the patient recover.
5 Suggest another advantage of adding banana before drinking the mixture, especially in the case of children.

6 Sports drinks contain a high proportion of glucose to help replace that used during strenuous exercise. Explain in terms of water potential why these drinks are therefore not suitable to rehydrate those suffering from diarrhoea.

The development of oral rehydration solutions resulted from a long process of scientific experimentation.

Early rehydration solutions led to side effects, especially in children. These were caused by excess sodium and so mixtures with lower sodium content, but more glucose, were tested. Unfortunately the additional glucose lowered the water potential in the lumen of the ileum so much that it started to draw out even more water from the epithelial cells. This made the dehydration even worse. Lowering the glucose content reduced this effect but, as glucose also acted as a respiratory substrate, it reduced the amount of energy being supplied to the patient. One answer was to use starch in place of some of the glucose.

7 Explain why using starch is better than using glucose.

Starch is broken down steadily by amylase and maltase in the ileum into its glucose **monomers**. By experimenting with different concentrations of starch, a rehydration solution was developed that released glucose at the optimum rate for it to be taken up as it was produced, without it adversely influencing the water potential. Further scientific research is being carried out to find the best source of starch.

Rice starch is a popular choice for two main reasons:

- It is readily available in many parts of the world, especially those where diarrhoeal diseases are common.

- It also provides other nutrients like amino acids. Not only are these nutrients nutritionally valuable but they also help the uptake of sodium ions from the ileum.

As rice flour produces a very viscous solution, it is not easy to swallow.

8 Suggest **one** possible method of reducing the viscosity of the rice flour solution and explain how it works.

We have seen how the development of an improved medicine takes place in a number of stages, each of which must be tested for its safety. While initial testing can be done on tissue cultures and animals, to be sure of a drug's effectiveness and safety, it must eventually be tested on humans.

9 Consider why drugs must ultimately be tested on humans.

Testing is normally carried out in four phases.

- A small number (20–80) of usually healthy people are given a tiny amount of the drug to test for side effects rather than to see if the drug is effective. The dose may be increased gradually in a series of such trials. This stage takes around six months.

- The drug is given to a slightly larger number of people (100–300) who have the condition the drug is designed to treat. This is to see that it works and to look at any safety issues. This stage takes up to two years.

- A large-scale trial of many thousands of patients takes place. Many are given a dummy drug called a **placebo**. Often, neither the scientists nor the patients know who has taken the real drug and who has taken the placebo until after the trial. This is known as a double-blind trial. These trials take many years.

- If the drug passes all these stages it may be granted a licence, but its use and effects are still monitored over many years to check on any long-term effects.

10 Suggest a reason why a placebo is necessary to ensure that the results of a drug trial can be relied upon.

11 Suggest why the results of a 'double-blind' trial might be more reliable than one in which the patients knew whether they were taking the real drug or a placebo.

1   A student investigated the effect of putting cylinders cut from a potato into sodium
    chloride solutions of different concentration. He cut cylinders from a potato and weighed
    each cylinder. He then placed each cylinder in a test tube. Each test tube contained a
    different concentration of sodium chloride solution. The tubes were left overnight. He
    then removed the cylinders from the solutions and reweighed them.

    (a) Before reweighing, the student blotted dry the outside of each cylinder.
         Explain why.   (2 marks)
         The student repeated the experiment several times at each concentration of
         sodium chloride solution. His results are shown in the graph.

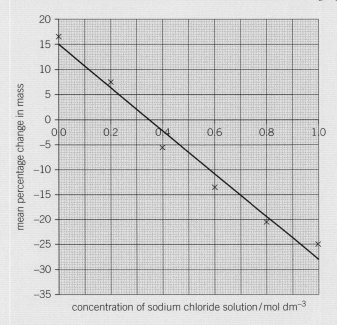

    (b) The student made up all the sodium chloride solutions using a 1.0 mol dm^{-3} sodium
         chloride solution and distilled water.
         Complete the table to show how he made 20 cm^3 of a 0.2 mol dm^{-3} sodium chloride
         solution.

Volume of 1.0 mol dm^{-3} sodium chloride solution	Volume of distilled water

                                                                                   (1 mark)
    (c) The student calculated the *percentage* change in mass rather than the change
         in mass.
         Explain the advantage of this.   (2 marks)
    (d) The student carried out several repeats at each concentration of sodium
         chloride solution. Explain why the repeats were important.   (2 marks)
    (e) Use the graph to find the concentration of sodium chloride solution that
         has the same water potential as the potato cylinders.

                                                                          mol dm^{-3}
                                                                                   (1 mark)
                                                                              AQA Jan 2011

2   Some substances can cross the cell-surface membrane of a cell by simple diffusion
    through the phospholipid bilayer. Describe other ways by which substances cross
    this membrane.   (5 marks)
                                                                              AQA Jan 2013

3 **(a)** Give **two** ways in which active transport is different from facilitated diffusion. (*2 marks*)

Scientists investigated the effect of a drug called a proton pump inhibitor. The drug is given as a tablet to people who produce too much acid in their stomach. It binds to a carrier protein in the surface membrane of cells lining the stomach. This carrier protein usually moves hydrogen ions into the stomach by active transport.

The scientists used two groups of people in their investigation. All the people produced too much acid in their stomach. People in group **P** were given the drug. Group **Q** was the control group.

The graph shows the results.

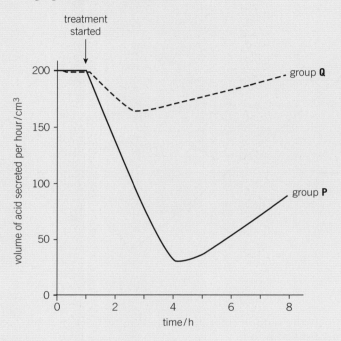

(b) **(i)** The scientists used a control group in this trial. Explain why. (*1 mark*)
  (ii) Suggest how the control group would have been treated. (*2 marks*)
(c) Describe the effect of taking the drug on acid secretion. (*1 mark*)

AQA June 2011

(d) Calculate the percentage decrease in acid secretion of group P compared to group Q after 8 hours. (*2 marks*)

4 Scientists investigated the percentages of different types of lipid in plasma membranes from different types of cell. **Table 2** shows some of their results.

▼ **Table 2**

Type of lipid	Percentage of lipid in plasma membrane by mass		
	Cell lining ileum of mammal	Red blood cell of mammal	The bacterium *Escherichia coli*
Cholesterol	17	23	0
Glycolipid	7	3	0
Phospholipid	54	60	70
Others	22	14	30

(a) The scientists expressed their results as **Percentage of lipid in plasma membrane by mass**. Explain how they would find these values. (*2 marks*)

(b) Cholesterol increases the stability of plasma membranes. Cholesterol does this by making membranes less flexible.
Suggest **one** advantage of the different percentage of cholesterol in red blood cells compared with cells lining the ileum. (*1 mark*)

(c) *E. coli* has no cholesterol in its cell-surface membrane. Despite this, the cell maintains a constant shape. Explain why. (*2 marks*)

AQA SAMS AS PAPER 1

Cell recognition and the immune system
5.1 Defence mechanisms

Learning objectives

→ Describe the main defence mechanisms of the body.

→ Explain how the body distinguishes between its own cells and foreign cells.

Specification reference: 3.2.4

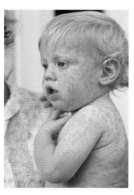

▲ **Figure 1** *Baby boy covered with measles rash. Measles is a highly infectious viral disease that mainly affects young children before they have acquired immunity to it*

Hint

The defensive mechanisms can be likened to the defences of a castle hundreds of years ago. The physical barrier is like the walls of the castle, the phagocytes are like the foot soldiers patrolling in the castle, who seek out and kill any intruders, and the lymphocytes are like specialised soldiers who respond to specific threats and use the intelligence gained from previous attacks to recognise, and quickly destroy, future intruders.

NB Don't use these descriptions in an examination!

Tens of millions of humans die each year from infectious diseases. Many more survive and others appear never to be affected in the first place. Why are there these differences?

Any infection is, in effect, an interaction between the **pathogen** and the body's various defence mechanisms. Sometimes the pathogen overwhelms the defences and the individual dies. Sometimes the body's defence mechanisms overwhelm the pathogen and the individual recovers from the disease. Having overwhelmed the pathogen, however, the body's defences seem to be better prepared for a second infection from the same pathogen and can kill it before it can cause any harm. This is known as **immunity** and is the main reason why some people are unaffected by certain pathogens.

There is a complete range of intermediates between the stages described above. Much depends on the overall state of health of an individual. A fit, healthy adult will rarely die of to an infection. Those in ill health, the young and the elderly are usually more vulnerable.

Defence mechanisms

The human body has a range of defences to protect itself from pathogens (Figure 2). Some are general and immediate defences like the skin forming a barrier to the entry of pathogens and phagocytosis (see Topic 5.2). Others are more specific, less rapid but longer-lasting. These responses involve a type of white blood cell called a lymphocyte and take two forms:

- cell-mediated responses involving T lymphocytes
- humoral responses involving B lymphocytes.

Before we look in detail at these defence mechanisms, let us first consider how the body distinguishes its own cells from foreign material.

Recognising your own cells

To defend the body from invasion by foreign material, lymphocytes must be able to distinguish the body's own cells and molecules (**self**) from those that are foreign (**non-self**). If they could not do this, the lymphocytes would destroy the organism's own tissues.

Each type of cell, self or non-self, has specific molecules on its surface that identify it. While these molecules can be of a variety of types, it is the proteins that are the most important. This is because proteins have enormous variety and a highly specific tertiary structure. It is this variety of specific 3-D structure that distinguishes one cell from another. It is these protein molecules which usually allow the immune system to identify:

- pathogens, for example the human immunodeficiency virus (see Topic 5.7).
- non-self material such as cells from other organisms of the same species.

- toxins including those produced by certain pathogens like the bacterium that causes cholera.
- abnormal body cells such as cancer cells.

All of the above are potentially harmful and their identification is the first stage in removing the threat they pose. Although this response is clearly advantageous to the organism, it has implications for humans who have had tissue or organ transplants. The immune system recognises these as non-self even though they have come from individuals of the same species. It therefore attempts to destroy the transplant. To minimise the effect of this tissue rejection, donor tissues for transplant are normally matched as closely as possible to those of the recipient. The best matches often come from relatives that are genetically close. In addition, immunosuppressant drugs are often administered to reduce the level of the immune response that still occurs.

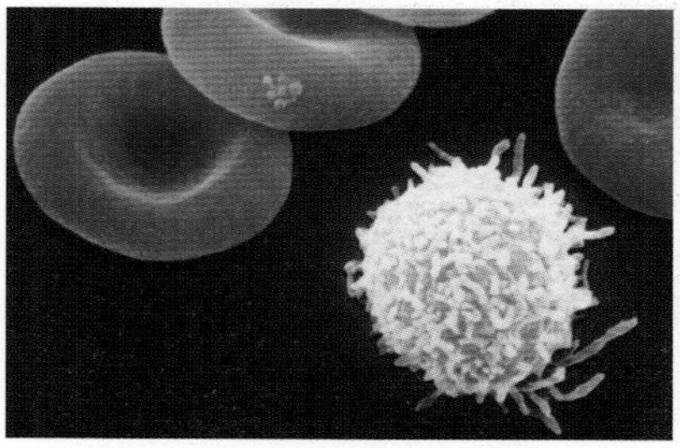

It is important to remember that specific lymphocytes are not produced in response to an infection, but that they already exist – all ten million different types. Given that there are so many different types of lymphocytes, there is a high probability that, when a pathogen gets into the body, one of these lymphocytes will have a protein on its surface that is complementary to one of the proteins of the pathogen. In other words, the lymphocyte will 'recognise' the pathogen. Not surprisingly with so many different lymphocytes, there are very few of each type. When an infection occurs, the one type already present that has the complementary proteins to those of the pathogen is stimulated to divide to build up its numbers to a level where it can be effective in destroying it. This is called clonal selection and you will learn more about it in Topic 5.4. This explains why there is a time lag between exposure to the pathogen and body's defences bringing it under control.

How lymphocytes recognise cells belonging to the body

- There are probably around ten million different lymphocytes present at any time, each capable of recognising a different chemical shape.
- In the fetus, these lymphocytes are constantly colliding with other cells.
- Infection in the fetus is rare because it is protected from the outside world by the mother and, in particular, the placenta.
- Lymphocytes will therefore collide almost exclusively with the body's own material (self).
- Some of the lymphocytes will have receptors that exactly fit those of the body's own cells.
- These lymphocytes either die or are suppressed.
- The only remaining lymphocytes are those that might fit foreign material (non-self), and therefore only respond to foreign material.
- In adults, lymphocytes produced in the bone marrow initially only encounter self-antigens.
- Any lymphocytes that show an immune response to these self-antigens undergo programmed cell death (apoptosis) before they can differentiate into mature lymphocytes.
- No clones of these anti-self lymphocytes will appear in the blood, leaving only those that might respond to non-self antigens.

▲ **Figure 3** *False-colour SEM of a single human lymphocyte (blue) and red blood cells (red)*

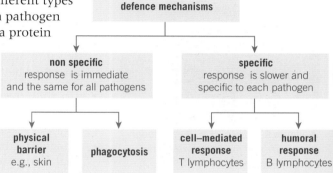

▲ **Figure 2** *Summary of defence mechanisms*

Summary questions

1 State *two* differences between a specific and a non-specific defence mechanism.

2 After a pathogen gains entry to the body it is often a number of days before the body's immune system begins to control it. Suggest a possible reason why this is so.

3 In the above case, suggest why it would be inaccurate to say that the body takes days to 'respond' to the pathogen.

If a pathogen is to infect the body it must first gain entry. Clearly then, the body's first line of defence is to form a physical or chemical barrier to entry. Should this fail, the next line of defence is the white blood cells. There are two types of white blood cell: **phagocytes** and **lymphocytes**. Phagocytes ingest and destroy the pathogen by a process called phagocytosis before it can cause harm. Lymphocytes are involved in immune responses (Topics 5.3 and 5.5).

Despite various barriers pathogens still frequently gain entry and the next line of defence is then phagocytosis.

Phagocytosis

Large particles, such as some types of bacteria, can be engulfed by cells in the vesicles formed from the cell-surface membrane. This process is called phagocytosis. In the blood, the types of white blood cells that carry out phagocytosis are known as **phagocytes**. They provide an important defence against the pathogens that manage to enter the body. Some phagocytes travel in the blood but can move out of blood vessels into other tissues. Phagocytosis is illustrated in Figure 2 and is summarised below and in Figure 3.

- Chemical products of pathogens or dead, damaged and abnormal cells act as attractants, causing phagocytes to move towards the pathogen (e.g., a bacterium).
- Phagocytes have several receptors on their cell-surface membrane that recognise, and attach to, chemicals on the surface of the pathogen.
- They engulf the pathogen to form a vesicle, known as a **phagosome**.
- Lysosomes move towards the vesicle and fuse with it.
- Enzymes called **lysozymes** are present within the lysosome. These lysozymes destroy ingested bacteria by hydrolysis of their cell walls. The process is the same as that for the digestion of food in the intestines, namely the hydrolysis of larger, insoluble molecules into smaller, soluble ones.
- The soluble products from the breakdown of the pathogen are absorbed into the cytoplasm of the phagocyte.

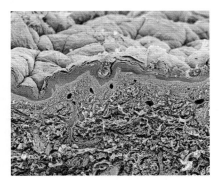

▲ **Figure 1** *Human skin forms a tough outer layer that forms a barrier to the entry of pathogens*

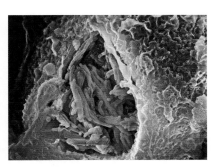

▲ **Figure 2** *False-colour SEM of a phagocyte (red) engulfing tuberculosis bacteria (yellow), a process known as phagocytosis*

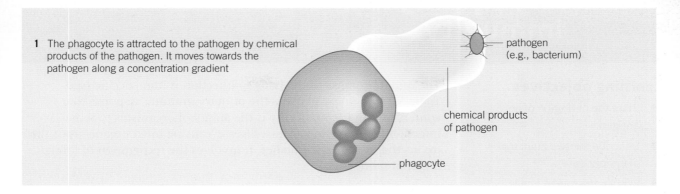

1   The phagocyte is attracted to the pathogen by chemical products of the pathogen. It moves towards the pathogen along a concentration gradient

pathogen (e.g., bacterium)

chemical products of pathogen

phagocyte

2   The phagocyte has several receptors on its cell-surface membrane that attach to chemicals on the surface of the pathogen

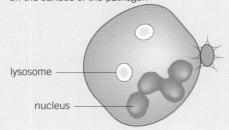

lysosome

nucleus

3   Lysosomes within the phagocyte migrate towards the phagosome formed by engulfing the bacterium

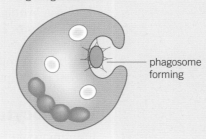

phagosome forming

4   The lysosomes release their lysozymes into the phagosome, where they hydrolyse the bacterium

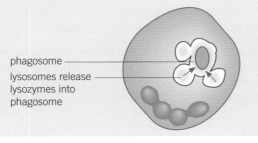

phagosome

lysosomes release lysozymes into phagosome

5   The hydrolysis products of the bacterium are absorbed by the phagocyte

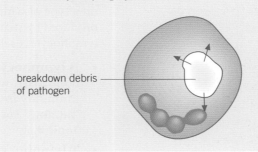

breakdown debris of pathogen

▲ **Figure 3** *Summary of phagocytosis*

Summary questions

1   In the following passage, state the missing word indicated by each letter **a**–**d**.

Pathogens that invade the body may be engulfed by cells which carry out **a**. The engulfed pathogen forms a vesicle known as a **b**. Once engulfed the pathogen is broken down by enzymes called **c** released from organelles called **d**.

2   Among other places, lysozymes are found in tears. Suggest a reason why this is so.

The initial response of the body to infection is non-specific (see Topic 5.2). The next phase is the primary immune response that confers immunity. Immunity is the ability of organisms to resist infection by protecting against disease-causing microorganisms or their toxins that invade their bodies. It involves the recognition of foreign material (antigens).

Antigens

An antigen is any part of an organism or substance that is recognised as non-self (foreign) by the immune system and stimulates an immune response. Antigens are usually proteins that are part of the cell-surface membranes or cell walls of invading cells, such as microorganisms, or abnormal body cells, such as cancer cells. The presence of an antigen triggers the production of an antibody as part of the body's defence system (see Topic 5.4).

Lymphocytes

Immune responses such as phagocytosis are **non-specific** (see Topic 5.2) and occur whatever the infection. The body also has **specific** responses that react to specific antigens. These are slower in action at first, but they can provide long-term immunity. This specific immune response depends on a type of white blood cell called a **lymphocyte**. Lymphocytes are produced by stem cells in the bone marrow. There are two types of lymphocyte, each with its own role in the immune response:

* **B lymphocytes (B cells)** are so called because they mature in the bone marrow. They are associated with humoral immunity, that is, immunity involving antibodies that are present in body fluids, or 'humour' such as blood plasma. This is described in more detail in Topic 5.5.
* **T lymphocytes (T cells)** are so called because they mature in the thymus gland. They are associated with cell-mediated immunity, that is immunity involving body cells.

Cell-mediated immunity

Lymphocytes respond to an organism's own cells that have been infected by non-self material from a different species, for example a virus. They also respond to cells from other individuals of the same species because these are genetically different. These therefore have different antigens on their cell-surface membrane from the antigens on the organism's own cells. T lymphocytes can distinguish these invader cells from normal cells because:

* phagocytes that have engulfed and hydrolysed a pathogen present some of a pathogen's antigens on their own cell-surface membrane
* body cells invaded by a virus present some of the viral antigens on their own cell-surface membrane.

- transplanted cells from individuals of the same species have different antigens on their cell-surface membrane
- cancer cells are different from normal body cells and present antigens on their cell-surface membranes.

Cells that display foreign antigens on their surface are called **antigen-presenting cells** because they can present antigens of other cells on their own cell-surface membrane.

T lymphocytes will only respond to antigens that are presented on a body cell (rather than to antigens within the body fluids). This type of response is called **cell-mediated immunity** or the **cellular response**. The role of the receptors on T cells is important. The receptors on each T cell respond to a single antigen. It follows that there is a vast number of different types of T cell, each one responding to a different antigen (see Topic 5.1). The stages in the response of T lymphocytes to infection by a pathogen are summarised in Figure 1 and explained below.

1   Pathogens invade body cells or are taken in by phagocytes.

2   The phagocyte places antigens from the pathogen on its cell-surface membrane.

3   Receptors on a specific helper T cell (T_H cell) fit exactly onto these antigens.

4   This attachment activates the T cell to divide rapidly by mitosis and form a clone of genetically identical cells.

5   The cloned T cells:

    a   develop into memory cells that enable a rapid response to future infections by the same pathogen

    b   stimulate phagocytes to engulf pathogens by phagocytosis

    c   stimulate B cells to divide and secrete their antibody

    d   activate cytotoxic T cells (T_C cells).

Hint

Three terms that are frequently confused are *antigen*, *antibody* and *antibiotic*. When dealing with immunity put *antibiotic* out of your mind – it has nothing to do with immunity.

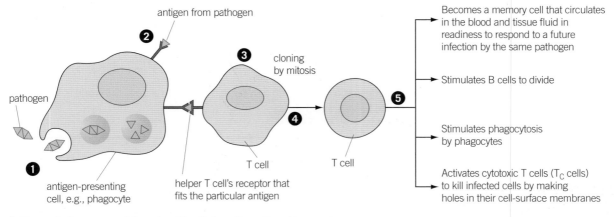

▲ **Figure 1** *Summary of the role of T cells in cell-mediated immunity*

How cytotoxic T cells kill infected cells

Cytotoxic T cells (T_C cells) kill abnormal cells and body cells that are infected by pathogens, by producing a protein called perforin that makes holes in the cell-surface membrane. These holes mean the cell membrane becomes freely permeable to all substances and the cell dies as a result. This illustrates the vital importance of cell-surface membranes in maintaining the integrity of cells and hence their survival. The action of T cells is most effective against viruses because viruses replicate inside cells. As viruses use living cells in which to replicate, this sacrifice of body cells prevents viruses multiplying and infecting more cells.

Summary questions

1 Define an antigen

2 State *two* similarities between T cells and B cells.

3 State *two* differences between T cells and B cells.

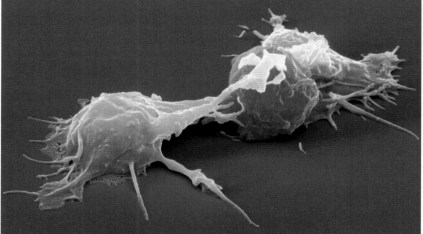

▲ **Figure 2** *False-colour SEM of two human cytotoxic T cells (yellow) attacking a cancer cell (red)*

Bird flu

Avian (bird) flu is caused by one of many strains of the influenza virus. Although it is adapted primarily to infect birds, the H5N1 strain of the virus can infect other species, including humans. Avian flu affects the lungs and can cause the immune system to go into overdrive. This results in a massive overproduction of T cells.

1 From your knowledge of cell-mediated immunity and lung structure suggest why humans infected with the H5N1 virus may sometimes die from suffocation.

2 Suggest a reason why any spread of bird flu across the world is likely to be very rapid.

We saw in Topic 5.3 that the first phase of the specific response to infection is the mitotic division of specific T cells to form a clone of the relevant T cells to build up their numbers. Some of these T cells produce factors that stimulate B cells to divide. It is these B cells that are involved in the next phase of the immune response: humoral immunity.

Humoral immunity

Humoral immunity is so called because it involves antibodies (see Topic 5.5), and antibodies are soluble in the blood and tissue fluid of the body. An old-fashioned word for body fluids is 'humour'. There are many different types of B cell, possibly as many as ten million, and each B cell starts to produce a specific antibody that responds to one specific antigen. When an antigen, for example, a protein on the surface of a pathogen, foreign cell, toxin, damaged or abnormal cell, enters the blood or tissue fluid, there will be one B cell that has an antibody on its surface whose shape exactly fits the antigen, that is, they are complementary. The antibody therefore attaches to this complementary antigen. The antigen enters the B cell by endocytosis and gets presented on its surface (processed). T_H cells bind to these processed antigens and stimulate this B cell to divide by mitosis (see Topic 3.7) to form a clone of identical B cells, all of which produce the antibody that is specific to the foreign antigen. This is called **clonal selection** and accounts for the body's ability to respond rapidly to any of a vast number of antigens.

In practice, a typical pathogen has many different proteins on its surface, all of which act as antigens. Some pathogens, such as the bacterium that causes cholera, also produce toxins. Each toxin molecule also acts as an antigen. Therefore many different B cells make clones, each of which produces its own type of antibody. As each clone produces one specific antibody these antibodies are referred to as monoclonal antibodies (see Topic 5.5). In each clone, the cells produced develop into one of two types of cell:

- **Plasma cells** secrete antibodies usually into blood plasma. These cells survive for only a few days, but each can make around 2000 antibodies every second during its brief lifespan. These antibodies lead to the destruction of the antigen. The plasma cells are therefore responsible for the immediate defence of the body against infection. The production of antibodies and memory cells (see below) is known as the **primary immune response**.

- **Memory cells** are responsible for the **secondary immune response.** Memory cells live considerably longer than plasma cells, often for decades. These cells do not produce antibodies directly, but circulate in the blood and tissue fluid. When they encounter the same antigen at a later date, they divide rapidly and develop into plasma cells and more memory cells. The plasma cells produce the antibodies needed to destroy the pathogen, while the new memory cells circulate in readiness for any future infection. In this way, memory cells provide long-term immunity against the original infection. An increased quantity of antibodies is secreted at a faster rate than in the primary immune response. It ensures that a new infection is destroyed before it can cause any harm – and individuals

Learning objectives

→ Explain the role of B cells (B lymphocytes) in humoral immunity.

→ Explain the roles of plasma cells and antibodies in the primary immune response.

→ Explain the role of memory cells in the secondary immune response.

→ Explain how antigenic variation affects the body's response to infection.

Specification reference: 3.2.4

Hint

Remember that B cells with the appropriate antibody to bind to antigens of a pathogen are not produced in response to the pathogen. They are present from birth. Being present, they simply *multiply* in response to the pathogen.

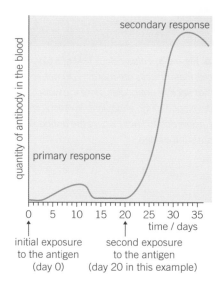

▲ **Figure 1** *Primary and secondary responses to an antigen*

are often totally unaware that they have ever been infected. Figure 1 illustrates the relative amounts of antibody produced in the primary and secondary immune responses.

The role of B cells in immunity is explained below and summarised in Figure 2.

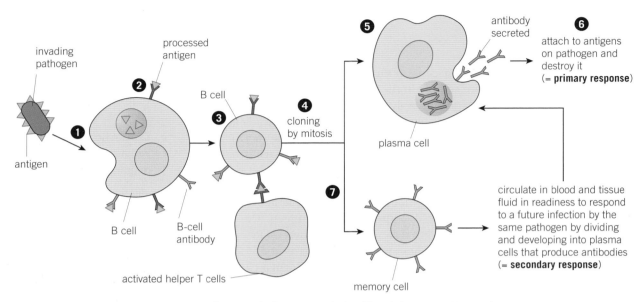

▲ **Figure 2** *Summary of role of B cells in humoral immunity*

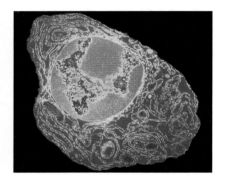

▲ **Figure 3** *False-colour TEM of a plasma cell. Plasma cells are mature B lymphocytes that secrete antibodies. Note the well-developed rough endoplasmic reticulum (yellow dotted lines) where the antibodies are synthesised.*

1 The surface antigens of an invading pathogen are taken up by a B cell.

2 The B cell processes the antigens and presents them on its surface.

3 Helper T cells (activated in the process described in Topic 5.3) attach to the processed antigens on the B cell thereby activating the B cell.

4 The B cell is now activated to divide by **mitosis** to give a clone of plasma cells.

5 The cloned plasma cells produce and secrete the specific antibody that exactly fits the antigen on the pathogen's surface.

6 The antibody attaches to antigens on the pathogen and destroys them (see Topic 5.5).

7 Some B cells develop into memory cells. These can respond to future infections by the same pathogen by dividing rapidly and developing into plasma cells that produce antibodies. This is the secondary immune response.

Summary questions

1 Explain why the secondary immune response is much more rapid than the primary one.

2 Contrast the cell-mediated and humoral responses to a pathogen.

3 Plasma cells can produce around 2000 protein antibodies each second. Suggest **three** cell organelles that you might expect to find in large quantities in a plasma cell, and explain why.

5.5 Antibodies

In Topic 5.4 we saw how B cells respond to **antigens** by producing antibodies. Let us now look at antibodies and how they work in more detail.

Antibodies

Antibodies are proteins with specific binding sites synthesised by B cells. When the body is infected by non-self material, a B cell produces a specific antibody. This specific antibody reacts with an antigen on the surface of the non-self material by binding to them. Each antibody has two identical binding sites. The antibody binding sites are complementary to a specific antigen. The massive variety of antibodies is possible because they are made of proteins – molecules that occur in an almost infinite number of forms.

Antibodies are made up of four polypeptide chains. The chains of one pair are long and are called **heavy chains**, while the chains of the other pair are shorter and are known as **light chains**. Each antibody has a specific binding site that fits very precisely onto a specific antigen to form what is known as an **antigen–antibody complex**. The binding site is different on different antibodies and is therefore called the **variable region**. Each binding site consists of a sequence of amino acids that form a specific 3-D shape that binds directly to a specific antigen. The rest of the antibody is known as the **constant region**. This binds to receptors on cells such as B cells. The structure of an antibody is illustrated in Figure 1.

How the antibody leads to the destruction of the antigen

It is important to understand that antibodies do not destroy antigens directly but rather prepare the antigen for destruction. Different antibodies lead to the destruction of an antigen in a range of ways. Take the example of when the antigen is a bacterial cell – antibodies assist in its destruction in two ways:

- They cause agglutination of the bacterial cells (Figure 2). In this way clumps of bacterial cells are formed, making it easier for the phagocytes to locate them as they are less spread-out within the body.
- They then serve as markers that stimulate phagocytes to engulf the bacterial cells to which they are attached.

Hint

One molecule fitting neatly with another is a recurring theme throughout biology. We met it with enzymes (see Topic 1.7) and with T cells (see Topic 5.3) and it features again here. While the 'lock and key' image is helpful, remember that, with the induced fit model of enzyme action, the molecules are flexible rather than rigid. This is the same for antibodies. The image of a hand fitting a glove is therefore perhaps a better one when it comes to understanding the process.

Study tip

Agglutination is possible because each antibody has two antigen binding sites.

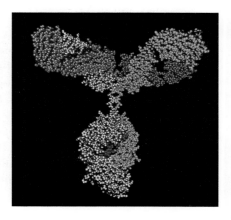

antigen-binding sites

light chain

heavy chain

receptor binding site

variable region (different in different antibodies)

constant region

◀ **Figure 1** *Structure of an antibody (left); molecular model of an antibody (right). This Y-shaped protein is produced by B lymphocytes as part of the immune response*

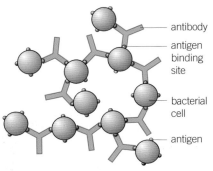

antibody

antigen binding site

bacterial cell

antigen

Each antibody attaches to two bacterial cell, causing them to clump together

▲ **Figure 2** *Agglutination*

Monoclonal antibodies

We have seen that a bacterium or other microorganism entering the body is likely to have many hundreds of different antigens on its surface. Each antigen will induce a different B cell to multiply and form a clone of itself. Each of these clones will produce a different antibody. It is of considerable medical value to be able to produce antibodies outside the body. It is even better if a single type of antibody can be isolated and cloned. Such antibodies are known as **monoclonal antibodies**.

Monoclonal antibodies have a number of useful functions in science and medicine.

Targeting medication to specific cell types by attaching a therapeutic drug to an antibody

As an antibody is very specific to particular antigen (protein), monoclonal antibodies can be used to target specific substances and specific cells. One type of cell they can target is cancer cells. Monoclonal antibodies can be used to treat cancer in a number of ways. By far the most successful so far is direct monoclonal antibody therapy.

- Monoclonal antibodies are produced that are specific to antigens on cancer cells.
- These antibodies are given to a patient and attach themselves to the receptors on their cancer cells.
- They attach to the surface of their cancer cells and block the chemical signals that stimulate their uncontrolled growth.

An example is herceptin, a monoclonal antibody used to treat breast cancer. The advantage of direct monoclonal antibody therapy is that since the antibodies are not toxic and are highly specific, they lead to fewer side effects than other forms of therapy.

Another method, called indirect monoclonal antibody therapy, involves attaching a radioactive or cytotoxic drug (a drug that kills cells) to the monoclonal antibody. When the antibody attaches to the cancer cells, it kills them.

For obvious reasons, monoclonal antibodies used in this way are referred to as 'magic bullets' and can be used in smaller doses, as they are targeted on specific sites. Using them in smaller doses is not only cheaper but also reduces any side effects the drug might have.

Medical diagnosis

Monoclonal antibodies are an invaluable tool in diagnosing disease with over a hundred different diagnostic products based on them. They are used for the diagnosis of influenza, hepatitis and chlamydia infections where they produce a much more rapid result than conventional methods of diagnosis. They are important in diagnosing certain cancers. For example, men with prostate cancer often produce more of a protein called prostate specific antigen (PSA) leading to unusually high levels of it in the blood. By using a monoclonal antibody that interacts with this antigen, it is possible to obtain a measure of the level of PSA in a sample of blood. While a higher than normal level of PSA is not itself diagnostic of the disease, it gives an early warning of its possibility and the need for further tests. The use of antibodies in the ELISA test is discussed in Topic 5.7.

Pregnancy testing

It is important that a mother knows as early as possible that she is pregnant, not least because there are certain actions she can take to ensure the welfare of herself and her unborn baby. The use of pregnancy testing kits that can easily be used at home has made possible the early detection of a pregnancy. These kits rely on the fact that the placenta produces a hormone called human chorionic gonadatrophin (hCG) and that this is found in the mother's urine. Monoclonal antibodies present on the test strip of a home pregnancy testing kit are linked to coloured particles. If hCG is present in the urine it binds to these antibodies. The hCG-antibody-colour complex moves along the strip until it is trapped by a different type of antibody creating a coloured line.

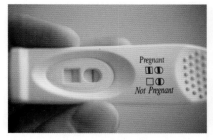

▲ **Figure 3** *Home pregnancy testing kit showing a positive result. These kits use monoclonal antibodies*

Ethical use of monoclonal antibodies

The development of monoclonal antibodies has provided society with the power and opportunity to treat diseases in hitherto unknown ways. However, with this power and opportunity comes responsibility. The use of monoclonal antibodies raises some ethical issues.

*   Production of monoclonal antibodies involves the use of mice. These mice are used to produce both antibodies and tumour cells. The production of tumour cells involves deliberately inducing cancer in mice. Despite the specific guidelines drawn up to minimise any suffering, some people still have reservations about using animals in this way.

*   Monoclonal antibodies have been used successfully to treat a number of diseases, including cancer and diabetes, saving many lives. There have also been some deaths associated with their use in the treatment of multiple sclerosis. It is important that patients have full knowledge of the risks and benefits of these drugs before giving permission for them to be used (=informed consent).

*   Testing for the safety of new drugs presents certain dangers. In March 2006, six healthy volunteers took part in the trial of a new monoclonal antibody (TGN1412) in London. Within minutes they suffered multiple organ failure, probably as a result of T cells overproducing chemicals that stimulate an immune response or attacking the body tissues. All the volunteers survived, but it raises issues about the conduct of drug trials.

Society must use the issues raised here, combined with current scientific knowledge about monoclonal antibodies, to make decisions about their use. We must balance the advantages that a new medicine provides with the dangers that its use might bring. Only then can we make informed decisions at individual, local, national and global levels about the ethical use of drugs such as monoclonal antibodies.

▲ **Figure 4** *A scientist adding monoclonal antibodies to human tissue samples in order to detect cancer*

Study tip

When discussing social and ethical issues such as these, do not resort to comments such as "Who are we to play God?". General arguments should be supported with sound biology.

Summary questions

1 Suggest why antibodies made of proteins, rather than carbohydrates or fats, are more likely to be effective against a wide range of diseases.

2 Distinguish between an antigen and an antibody.

3 Discuss whether drug trials should be limited to volunteers who are terminally ill with a condition that the monoclonal antibody is designed to treat.

 Producing monoclonal antibodies

The production of a large quantities of a specific antibody has long been recognised as useful. The problem had always been that B cells are short-lived and only divide inside a living organism. Nowadays, large quantities of a single antibody can be produced outside the body.

There was much competition among scientific research teams to overcome the problem of getting B cells to grow indefinitely outside of the body and a variety of methods was investigated. Cesar Milstein and Georges Kohler evaluated these methods in relation to the behaviour of cancer cells. In 1975, they produced a solution to the problem by developing the following procedure:

- A mouse is exposed to the non-self material against which an antibody is required.

- The B cells in the mouse then produce a mixture of antibodies, which are extracted from the spleen of the mouse.

- To enable these B cells to divide outside the body, they are mixed with cells that divide readily outside the body, for example, cells from a cancer tumour.

- Detergent is added to the mixture to break down the cell-surface membranes of both types of cell and enable them to fuse together. The fused cells are called **hybridoma cells**.

- The hybridoma cells are separated under a microscope and each single cell is cultured to form a clone. Each clone is tested to see whether it is producing the required antibody.

- Any clone producing the required antibody is grown on a large scale and the antibodies are extracted from the growth medium.

- Because these antibodies come from a clone formed from a single B cell, they are called **monoclonal antibodies**.

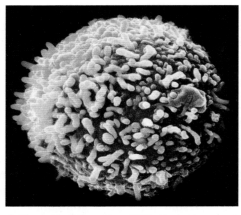

▲ **Figure 5** *False-colour SEM of hybridoma cell used to produce monoclonal antibodies*

As these monoclonal antibodies come from mouse tissue, they have to be modified to make them like human cells before they can be used. This process is called humanisation.

1 From your knowledge of membrane structure, suggest a reason why detergent might causes B cells and tumour cells to fuse.

2 When the detergent is added to the cells, the mixture is gently agitated. Suggest a reason why.

3 Explain why cells from cancer tumours are used to fuse with the B cells?

4 Some B cells and tumour cells fuse together. Suggest which other cells might also fuse together.

5 Explain why it is necessary to carry out 'humanisation' of the monoclonal antibodies.

6 One way to eliminate the need for humanisation would be to inject humans with an antigen and then extract the antibodies produced in response to it. Suggest reasons why this is considered unethical.

Immunity is the ability of an organism to resist infection. This immunity takes two forms.

- **Passive immunity** is produced by the introduction of antibodies into individuals from an outside source. No direct contact with the pathogen or its antigen is necessary to induce immunity. Immunity is acquired immediately. As the antibodies are not being produced by the individuals themselves, the antibodies are not replaced when they are broken down, no memory cells are formed and so there is no lasting immunity. Examples of passive immunity include anti-venom given to the victims of snake bites and the immunity acquired by the fetus when antibodies pass across the placenta from the mother.

- **Active immunity** is produced by stimulating the production of antibodies by the individuals' own immune system. Direct contact with the pathogen or its antigen is necessary. Immunity takes time to develop. It is generally long-lasting and is of two types:

  - **Natural active immunity** results from an individual becoming infected with a disease under normal circumstances. The body produces its own antibodies and may continue to do so for many years.

  - **Artificial active immunity** forms the basis of vaccination (immunisation). It involves inducing an immune response in an individual, without them suffering the symptoms of the disease.

Vaccination is the introduction of the appropriate disease antigens into the body, either by injection or by mouth. The intention is to stimulate an immune response against a particular disease. The material introduced is called **vaccine** and, in whatever form (see below), it contains one or more types of antigen from the pathogen. These antigens stimulate the immune response as described in Topics 5.3 and 5.4. The response is slight because only a small amount of antigen has been introduced, However, the crucial factor is that memory cells (see Topic 5.4) are produced. These remain in the blood and allow a greater, and more immediate, response to a future infection with the pathogen. The result is that there is a rapid production of antibodies and the new infection is rapidly overcome before it can cause any harm and with few, if any, symptoms.

When carried out on a large scale, this provides protection against disease not only for individuals, but also for whole populations.

Features of a successful vaccination programme

It is important to understand that vaccination is used as a precautionary measure to prevent individuals contracting a disease. It is not a means of treating individuals who already have the disease. Some programmes of vaccination against diseases have had considerable success. Yet, in other instances, similar measures have been less successful. The success of a vaccination programme depends on a number of factors:

- A suitable vaccine must be economically available in sufficient quantities to immunise most of the vulnerable population.

Learning objectives

→ Describe the nature of vaccines.

→ Describe the features of an effective vaccination programme.

→ Explain why vaccination rarely eliminates a disease.

→ Discuss the ethical issues associated with vaccination programmes.

Specification reference: 3.2.4

▲ **Figure 1** *The development of new vaccines is a highly technological process requiring sterile conditions*

- There must be few side-effects, if any, from vaccination. Unpleasant side-effects may discourage individuals in the population from being vaccinated.
- Means of producing, storing and transporting the vaccine must be available. This usually involves technologically advanced equipment, hygienic conditions and refrigerated transport.
- There must be the means of administering the vaccine properly at the appropriate time. This involves training staff with appropriate skills at different centres throughout the population.
- It must be possible to vaccinate the vast majority of the vulnerable population to produce **herd immunity**.

Herd immunity

Herd immunity arises when a sufficiently large proportion of the population has been vaccinated to make it difficult for a pathogen to spread within that population. The concept is based on the idea that pathogens are passed from individual to individual when in close contact. Where the vast majority of the population is immune, it is highly improbable that a susceptible individual will come in contact with an infected person. In this way those individuals who are not immune to the disease are nevertheless protected.

Herd immunity is important because it is never possible to vaccinate everyone in a large population. For example, babies and very young children are not vaccinated because their immune system is not yet fully functional. It could also be dangerous to vaccinate those who are ill or have compromised immune systems. The percentage of the population that must be vaccinated in order to achieve herd immunity is different for each disease. To achieve herd immunity, vaccination is best carried out at one time. This means that, for a certain period, there are very few individuals in the population with the disease and the transmission of the pathogen is interrupted.

Why vaccination may not eliminate a disease

Even when these criteria for successful vaccination are met, it can still prove extremely difficult to eradicate a disease. The reasons are as follows:

- Vaccination fails to induce immunity in certain individuals, for example people with defective immune systems.
- Individuals may develop the disease immediately after vaccination but before their immunity levels are high enough to prevent it. These individuals may harbour the pathogen and reinfect others.
- The pathogen may mutate frequently, so that its antigens change suddenly rather than gradually. This means that vaccines suddenly become ineffective because the new antigens on the pathogen are no longer recognised by the immune system. As a result the immune system does not produce the antibodies to destroy the pathogen. This **antigenic variability** happens with the influenza virus, which changes its antigens frequently. Immunity is therefore short-lived and individuals may develop repeated bouts of influenza during their lifetime.

- There may be so many varieties of a particular pathogen that it is almost impossible to develop a vaccine that is effective against them all. For example, there are over 100 varieties of the common cold virus and new ones are constantly evolving.

- Certain pathogens 'hide' from the body's immune system, either by concealing themselves inside cells, or by living in places out of reach, such as within the intestines, for example, the cholera pathogen.

- Individuals may have objections to vaccination for religious, ethical or medical reasons. For example, unfounded concerns over the measles, mumps and rubella (MMR) triple vaccine has led a number of parents to opt for separate vaccinations for their children, or to avoid vaccination altogether.

▲ **Figure 2** *Vaccination programmes for children have considerably reduced deaths from infectious diseases*

The ethics of using vaccines

As vaccinations have saved millions of lives, it is easy to accept vaccination programmes without question. However, they do raise ethical issues that need to be addressed if such programmes are to command widespread support. The production and use of vaccines raises the following questions:

- The production of existing vaccines, and the development of new ones, often involves the use of animals. How acceptable is this?

- Vaccines have side-effects that may sometimes cause long-term harm. How can the risk of side-effects be balanced against the risk of developing a disease that causes even greater harm?

- On whom should vaccines be tested? How should such trials be carried out? To what extent should individuals be asked to accept risk in the interests of the public health?

- Is it acceptable to trial a new vaccine with unknown health risks only in a country where the targeted disease is common, on the basis that the population there has most to gain if it proves successful?

- To be fully effective the majority, and preferably all, of the population should be vaccinated. Is it right, in the interests of everyone's health, that vaccination should be compulsory? If so, should this be at any time, or just when there is a potential epidemic? Can people opt out? If so, on what grounds: religious belief, medical circumstances, personal belief?

- Should expensive vaccination programmes continue when a disease is almost eradicated, even though this might mean less money for the treatment of other diseases?

- How can any individual health risks from vaccination be balanced against the advantages of controlling a disease for the benefit of the population at large?

> ### Study tip
> When discussing ethical issues, always present a balanced view that reflects both sides of the debate and support your arguments with relevant biological information.

Summary questions

1 Distinguish between active immunity and passive immunity.

2 Explain why vaccinating against influenza is not always effective.

MMR vaccine

In 1988, a combined vaccine for measles, mumps and rubella (MMR) was introduced into the UK to replace three separate vaccines. All three diseases are potentially disabling. Mumps can lead to orchitis in men possibly causing sterility and measles is potentially lethal. Ten years later a study was published in a well-respected medical journal. This suggested that there was a higher incidence of autism among children who had received the triple MMR vaccine than those who had received separate vaccinations. Autism is a condition in which individuals have impaired social interaction and communication skills.

In the wake of the media furore that followed, many parents decided to have their children vaccinated separately for the three diseases, while others opted for no vaccination at all. Parents of autistic children recalled that symptoms of the disorder emerged at around 14 months of age – shortly after the children had been given the MMR vaccination, adding to public concern about the MMR vaccine. The incidence of measles, mumps and rubella rose.

The vast majority of scientists now think that the vaccine is safe. A number of facts have emerged since the first research linking the MMR vaccine to autism.

- The author of the research had a conflict of interests. He was also being paid by the Legal Aid Board to discover whether parents who claimed their children had been damaged by MMR had a case. Some children were included in both studies.
- Further studies, including one in Japan involving over 30 000 children, have found no link between the MMR vaccine and autism.
- The sample size of the initial research was very small relative to later studies.
- The journal that published the initial research has publicly declared that, had it known all the facts, it would not have published the work.

1 Autism experts point out that many of the symptoms of autism first occur around the age of 14 months. Explain why this information is relevant to the debate on whether MMR vaccine and autism are linked.

2 Discuss how an organisation funding research might influence the outcome of that research without dishonestly altering the findings.

Even without this additional evidence, care has to be exercised when looking at data, specially where there are correlations between two factors, In this example, almost all the population had been vaccinated with the MMR vaccine. There would therefore be a correlation between people who had been vaccinated and almost everything – what they

▲ **Figure 3** *MMR vaccination phial*

ate, where they lived etc. For example, data would have shown that that the majority of children who died in road accidents had been given the MMR vaccine. It does not follow that MMR causes road accidents. It was clearly a difficult choice for parents. Some parents, understandably, opted for separate vaccinations. Others mistrusted vaccinations in general and left their children unprotected. As a result, some children have developed disabilities that could have been avoided. On the other hand, had the research proved valid, it would have been those who held faith with the MMR vaccine who would have been putting their children's health at risk. It was a real dilemma.

The public sometimes believe that all such evidence must be true and accept it uncritically. However, all scientific evidence should be initially treated with caution – after all it is fellow scientists who are often quickest to criticise. There are various reasons for this caution:

- To be universally accepted, a scientific theory must first be critically appraised and confirmed by other scientists in the field. The confirmation of a theory takes time.
- Some scientists may not be acting totally independently but may be funded by other people or organisations who are anticipating a particular outcome from the research.
- Scientists' personal beliefs, views and opinions may influence the way they approach or represent their research.
- The facts, as presented by media headline writers, companies, governments and other organisations, may have been biased or distorted to suit their own interests.
- New knowledge may challenge accepted scientific beliefs; theories are being modified all the time.

The human immunodeficiency virus (HIV)

The human immunodeficiency virus causes the disease **acquired immune deficiency syndrome (AIDS)**. Among contagious diseases it is a relative newcomer, having been first diagnosed in 1981. In this topic we will look at the structure of HIV and how it leads to the symptoms of AIDS.

Structure of the human immunodeficiency virus

The structure of HIV is shown in Figure 1. On the outside is a **lipid envelope**, embedded in which are peg-like **attachment proteins**. Inside the envelope is a protein layer called the **capsid** that encloses two single strands of **RNA** and some enzymes. One of these enzymes is **reverse transcriptase**, so-called because it catalyses the production of DNA from RNA – the reverse reaction to that carried out by transcriptase. The presence of reverse transcriptase, and consequent ability to make DNA from RNA, means that HIV belongs to a group of viruses called **retroviruses**.

Learning objectives

→ Describe the structure of the human immunodeficiency virus.

→ Explain how the human immunodeficiency virus replicates.

→ Explain how the human immunodeficiency virus causes AIDS.

→ Describe the treatment and control of AIDS.

→ Explain how the ELISA test works.

→ Explain why antibiotics are ineffective against viruses.

Specification reference: 3.2.4

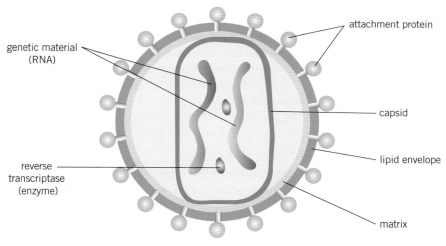

genetic material (RNA)

attachment protein

capsid

lipid envelope

reverse transcriptase (enzyme)

matrix

▲ **Figure 1** *Structure of HIV*

Replication of the human immunodeficiency virus

Being a virus, HIV cannot replicate itself. Instead it uses its genetic material to instruct the host cell's biochemical mechanisms to produce the components required to make new HIV. It does so as follows:

- Following infection HIV enters the bloodstream and circulates around the body.
- A protein on the HIV readily binds to a protein called CD4. While this protein occurs on a number of different cells, HIV most frequently attaches to helper T cells (see Topic 5.3).
- The protein capsid fuses with the cell-surface membrane. The RNA and enzymes of HIV enter the helper T cell.
- The HIV reverse transcriptase converts the virus's RNA into DNA.

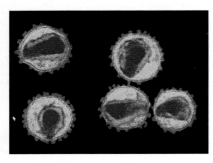

▲ **Figure 2** *Colourised TEM of HIV*

- The newly made DNA is moved into the helper T cell's nucleus where it is inserted into the cell's DNA.
- The HIV DNA in the nucleus creates **messenger RNA** (mRNA), using the cell's enzymes. This mRNA contains the instructions for making new viral proteins and the RNA to go into the new HIV.
- The mRNA passes out of the nucleus through a nuclear pore and uses the cell's protein synthesis mechanisms to make HIV particles.
- The HIV particles break away from the helper T cell with a piece of its cell-surface membrane surrounding them which forms their lipid envelope.

Once infected with HIV a person is said to be **HIV positive**. However, the replication of HIV often goes into dormancy and only recommences, leading to AIDS, many years later.

How HIV causes the symptoms of AIDS

The human immunodeficiency virus specifically attacks helper T cells. HIV causes AIDS by killing or interfering with the normal functioning of helper T cells. An uninfected person normally has between 800 and 1200 helper T cells in each mm^3 of blood. In a person suffering from AIDS this number can be as low as $200\,mm^{-3}$. We have seen (Topic 5.3) that helper T cells are important in cell-mediated immunity. Without a sufficient number of helper T cells, the immune system cannot stimulate B cells to produce antibodies or the cytotoxic T cells that kill cells infected by pathogens. Memory cells may also become infected and destroyed. As a result, the body is unable to produce an adequate immune response and becomes susceptible to other infections and cancers. Many AIDS sufferers develop infections of the lungs, intestines, brain and eyes, as well as experiencing weight loss and diarrhoea. It is these secondary, diseases that ultimately cause death.

HIV does not kill individuals directly. By infecting the immune system, HIV prevents it from functioning normally. As a result those infected by HIV are unable to respond effectively to other pathogens. It is these infections, rather than HIV, that ultimately cause ill health and eventual death.

The ELISA test

ELISA stands for **enzyme linked immunosorbant assay**. It uses antibodies to not only detect the presence of a protein in a sample but also the quantity. It is extremely sensitive and so can detect very small amounts of a molecule. To understand how the test works, imagine that we are trying to find whether a particular protein, in this case an antigen, is present in a sample. The procedure is as follows:

- Apply the sample to a surface, for example a slide, to which all the antigens in the sample will attach.
- Wash the surface several times to remove any unattached antigens.
- Add the antibody that is specific to the antigen we are trying to detect and leave the two to bind together.
- Wash the surface to remove excess antibody.

- Add a second antibody that binds with the first antibody. This second antibody has an enzyme attached to it.
- Add the colourless substrate of the enzyme. The enzyme acts on the substrate to change it into a coloured product.
- The amount of the antigen present is relative to the intensity of colour that develops.

This basic technique can be used to detect HIV and the pathogens of diseases including tuberculosis and hepatitis. ELISA is especially useful where the quantity of an antigen needs to be measured. In testing for particular drugs in the body for example. The mere presence of a drug is often less important than its quantity as many drugs are found naturally in low concentrations. ELISA is therefore very useful in both drug and allergen tests.

Why antibiotics are ineffective against viral diseases like AIDS

Antibiotics work in a number of different ways. One is by preventing bacteria from making normal cell walls.

In bacterial cells, as in plant cells, water constantly enters by osmosis. This entry of water would normally cause the cell to burst. That it doesn't burst is due to the wall that surrounds all bacterial cells. This wall is made of **murein** (peptidoglycan) a tough material that is not easily stretched. As water enters the cell by osmosis, the cell expands and pushes against the cell wall. Being relatively inelastic, the cell wall resists expansion and so halts further entry of water. Antibiotics like penicillin inhibit certain enzymes required for the synthesis and assembly of the peptide cross-linkages in bacterial cell walls. This weakens the walls, making them unable to withstand pressure. As water enters naturally by osmosis, the cell bursts and the bacterium dies.

Viruses rely on the host cells to carry out their metabolic activities and therefore lack their own metabolic pathways and cell structures. As a result antibiotics are ineffective because there are no metabolic mechanisms or cell structures for them to disrupt. Viruses also have a protein coat rather than a murein cell wall and so do not have sites where antibiotics can work. In any case, when viruses are within an organism's own cells, antibiotics cannot reach them.

Summary questions

1 Explain why is HIV called a retrovirus.
2 Distinguish between HIV and AIDS.
3 Tuberculosis (TB) is a lung disease spread through the air. Suggest a possible reason why the widespread use of condoms might help reduce the incidence of TB in a population.

1   (a)   What is a pathogen?   (*1 mark*)
    (b)   When a pathogen enters the body it may be destroyed by phagocytosis.
          Describe how.   (*4 marks*)
    (c)   When a pathogen causes an infection, plasma cells secrete antibodies
          which destroy this pathogen. Explain why these antibodies are only
          effective against a specific pathogen.   (*2 marks*)
                                                                                       AQA June 2012

2   The diagram shows an antibody molecule.

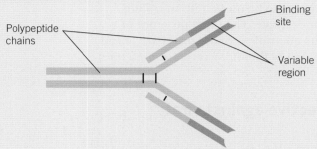

Polypeptide chains — Binding site — Variable region

    (a)   What is the evidence from the diagram that this antibody has
          a quaternary structure?   (*1 mark*)
    (b)   Scientists use this antibody to detect an antigen on the bacterium
          that causes stomach ulcers. Explain why the antibody will only
          detect this antigen.   (*3 marks*)
                                                                                       AQA Jan 2012

3   The table shows the cumulative rise in cases of the
    infectious disease Ebola over a five week
    period in 2014.
    (a)   Plot a graph of the above information
          with the number of weeks on the X axis   (*1 mark*)
    (b)   Calculate the rate of increase in number
          of cases of Ebola in the time period shown
          on the graph.   (*2 marks*)
    (c)   A scientist suggests that the increase in the
          number of cases in the following six months
          will be exponential. Explain how plotting the
          next six months' data on a log scale would
          show whether the increase is exponential.   (*1 mark*)
    (d)   Ebola is a rare disease in the human population, but can be passed
          on to humans from wild animals. Suggest, using your knowledge of
          the immune system, why the disease spreads fast once it is present
          in one human in an urban area.   (*4 marks*)

Week	Number of cases	Number of deaths
1	70	20
2	112	40
3	168	95
4	200	119
5	230	134
6	250	148

4   Read the following passage.

    Microfold cells are found in the epithelium of the small intestine. Unlike other
    epithelial cells in the small intestine, microfold cells do not have adaptations for the
    absorption of food.

    Microfold cells help to protect against pathogens that enter the intestine. They have
    receptor proteins on their cell-surface membranes that bind to antigens on the   5
    surface of pathogens. The microfold cells take up the antigens and transport them to
    cells of the immune system. Antibodies are then produced which give protection
    against the pathogen.

    Scientists believe that it may be possible to develop vaccines that make use of
    microfold cells. These vaccines could be swallowed in tablet form.   10

    Use information from the passage and your own knowledge to answer the
    following questions.

    (a)   Microfold cells do not have adaptations for the absorption of food (lines 2–3).
          Give two adaptations that other epithelial cells have for the absorption of food. (*2 marks*)

(b) (i) Microfold cells have receptor proteins on their cell-surface membranes that bind to antigens (line 5). What is an antigen? *(1 mark)*

(ii) Microfold cells take up the antigens and transport them to cells of the immune system (lines 6–7). Antigens are not able to pass through the cell-surface membranes of other epithelial cells. Suggest **two** reasons why. *(2 marks)*

(c) Scientists believe that it may be possible to develop vaccines that make use of microfold cells (lines 9–10). Explain how this sort of vaccine would lead to a person developing immunity to a pathogen. *(5 marks)*

AQA June 2013

5 Read the passage below.

Most cases of cervical cancer are caused by infection with Human Papilloma Virus (HPV). This virus can be spread by sexual contact. There are many types of HPV, each identified by a number. Most of these types are harmless but types 16 and 18 are most likely to cause cervical cancer.

A vaccine made from HPV types 16 and 18 is offered to girls aged 12 to 13. Three injections of the vaccine are given over six months. In clinical trials, the vaccine has proved very effective in protecting against HPV types 16 and 18. However, it will be many years before it can be shown that this vaccination programme has reduced cases of cervical cancer. Until then, smear tests will continue to be offered to women, even if they have been vaccinated. A smear test allows abnormal cells in the cervix to be identified so that they can be removed before cervical cancer develops.

The Department of Health has estimated that 80% of girls aged 12 to 13 need to be vaccinated to achieve herd immunity to HPV types 16 and 18. Herd immunity is where enough people have been vaccinated to reduce significantly the spread of HPV through the population.

Use information from this passage and your own knowledge to answer the following questions.

(a) HPV vaccine is offered to girls aged 12 to 13 (line 5). Suggest why it is offered to this age group. *(1 mark)*

(b) The vaccine is made from HPV types 16 and 18 (line 5). Explain why this vaccine may **not** protect against other types of this virus. *(2 marks)*

(c) Three injections of the vaccine are given (lines 5 to 6). Use your knowledge of immunity to suggest why. *(2 marks)*

(d) It will be many years before it can be shown that this vaccination programme has reduced cases of cervical cancer (lines 7 to 9). Suggest two reasons why. *(2 marks)*

(e) Smear tests will continue to be offered to women, even if they have been vaccinated (lines 9 to 10). Suggest why women who have been vaccinated still need to be offered smear tests. *(1 mark)*

AQA Jan 2011

The table shows the uptake of the vaccine in one year in four health authority areas.

health authority area	number of girls aged 12–13	number of girls vaccinated
A	14 053	11 151
B	12 789	10 743
C	11 892	8 662
D	8 054	6 524

(f) Analyse mathematically which of these areas would be most likely to show a reduction in the spread of HPV through the population (lines 14 to 16). Use calculations to support your answer. *(2 marks)*

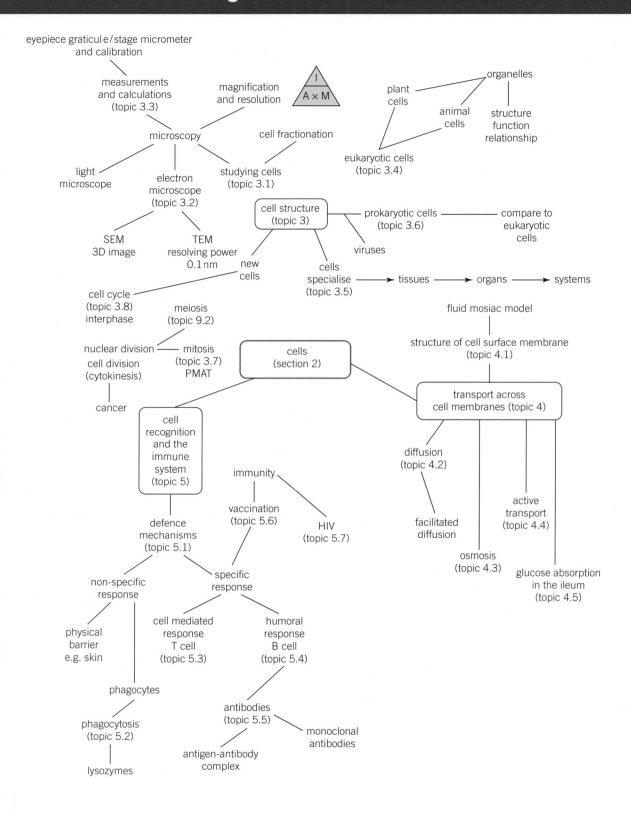

Practical skills

In this section you have met the following practical skills:

- How to use an optical microscope to measure the size of an object using an eyepiece graticule
- How to calibrate an eyepiece graticule
- Understanding the processes of cell fractionation, homogenation and ultracentrifugation
- Using scientific knowledge to solve practical problems
- Considering ethical issues when carrying out experiments.

Maths skills

In this section you have met the following maths skills:

- Using percentages in calculating a mitotic index
- Calculating magnifications and altering the magnification formula to determine the size of the image and/or real object
- Making use of appropriate units in calculations
- Changing the subject of an equation
- Calculating arithmetical means
- Interpreting graphs and finding values using the intercept of a graph.

Extension task

You are a research scientist working for a major pharmaceutical company that is developing immunosuppressant drugs to prevent rejection of newly transplanted tissue. Your section leader has asked you to consider the features that any new immunosuppressant drug should have. She has asked you to produce a report under the following headings:

Possible method(s) of transporting the drug across cell membranes.

Cells of the immune system that the drug should be targeted at.

How the effects of the drug on the targeted cells will reduce the likelihood of the transplanted tissue being rejected.

Possible side effects on other cells.

Likely detrimental effects of the immunosuppressant drug on the recipients and how these might be overcome.

Using the information given in the section of the book you have just completed, and additional information from textbooks, journals and the internet, write a report in around 1000 words. Use clear scientific terminology and specific biological terms.

Section 2 Practice questions

1 **Figure 6** represents part of a DNA molecule.

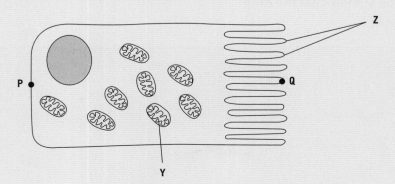

(i)   Draw a box around a single nucleotide.   (*1 mark*)

Table 4 shows the percentage of bases in each of the strands of a DNA molecule.

DNA strand	% of each base			
	A	C	G	T
strand 1	16			
strand 2		21	34	

(ii)   Complete **Table 4** by adding the missing values.   (*2 marks*)
(iii) During replication, the two DNA strands separate and each acts as a template
       for the production of a new strand. As new DNA strands are produced,
       nucleotides can only be added in the 5′ to 3′ direction.
       Use **Figure 6** and your knowledge of enzyme action and DNA replication
       to explain why new nucleotides can only be added in a 5′ to 3′ direction.   (*4 marks*)

AQA SAMS A LEVEL PAPER 1

2 The diagram shows an epithelial cell from the small intestine.

(a) (i)   Name organelle **Y**.   (*1 mark*)
    (ii)   There are large numbers of organelle Y in this cell. Explain how these
        organelles help the cell to absorb the products of digestion.   (*2 marks*)
(b)   This diagram shows the cell magnified 1000 times. Calculate the actual
    length of the cell between points **P** and **Q**. Give your answer in μm.
    Show your working.

                                                            (*2 marks*)

(c)   Coeliac disease is a disease of the human digestive system. In coeliac disease,
    the structures labelled **Z** are damaged.

    Although people with coeliac disease can digest proteins they have low
    concentrations of amino acids in their blood.

    Explain why they have low concentrations of amino acids in their blood.   (*2 marks*)

AQA Jan 2010

3   The human immunodeficiency virus (HIV) leads to the development of
       acquired immunodeficiency syndrome (AIDS). Eventually, people with
       AIDS die because they are unable to produce an immune response
       to pathogens.

       Scientists are trying to develop an effective vaccine to protect people   5
       against HIV. There are three main problems. HIV rapidly enters host
       cells. HIV causes the death of T cells that activate B cells. HIV shows
       a lot of antigenic variability.

       Scientists have experimented with different types of vaccine for HIV.
       One type contains HIV in an inactivated form. A second type contains   10
       attenuated HIV which replicates in the body but does not kill host cells.
       A third type uses a different, non-pathogenic virus to carry genetic
       information from HIV into the person's cells. This makes the person's
       cells produce HIV proteins. So far, these types of vaccine have not
       been considered safe to use in a mass vaccination programme.   15

Use the information in the passage and your own knowledge to answer the following
questions.

(a)   People with AIDS die because they are unable to produce an immune
    response to pathogens (lines 2–4). Explain why this leads to death.   (*3 marks*)
(b)   Explain why each of the following means that a vaccine might **not** be
    effective against HIV.
    (i)   HIV rapidly enters host cells (lines 6–7).   (*2 marks*)
    (ii)   HIV shows a lot of antigenic variability (lines 7–8).   (*2 marks*)
(c)   So far, these types of vaccine have not been considered safe to use in
    a mass vaccination programme (lines 14–15). Suggest why they have
    not been considered safe.   (*3 marks*)

AQA Jan 2013

HAMPTON SCHOOL
BIOLOGY DEPARTMENT

Section 3
Organisms exchange substances with their environment

Introduction

All cells and organisms exchange material between themselves and their environment. To enter or leave an organism, substances must pass across a plasma membrane. Single-celled and small multicellular organisms can satisfactorily exchange materials over their body surfaces using diffusion alone, especially if their metabolic rate is low. As organisms evolved and became larger, their surface area to volume ratios decreased and specialised respiratory surfaces evolved to meet the increasing requirement to exchange ever larger quantities of materials.

Where large size is combined with a high metabolic rate there is a requirement for a mass transport system to move substances between the exchange surface and the cells of which the organism is composed. In animals these systems often involve circulating a specialised transport medium (blood) through vessels using a pump (heart).

Plants do not move from place to place and have a relatively low metabolic rate and consequently reduced demand for oxygen and glucose. Coupled with their large surface area, essential for obtaining light for photosynthesis, they have not evolved a pumped circulatory system. Plants do, however, transport water up from their roots to the leaves and distribute the products of photosynthesis. Their mass transport system comprises vessels too – xylem and phloem, but the movement of fluid within them is largely a passive process.

The internal environment of a cell or organism differs from the environment around it. The cells of large multicellular animals are surrounded by tissue fluid, the composition of which is kept within a suitable metabolic range. In both plants and animals, it is the mass transport system that maintains the final diffusion gradients which allows substances to be exchanged across cell-surface membranes.

Working scientifically

Studying exchange between organisms and the environment allows you to carry out practical work and to develop practical skills. A required practical activity is the dissection of an animal or plant gas exchange system or mass transport system or of an organ within such a system.

You will require a range of mathematical skills; in particular the ability to change the subject of an equation and calculate the surface areas and volumes of various shapes.

What you already know

The material in this unit is intended to be self-explanatory. However, there is some knowledge from GCSE that will aid your understanding of this section. This information includes:

- ☐ The effectiveness of a gas-exchange surface is increased by having a large surface area, being thin, having an efficient blood supply and being ventilated.

- ☐ In humans the surface area of the lungs is increased by alveoli and that of the small intestine by villi. The villi provide a large surface area with an extensive network of capillaries to absorb the products of digestion by diffusion and active transport.

- ☐ Breathing in involves the ribcage moving out and up and the diaphragm becoming flatter. Breathing out involves these changes being reversed.

- ☐ In plants, water and mineral ions are absorbed by roots, the surface area of which is increased by root hairs.

- ☐ Plants have stomata in their leaves through which carbon dioxide and oxygen are exchanged with the atmosphere by diffusion. The size of stomata is controlled by guard cells that surround them and help control water loss.

- ☐ In flowering plants, xylem tissue transports water and mineral ions from the roots to the stem and leaves and phloem tissue carries dissolved sugars from the leaves to the rest of the plant.

- ☐ In animals a circulatory system transports substances using a heart, which is a muscular organ with four main chambers – left and right atria and ventricles.

- ☐ Blood flows from the heart to the organs through arteries and returns through veins. Arteries have thick walls containing muscle and elastic fibres. Veins have thinner walls and often have valves to prevent back-flow of blood.

- ☐ Blood is a tissue and consists of plasma in which red blood cells, white blood cells and platelets are suspended.

- ☐ Red blood cells have no nucleus and are packed with haemoglobin. In the lungs haemoglobin combines with oxygen to form oxyhaemoglobin. In other organs oxyhaemoglobin splits up into haemoglobin and oxygen.

- ☐ White blood cells have a nucleus and form part of the body's defence system against microorganisms.

6.1 Exchange between organisms and their environment

The external environment is different from the internal environment found within an organism and within its cells. To survive, organisms transfer materials between the two environments. This transfer takes place at exchange surfaces and always involves crossing cell plasma membranes. The environment around the cells of multicellular organisms is called **tissue fluid**. The majority of cells are too far from exchange surfaces for diffusion alone to supply or remove their tissue fluid with the various materials needed to keep its composition relatively constant. Therefore, once absorbed, materials are rapidly distributed to the tissue fluid and the waste products returned to the exchange surface for removal. This involves a mass transport system. It is this mass transport system that maintains the diffusion gradients that bring materials to and from the cell-surface membranes.

The size and metabolic rate of an organism will affect the amount of each material that is exchanged. For example, organisms with a high metabolic rate exchange more materials and so require a larger surface area to volume ratio. In turn this is reflected in the type of exchange surface and transport system that evolved to meet the requirements of each organism. In this chapter we will investigate the adaptations of exchange surfaces and transport systems in a variety of organisms.

Examples of things that need to be interchanged between an organism and its environment include: respiratory gases (oxygen and carbon dioxide); nutrients (glucose, fatty acids, amino acids, vitamins, minerals); excretory products (urea and carbon dioxide); and heat.

Except for heat, these exchanges can take place in two ways:

- passively (no metabolic energy is required), by **diffusion** and **osmosis**
- actively (metabolic energy is required), by **active transport**.

Surface area to volume ratio

Exchange takes place at the surface of an organism, but the materials absorbed are used by the cells that mostly make up its volume. For exchange to be effective, the exchange surface(s) of the organism must be large compared with its volume.

Small organisms have a surface area that is large enough, compared with their volume, to allow efficient exchange across their body surface. However, as organisms become larger, their volume increases at a faster rate than their surface area (Table 1). Because of this, simple diffusion of substances across the outer surface can only meet the needs of relatively inactive organisms. Even if the outer surface could supply enough of a substance, it would still take too long for it to reach the middle of the organism if diffusion alone was the method of transport. Organisms have evolved one or more of the following features:

▼ Table 1 *How the surface area to volume ratio gets smaller as an object becomes larger*

Length of edge of a cube / cm	Surface area of whole cube (area of one side × 6 sides) / cm²	Volume of cube (length × width × height) / cm³	Ratio of surface area to volume (surface area ÷ volume)
1	1 × 6 = 6	1 × 1 × 1 = 1	$\frac{6}{1}$ = 6.0 : 1
2	4 × 6 = 24	2 × 2 × 2 = 8	$\frac{24}{8}$ = 3.0 : 1
3	9 × 6 = 54	3 × 3 × 3 = 27	$\frac{54}{27}$ = 2.0 : 1
4	16 × 6 = 96	4 × 4 × 4 = 64	$\frac{96}{64}$ = 1.5 : 1
5	25 × 6 = 150	5 × 5 × 5 = 125	$\frac{150}{125}$ = 1.2 : 1
6	36 × 6 = 216	6 × 6 × 6 = 216	$\frac{216}{216}$ = 1.0 : 1

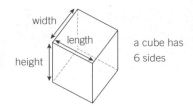

▲ Figure 1 *Calculating volume*

- a flattened shape so that no cell is ever far from the surface (e.g. a flatworm or a leaf)
- specialised exchange surfaces with large areas to increase the surface area to volume ratio (e.g., lungs in mammals, gills in fish).

You may be asked to calculate the surface area to volume ratio of cells with different shapes. To make these calculations reasonably straightforward, cells or organisms may have to be assumed to have a uniform shape although in practice they almost never do.

Maths link √x̄

MS 0.3 and 4.1, see Chapter 11.

Calculating the surface area to volume ratio of cells with different shapes

For example, let us assume a cell has the shape of a sphere that is 10 μm in diameter. The surface area of a sphere is calculated using the formula: $4\pi r^2$

In our example: r = 5 μm (radius = half the diameter) and we will use the value of π as 3.14.

Therefore the surface area of the cell = 4 × 3.14 × (5 × 5) = 314 μm²

The volume of a sphere is calculated using the formula: $\frac{4}{3}\pi r^3$

Therefore the volume of the cell = $\frac{4}{3}$ × 3.14 × (5 × 5 × 5) = 523.33 μm³

The surface area to volume ratio is therefore 314 ÷ 523.33 = 0.6 : 1

Features of specialised exchange surfaces

To allow effective transfer of materials across specialised exchange surfaces by diffusion or active transport, exchange surfaces show the following characteristics:

- a large surface area relative to the volume of the organism which increases the rate of exchange
- very thin so that the diffusion distance is short and therefore materials cross the exchange surface rapidly
- selectively permeable to allow selected materials to cross

Hint

Remember that substances not only have to move into cells through the cell-surface membrane but also into organelles like mitochondria through the plasma membrane that surrounds them. All plasma membranes are therefore thin not just cell-surface membranes.

Summary questions

1 Name **four** general things that need to be exchanged between organisms and their environment.

2 Calculate the surface area to volume ratio of a cube that has sides 10 mm long.

3 Name **three** factors that affect the rate of diffusion of substances into cells.

- movement of the environmental medium, for example, air, to maintain a diffusion gradient
- A transport system to ensure the movement of the internal medium, for example blood, in order to maintain a diffusion gradient.

We saw in Topic 4.2 that the relationship between certain of these factors can be expressed as:

$$\text{diffusion} \propto \frac{\text{surface area} \times \text{difference in concentration}}{\text{length of diffusion path}}$$

Being thin, specialised exchange surfaces are easily damaged and dehydrated. They are therefore often located inside an organism. Where an exchange surface is located inside the body, the organism needs to have a means of moving the external medium over the surface, e.g. a means of ventilating the lungs in a mammal.

Significance of the surface area to volume ratio in organisms

The graph in Figure 2 shows the surface area to volume ratios of different-sized cubes. The ratios are actually 1:1, 2:1, 3:1 etc. but are shown as single numbers for ease of plotting.

1 Microscopic organisms obtain their oxygen by diffusion in across their body surface. Using the graph, explain how they are able to obtain sufficient oxygen for their needs.

2 The blue whale (Figure 3) is the largest organism on the planet. It spends much of its life in cold waters with temperatures between 0 °C and 6 °C. Use the graph to explain one way in which large size is an advantage to blue whales.

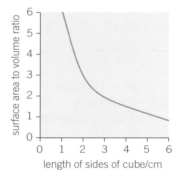

▲ **Figure 2** *Surface area to volume ratios*

Maths link

MS 0.3 and 4.1, see Chapter 11.

Calculating a surface area to volume ratio

Consider the shape shown in Figure 4, which has dimensions marked on it. Use the information below to calculate the ratio of surface area to volume of this shape (to two decimal places).

The area of a disc (like those at the ends of an enclosed cylinder) is calculated using the formula πr^2

The external surface area of an enclosed cylinder is calculated using the formula $2\pi rh + 2\pi r^2$.

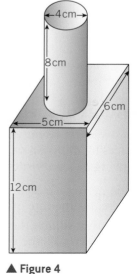

▲ **Figure 4**

Gas exchange in single-celled organisms

Single-celled organisms are small and therefore have a large surface area to volume ratio. Oxygen is absorbed by diffusion across their body surface, which is covered only by a cell-surface membrane. In the same way, carbon dioxide from respiration diffuses out across their body surface. Where a living cell is surrounded by a cell wall, this is no additional barrier to the diffusion of gases.

Gas exchange in insects

As with all terrestrial organisms, insects have evolved mechanisms to conserve water. The increase in surface area required for gas exchange conflicts with conserving water because water will evaporate from it. How insects overcome water loss is discussed in Topic 6.5. For gas exchange, insects have evolved an internal network of tubes called **tracheae**. The tracheae are supported by strengthened rings to prevent them from collapsing. The tracheae divide into smaller dead-end tubes called **tracheoles**. The tracheoles extend throughout all the body tissues of the insect. In this way atmospheric air, with the oxygen it contains, is brought directly to the respiring tissues, as there is a short diffusion pathway from a tracheole to any body cell.

Respiratory gases move in and out of the tracheal system in three ways.

- **Along a diffusion gradient**. When cells are respiring, oxygen is used up and so its concentration towards the ends of the tracheoles falls. This creates a diffusion gradient that causes gaseous oxygen to diffuse from the atmosphere along the tracheae and tracheoles to the cells. Carbon dioxide is produced by cells during respiration. This creates a diffusion gradient in the opposite direction. This causes gaseous carbon dioxide to diffuse along the tracheoles and tracheae from the cells to the atmosphere. As diffusion in air is much more rapid than in water, respiratory gases are exchanged quickly by this method.

- **Mass transport**. The contraction of muscles in insects can squeeze the trachea enabling mass movements of air in and out. This further speeds up the exchange of respiratory gases.

- **The ends of the tracheoles are filled with water**. During periods of major activity, the muscle cells around the tracheoles respire carry out some anaerobic respiration. This produces lactate, which is soluble and lowers the water potential of the muscle cells. Water therefore moves into the cells from the tracheoles by osmosis. The water in the ends of the tracheoles decreases in volume and in doing so draws air further into them. This means the final diffusion pathway is in a gas rather than a liquid phase, and therefore diffusion is more rapid. This increases the rate at which air is moved in the tracheoles but leads to greater water evaporation.

Gases enter and leave tracheae through tiny pores, called **spiracles**, on the body surface. The spiracles may be opened and closed by a valve. When the spiracles are open, water vapour can evaporate from

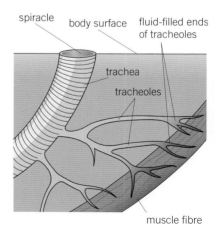

▲ **Figure 1** *Part of an insect tracheal system*

Practical link

Required practical 5. Dissection of animal or plant gas exchange system or mass transport system or of an organ within such a system.

the insect. For much of the time insects keep their spiracles closed to prevent this water loss. Periodically they open the spiracles to allow gas exchange. Part of an insect tracheal system is illustrated in Figure 1.

The tracheal system is an efficient method of gas exchange. It does, however, have some limitations. It relies mostly on diffusion to exchange gases between the environment and the cells. For diffusion to be effective, the diffusion pathway needs to be short which is why insects are of a small size. As a result the length of the diffusion pathway limits the size that insects can attain. Not that being small has hindered insects. They are one of the most successful groups of organisms on Earth.

Summary questions

1   Name the process by which carbon dioxide is removed from a single-celled organism.

2   Explain why there is a conflict in terrestrial insects between gas exchange and conserving water.

3   Explain how the tracheal system limits the size of insects.

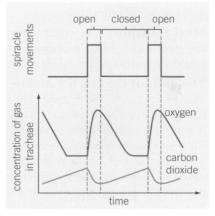

▲ **Figure 2** *Scanning electron micrograph (SEM) of a spiracle of an insect*

Spiracle movements

An experiment was carried out to measure the concentration of oxygen and carbon dioxide in the tracheal system of an insect over a period of time. During the experiment the opening and closing of the insect's spiracles was observed and recorded. The results are shown in Figure 3.

1   Describe what happens to the concentration of oxygen in the tracheae when the spiracles are closed.

2   Suggest an explanation for this change in the concentration of oxygen when the spiracles are closed.

3   Use the information provided by the graph to suggest what causes the spiracles to open.

4   Suggest an advantage of these spiracle movements to a terrestrial insect.

5   Fossil insects have been discovered that are larger than insects that occur on Earth today. What does this suggest about the composition of the atmosphere at the time when these fossil insects lived.

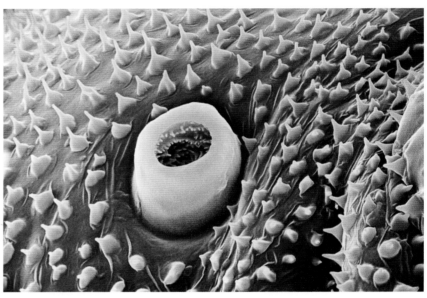

▲ **Figure 3**

Fish have a waterproof, and therefore a gas-tight, outer covering. Being relatively large they also have a small surface area to volume ratio. Their body surface is therefore not adequate to supply and remove their respiratory gases and so, like insects and humans, they have evolved a specialised internal gas exchange surface: the gills.

Structure of the gills

The gills are located within the body of the fish, behind the head. They are made up of **gill filaments**. The gill filaments are stacked up in a pile, rather like the pages in a book. At right angles to the filaments are **gill lamellae**, which increase the surface area of the gills. Water is taken in through the mouth and forced over the gills and out through an opening on each side of the body. The position and arrangement of the gill filaments and gill lamellae are shown in Figure 1. From this figure you will notice that the flow of water over the gill lamellae and the flow of blood within them are in opposite directions. This is known as a **countercurrent flow**.

Learning objectives

→ Describe the structure of fish gills.

→ Describe how water is passed along fish gills.

→ Explain the difference between parallel flow and countercurrent flow.

→ Explain how countercurrent flow increases the rate of gas exchange.

Specification reference: 3.3.2

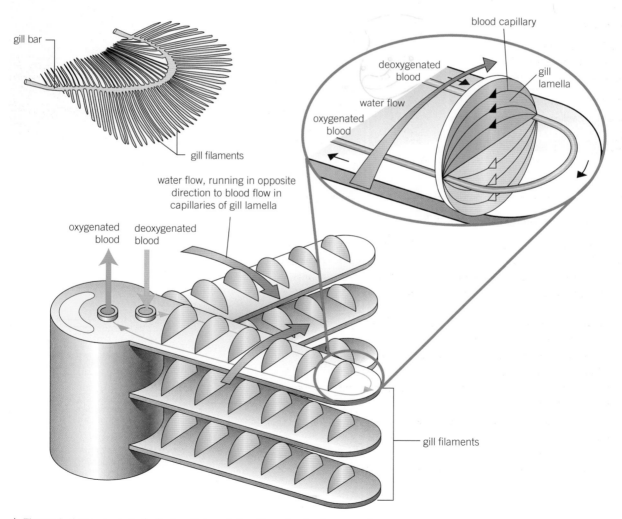

▲ **Figure 1** *Arrangement of gills in a fish and direction of water flow over them*

It is important for ensuring that the maximum possible gas exchange is achieved. If the water and blood flowed in the same direction, far less gas exchange would take place.

The countercurrent exchange principle

The essential feature of the countercurrent exchange system is that the blood and the water that flow over the gill lamellae do so in opposite directions. This arrangement means that:

- Blood that is already well loaded with oxygen meets water, which has its maximum concentration of oxygen. Therefore **diffusion** of oxygen from the water to the blood takes place.
- Blood with little oxygen in it meets water which has had most, but not all, of its oxygen removed. Again, diffusion of oxygen from the water to blood takes place.

As a result, a diffusion gradient for oxygen uptake is maintained across the entire width of the gill lamellae. In this way, about 80% of the oxygen available in the water is absorbed into the blood of the fish. If the flow of water and blood had been in the same direction (parallel flow), the diffusion gradient would only be maintained across part of the length of the gill lamellae and only 50% of the available oxygen would be absorbed by the blood.

countercurrent flow

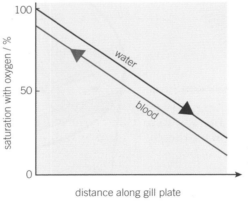

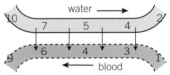

numbers represent relative oxygen concentrations

Diffusion of oxygen
A diffusion gradient is maintained all the way across the gill lamellae. Almost all the oxygen from the water diffuses into the blood.

parallel flow

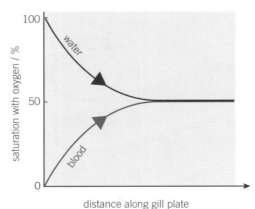

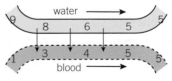

numbers represent relative oxygen concentrations

Diffusion of oxygen
A diffusion gradient is maintained for only half of the distance across the gill lamellae. Only 50% of the oxygen from the water diffuses into the blood.

▲ **Figure 2** *Parallel flow and countercurrent flow in the gills of a fish*

Like animal cells, all plant cells require oxygen and produce carbon dioxide during respiration. When it comes to gas exchange, however, plants show one important difference from animals. Some plant cells carry out photosynthesis. During photosynthesis, plant cells take in carbon dioxide and produce oxygen. At times the gases produced in one process can be used for the other. This reduces gas exchange with the external air. Overall, this means that the volumes and types of gases that are being exchanged by a plant leaf change. This depends on the balance between the rates of photosynthesis and respiration.

- When photosynthesis is taking place, although some carbon dioxide comes from respiration of cells, most of it is obtained from the external air. In the same way, some oxygen from photosynthesis is used in respiration but most of it **diffuses** out of the plant.

- When photosynthesis is not occurring, for example, in the dark, oxygen diffuses into the leaf because it is constantly being used by cells during respiration. In the same way, carbon dioxide produced during respiration diffuses out.

Structure of a plant leaf and gas exchange

In some ways, gas exchange in plants is similar to that of insects (see Topic 6.2).

- No living cell is far from the external air, and therefore a source of oxygen and carbon dioxide.

- Diffusion takes place in the gas phase (air), which makes it more rapid than if it were in water.

Overall, therefore, there is a short, fast diffusion pathway. In addition, the air spaces inside a leaf have a very large surface area compared with the volume of living tissue. There is no specific transport system for gases, which simply move in and through the plant by diffusion. Most gaseous exchange occurs in the leaves, which show the following adaptations for rapid diffusion:

- many small pores, called stomata, and so no cell is far from a stoma and therefore the diffusion pathway is short (Figure 1)

- numerous interconnecting air-spaces that occur throughout the mesophyll so that gases can readily come in contact with mesophyll cells

- large surface area of mesophyll cells for rapid diffusion.

The structure of a leaf is shown in Figure 2.

Stomata

Stomata are minute pores that occur mainly, but not exclusively, on the leaves, especially the underside. Each stoma (singular) is surrounded by a pair of special cells (guard cells). These cells can open and close the stomatal pore (Figure 3). In this way they can control the rate of gaseous exchange. This is important because terrestrial organisms lose water by evaporation. Plants have evolved to balance the conflicting needs of gas exchange and control of water loss. They do this by closing stomata at times when water loss would be excessive.

Learning objectives

→ Describe how plants exchange gases.

→ Describe the structure of a dicotyledonous plant leaf.

→ Explain the adaptations of leaves for efficient gas exchange.

Specification reference: 3.3.2

Study tip

The diffusion gradients in and out of the leaf are maintained by mitochondria carrying out respiration and chloroplasts carrying out photosynthesis.

Hint

Remember that plant cells respire all the time, but only plant cells with chloroplasts photosynthesise — and then only when the conditions are right.

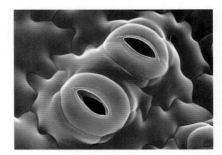

▲ **Figure 1** *False-colour SEM of open stomata on the surface a leaf*

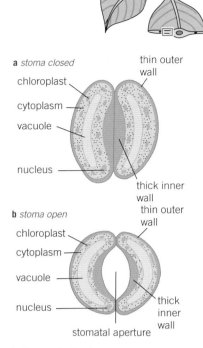

a *leaf structure*

leaf blade

leaf stalk

a *stoma closed*

thin outer wall

chloroplast

cytoplasm

vacuole

nucleus

thick inner wall

b *stoma open*

thin outer wall

chloroplast

cytoplasm

vacuole

nucleus

thick inner wall

stomatal aperture

▲ **Figure 3** *Surface view of a stoma closed and open*

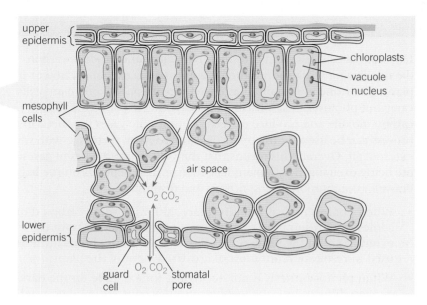

upper epidermis

chloroplasts

vacuole

nucleus

mesophyll cells

air space

O_2 CO_2

lower epidermis

guard cell

O_2 CO_2

stomatal pore

b *vertical section through a dicotyledonous leaf*

▲ **Figure 2** *Section through a leaf of a dicotyledonous plant showing gas exchange when photosynthesis is taking place*

Summary questions

1 State **two** similarities between gas exchange in a plant leaf and gas exchange in a terrestrial insect.

2 State **two** differences between gas exchange in a plant leaf and gas exchange in a terrestrial insect.

3 Explain the advantage to a plant of being able to control the opening and closing of stomata.

Maths link $\sqrt{x}$

MS 0.3, 1.1 and 3.4, see Chapter 11.

Exchange of carbon dioxide

The graph in Figure 4 shows the volume of carbon dioxide produced by a sample of tomato plants at different light intensities.

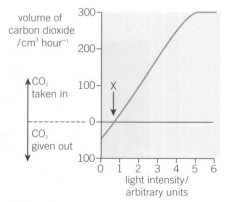

volume of carbon dioxide /cm^3 $hour^{-1}$

CO_2 taken in

CO_2 given out

X

light intensity/ arbitrary units

▲ **Figure 4**

1 Name the process which produces carbon dioxide in the tomato plants.

2 Name the process which uses up carbon dioxide in the tomato plants.

3 Explain why, at point X, carbon dioxide is neither taken up nor given out by the tomato plants.

4 $\sqrt{x}$ A plant at a light intensity of 10000 lux produced 115 $cm^3 hour^{-1}$ of carbon dioxide. When the light intensity was increased to 15000 lux the amount of carbon dioxide produced was 160 $cm^3 hour^{-1}$. Calculate the percentage increase in carbon dioxide at 15000 lux to four significant figures.

5 Some herbicides cause the stomata of plants to close. Suggest how these herbicides might lead to the death of a plant.

6 Suggest what information is provided by the point at which the line of the graph meets the y-axis.

In terrestrial organisms like insects and plants problems arise from the opposing needs of an efficient gas-exchange system and the requirement to conserve water. The features that make a good gas-exchange system are the same features that increase water loss. In order to survive, terrestrial organisms must limit their water loss without compromising the efficiency of their gas-exchange systems. The gas exchange surfaces of terrestrial organisms are inside the body. The air at the exchange surface is more or less 100% saturated with water vapour. This means there is less evaporation of water from the exchange surface.

Limiting water loss in insects

Most insects are terrestrial (live on land). The problem for all terrestrial organisms is that water easily evaporates from the surface of their bodies and they can become dehydrated. They have evolved adaptations to conserve water.

However, efficient gas exchange requires a thin, permeable surface with a large area. These features conflict with the need to conserve water. Overall, as a terrestrial organism, the insect has to balance the opposing needs of exchanging respiratory gases with limiting water loss.

Insects have evolved the following adaptations that reduce water loss:

- **Small surface area to volume ratio** to minimise the area over which water is lost.
- **Waterproof coverings** over their body surfaces. In the case of insects this covering is a rigid outer skeleton of chitin that is covered with a waterproof cuticle.
- **Spiracles** are the openings of the tracheae at the body surface and these can be closed to reduce water loss. This conflicts with the need for oxygen and so occurs largely when the insect is at rest.

These features mean that insects cannot use their body surface to diffuse respiratory gases in the way a single-celled organism does. Instead they have an internal network of tubes called **tracheae** that carry air containing oxygen directly to the tissues (see Topic 6.2).

Limiting water loss in plants

While plants also have waterproof coverings, they cannot have a small surface area to volume ratio. This is because they photosynthesise, and photosynthesis requires a large leaf surface area for the capture of light and for the exchange of gases. So how do plants limit water loss?

To reduce water loss, terrestrial plants have a waterproof covering over parts of the leaves and the ability to close stomata when necessary. Certain plants with a restricted supply of water, have also evolved a range of other adaptations to limit water loss through transpiration. These plants are called **xerophytes**.

Xerophytes are plants that are adapted to living in areas where water is in short supply. Without these adaptations these plants would become desiccated and die.

Learning objectives

→ Explain how terrestrial plants and insects balance the need for gas-exchange and the need to conserve water.

Specification reference: 3.3.2

▲ **Figure 1** *Conifers have needle-like leaves to reduce water loss*

▲ **Figure 2** *Holly has leaves with a thick waxy cuticle that reduces water loss*

Hint

Climate change affects rainfall and rate of evaporation of water. As a result, the distribution of plant species changes. As regions become drier, so the number of xerophytic plants in them increases.

The main way of surviving in habitats where there is a high rate of water loss and a limited water supply is to reduce the rate at which water can be lost through evaporation. As the vast majority of water loss occurs through the leaves, it is these organs that usually show most modifications. Examples of these modifications include:

- **a thick cuticle**. Although the waxy cuticle on leaves forms a waterproof barrier, up to 10% of water loss can still occur by this route. The thicker the cuticle, the less water can escape by this means, for example holly.

- **rolling up of leaves**. Most leaves have their stomata largely, or entirely, confined to the lower epidermis. The rolling of leaves in a way that protects the lower epidermis from the outside helps to trap a region of still air within the rolled leaf. This region becomes saturated with water vapour and so has a very high water potential. There is no water potential gradient between the inside and outside of the leaf and therefore no water loss. Marram grass rolls its leaves.

- **hairy leaves**. A thick layer of hairs on leaves, especially on the lower epidermis, traps still, moist air next to the leaf surface. The water potential gradient between the inside and the outside of the leaves is reduced and therefore less water is lost by evaporation. One type of heather plant has this modification.

- **stomata in pits or grooves**. These again trap still, moist air next to the leaf and reduce the water potential gradient. Examples of plants using this mechanism include pine trees.

- **a reduced surface area to volume ratio of the leaves**. We saw in Topic 6.1 that the smaller the surface area to volume ratio, the slower the rate of diffusion. By having leaves that are small and roughly circular in cross-section, as in pine needles, rather than leaves that are broad and flat, the rate of water loss can be considerably reduced. This reduction in surface area is balanced against the need for a sufficient area for photosynthesis to meet the requirements of the plant.

▲ **Figure 3** *This cactus stores water in its swollen stem. The leaves are needle-like to reduce their surface area and hence water loss*

Summary questions

1 Insects and plants face the same problems when it comes to living on land. What is the main problem they share?

2 State **one** modification to reduce water loss that is shared by plants and insects.

3 Insects limit water loss by having a small surface area to volume ratio. Why is this not a feasible way of limiting water loss in plants?

4 Plants such as marram grass roll up their leaves, with the lower epidermis on the inside, to reduce water loss.

   a Explain how rolling up their leaves helps to reduce water loss.

   b Why would rolling the leaf the other way (with the upper epidermis on the inside) not be effective in reducing water loss?

Study tip

When explaining adaptations of xerophytic plants to reduce water loss always relate these adaptations to reducing the water potential gradient and therefore slower diffusion, less water loss from air spaces and hence reduced evaporation of water.

Not only desert plants have problems obtaining water

Xerophytes are typically thought of as desert plants, which show a wide range of adaptations for coping with hot, dry conditions. However, similar adaptations may also be seen in plants found in sand dunes or other dry, windy places in temperate climates where rainfall is high and temperature relatively low. These adaptations are essential because the rain quickly drains away through the sand and out of the reach of the roots, making it difficult for these plants to obtain water. Plants living on salt marshes near the coast may have their roots drenched in water but find it difficult to absorb it. In addition, coastal regions are exposed to high wind speeds, which increase transpiration rates. Plants living in cold regions often have difficulty obtaining water for much of the year. Most plants living in these habitats show xerophytic modifications to enable them to reduce transpiration and so survive.

▲ **Figure 4** *Sand dunes*

1   List **two** reasons why plants growing on sand dunes (Figure 4) need to have xerophytic features even though there is plentiful rainfall.
2   Explain in terms of water potential why salt marsh plants have difficulty absorbing water, despite having plenty around their roots.
3   Explain why plants in cold regions 'have difficulty obtaining water from the soil for much of the year'.
4   Plants living in cold regions often reduce water loss by having leaves with a small surface area to volume ratio. This reduces the surface area available to capture light for photosynthesis. Photosynthesis is, in part, an enzyme-controlled process. Suggest a reason why having a smaller leaf area does not reduce the rate of photosynthesis in the same way as it would for plants in warmer climates.

Learning objectives

→ Describe how the human gas-exchange system is arranged.

→ Explain the functions of the human gas-exchange system.

Specification reference: 3.3.2

All aerobic organisms require a constant supply of oxygen to release energy in the form of **ATP** during respiration. The carbon dioxide produced in the process needs to be removed as its build-up could be harmful to the body.

The volume of oxygen that has to be absorbed and the volume of carbon dioxide that must be removed are large in mammals because:

- they are relatively large organisms with a large volume of living cells
- they maintain a high body temperature which is related to them having high metabolic and respiratory rates.

As a result mammals have evolved specialised surfaces, called **lungs**, to ensure efficient gas exchange between the air and their blood.

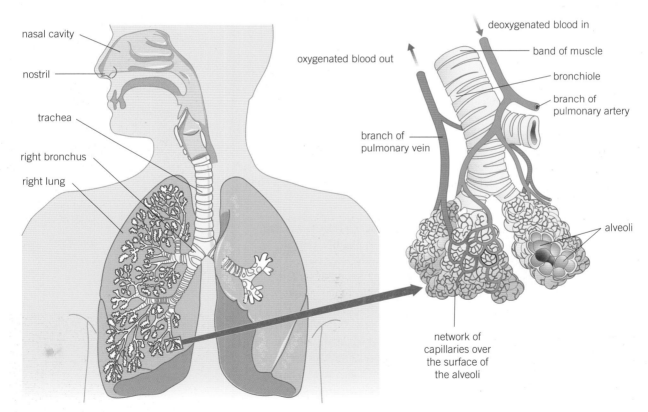

▲ **Figure 1** *The gross structure of the human gas-exchange system*

Mammalian lungs

The lungs are the site of gas exchange in mammals. They are located inside the body because:

- air is not dense enough to support and protect these delicate structures
- the body as a whole would otherwise lose a great deal of water and dry out.

The lungs are supported and protected by a bony box called the **ribcage**. The ribs can be moved by the muscles between them. The lungs are ventilated by a tidal stream of air, thereby ensuring that the air within them is constantly replenished. The main parts of the human gas-exchange system and their structure and functions are described below.

- The **lungs** are a pair of lobed structures made up of a series of highly branched tubules, called bronchioles, which end in tiny air sacs called alveoli.
- The **trachea** is a flexible airway that is supported by rings of cartilage. The cartilage prevents the trachea collapsing as the air pressure inside falls when breathing in. The tracheal walls are made up of muscle, lined with ciliated epithelium and goblet cells.
- The **bronchi** are two divisions of the trachea, each leading to one lung. They are similar in structure to the trachea and, like the trachea, they also produce mucus to trap dirt particles and have cilia that move the dirt-laden mucus towards the throat. The larger bronchi are supported by cartilage, although the amount of cartilage is reduced as the bronchi get smaller.
- The **bronchioles** are a series of branching subdivisions of the bronchi. Their walls are made of muscle lined with epithelial cells. This muscle allows them to constrict so that they can control the flow of air in and out of the alveoli.
- The **alveoli** are minute air-sacs, with a diameter of between 100 μm and 300 μm, at the end of the bronchioles. Between the alveoli there are some collagen and elastic fibres. The alveoli are lined with epithelium. The elastic fibres allow the alveoli to stretch as they fill with air when breathing in. They then spring back during breathing out in order to expel the carbon dioxide-rich air. The alveolar membrane is the gas-exchange surface.

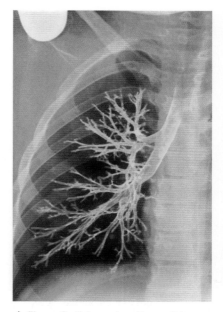

▲ **Figure 2** *False-colour X-ray of the bronchus and bronchioles of a healthy human lung*

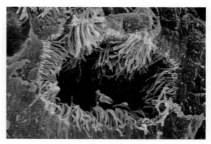

▲ **Figure 3** *False-colour SEM of a section of the epithelium of the trachea showing ciliated cells (green)*

Summary questions

1 State **two** reasons why humans need to absorb large volumes of oxygen from the lungs.

2 List in the correct sequence all the structures that air passes through on its journey from the gas-exchange surface of the lungs to the nose.

3 Explain how the cells lining the trachea and bronchus protect the alveoli from damage.

Hint

Do not write about 'respiration' when you mean 'breathing' and vice versa.

Hint

There are two basic physical laws that will help you to understand the movement of air during breathing:

Within a closed container, as the volume of a gas increases its pressure decreases. Similarly, as the volume of a gas decreases so the pressure increases.

Gases move from a region where their pressure is higher to a region where their pressure is lower.

To maintain diffusion of gases across the alveolar epithelium, air is constantly moved in and out of the lungs. This process is called breathing, or **ventilation**. When the air pressure of the atmosphere is greater than the air pressure inside the lungs, air is forced into the lungs. This is called **inspiration** (inhalation). When the air pressure in the lungs is greater than that of the atmosphere, air is forced out of the lungs. This is called **expiration** (exhalation). The pressure changes within the lungs are brought about by the movement of three sets of muscles:

- the diaphragm, which is a sheet of muscle that separates the thorax from the abdomen
- the intercostal muscles, which lie between the ribs. There are two sets of intercostal muscles:
  - the **internal intercostal muscles**, whose contraction leads to expiration
  - the **external intercostal muscles**, whose contraction leads to inspiration.

Figure 1 shows the arrangement of these various muscles.

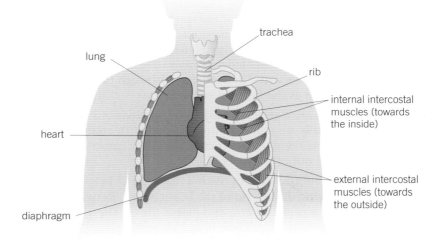

▲ **Figure 1** *The arrangement of the diaphragm and intercostal muscles*

Inspiration

Breathing in is an active process (it uses energy) and occurs as follows:

- The external intercostal muscles contract, while the internal intercostal muscles relax.
- The ribs are pulled upwards and outwards, increasing the volume of the thorax.
- The diaphragm muscles contract, causing it to flatten, which also increases the volume of the thorax.
- The increased volume of the thorax results in reduction of pressure in the lungs.
- Atmospheric pressure is now greater than pulmonary pressure, and so air is forced into the lungs.

Expiration

Breathing out is a largely passive process (it does not require much energy) and occurs as follows:

- The internal intercostal muscles contract, while the external intercostal muscles relax.
- The ribs move downwards and inwards, decreasing the volume of the thorax.
- The diaphragm muscles relax and so it is pushed up again by the contents of the abdomen that were compressed during inspiration. The volume of the thorax is therefore further decreased.
- The decreased volume of the thorax increases the pressure in the lungs.
- The pulmonary pressure is now greater than that of the atmosphere, and so air is forced out of the lungs.

During normal quiet breathing, the recoil of the elastic tissue in the lungs is the main cause of air being forced out (like air being expelled from a partly inflated balloon). Only under more strenuous conditions such as exercise do the various muscles play a major part.

Summary questions

1. From the graphs in Figure 3, calculate the rate of breathing of this person. Give your answer in breaths per minute. Show how you arrived at your answer.

2. If the volume of air in the lungs when the person inhaled was 3 000 cm³ calculate the volume of air in the lungs after the person had exhaled. Show your working.

3. Explain how muscles create the change of pressure in the alveoli over the period 0 to 0.5 s.

➕ Pulmonary ventilation √x̄

It is sometimes useful to know how much air is taken in and out of the lungs in a given time. To do this we use a measure called pulmonary ventilation rate. Pulmonary ventilation rate is the total volume of air that is moved into the lungs during 1 minute. To calculate it we multiply together two factors:

- tidal volume, which is the volume of air normally taken in at each breath when the body is at rest. This is usually around $0.5\ dm^3$.
- breathing (ventilation) rate, that is, the number of breaths taken in 1 minute. This is normally 12–20 breaths in a healthy adult.

Pulmonary ventilation rate is expressed as $dm^3\ min^{-1}$.

To summarise:

$$\text{pulmonary ventilation rate} = \text{tidal volume} \times \text{breathing rate}$$
$$(dm^3\ min^{-1}) \qquad (dm^3) \qquad (min^{-1})$$

1. A person has a pulmonary ventilation rate of $10.2\ dm^3\ min^{-1}$ and a tidal volume of $0.6\ dm^3$. Calculate the person's breathing rate.

BREATHING IN (inspiration)

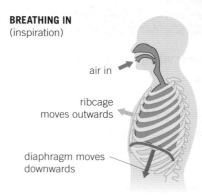

air in
ribcage moves outwards
diaphragm moves downwards

BREATHING OUT (expiration)

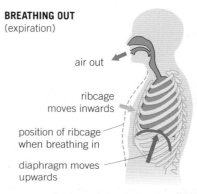

air out
ribcage moves inwards
position of ribcage when breathing in
diaphragm moves upwards

▲ **Figure 2** *Position of ribs and diaphragm during inspiration and expiration*

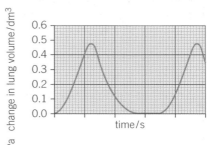

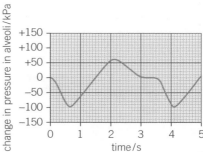

▲ **Figure 3** *The volume and pressure changes that occurred in the lungs of a person during breathing while at rest*

Maths link

MS 0.1, 2.2, 2.4 and 3.1, see Chapter 11.

The site of gas exchange in mammals is the epithelium of the alveoli. These alveoli are minute air sacs some 100–300 µm in diameter and situated in the lungs. To ensure a constant supply of oxygen to the body, a diffusion gradient must be maintained at the alveolar surface. We saw in Topic 6.1 that, to enable efficient transfer of materials across them, exchange surfaces are thin, partially permeable and have a large surface area. To maintain a diffusion gradient, there also has to be movement of both the environmental medium (for example, air) and the internal medium (for example, blood).

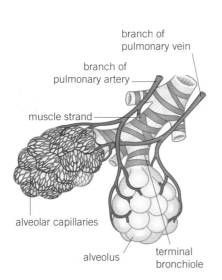

▲ **Figure 1** *Alveoli*

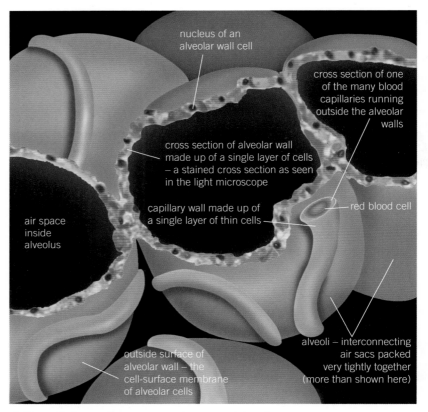

▲ **Figure 2** *External appearance of a group of alveoli*

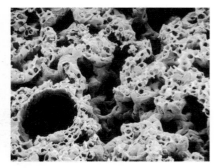

▲ **Figure 3** *False-colour SEM of a section of human lung tissue showing alveoli surrounded by blood capillaries*

Being thin, these specialised exchange surfaces are easily damaged and therefore are often located inside an organism for protection. Where an exchange surface, such as the lungs, is located inside the body, the organism has some means of moving the external medium over the surface, for example a means of ventilating the lungs in a mammal. This is because diffusion alone is not fast enough to maintain adequate transfer of oxygen and carbon dioxide along the trachea, bronchi and bronchioles. Breathing is basically a form of mass transport.

Role of the alveoli in gas exchange

There are about 300 million alveoli in each human lung. Their total surface area is around 70 m² – about half the area of a tennis court. Their structure is shown in Figures 1 and 2. Each alveolus is lined with epithelial cells only 0.05 μm to 0.3 μm thick. Around each alveolus is a network of pulmonary capillaries, so narrow (7–10 μm) that red blood cells are flattened against the thin capillary walls in order to squeeze through. These capillaries have walls that are only a single layer of cells thick (0.04–0.2 μm). Diffusion of gases between the alveoli and the blood will be very rapid because:

- red blood cells are slowed as they pass through pulmonary capillaries, allowing more time for diffusion
- the distance between the alveolar air and red blood cells is reduced as the red blood cells are flattened against the capillary walls
- the walls of both alveoli and capillaries are very thin and therefore the distance over which diffusion takes place is very short
- alveoli and pulmonary capillaries have a very large total surface area
- breathing movements constantly ventilate the lungs, and the action of the heart constantly circulates blood around the alveoli. Together, these ensure that a steep concentration gradient of the gases to be exchanged is maintained
- blood flow through the pulmonary capillaries maintains a concentration gradient.

The diffusion of gases in an alveolus is illustrated in Figure 4.

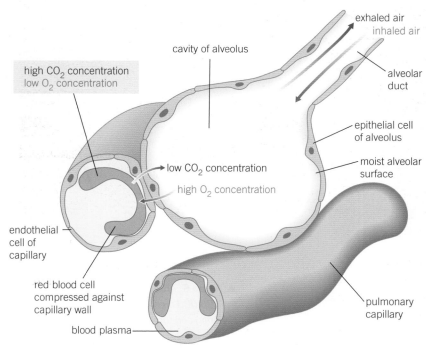

high CO_2 concentration
low O_2 concentration

exhaled air
inhaled air

cavity of alveolus

alveolar duct

epithelial cell of alveolus

low CO_2 concentration
high O_2 concentration

moist alveolar surface

endothelial cell of capillary

red blood cell compressed against capillary wall

blood plasma

pulmonary capillary

▲ **Figure 4** *Diffusion of gases in an alveolus*

Hint

The diffusion pathway is short because the alveoli have only a single layer of epithelial cells and the blood capillaries have only a single layer of endothelial cells. Don't say they have cells with thin membranes.

Summary questions

1 Explain how each of the following features contributes to the efficiency of gas exchange in alveoli.

   a The wall of each alveolus is not more than 0.3 μm thick.

   b There are 300 million alveoli in each lung.

   c Each alveolus is covered by a dense network of pulmonary blood capillaries.

   d Each pulmonary capillary is very narrow.

2 √x̄ If the number of alveoli in each lung was increased to 600 million and the pulmonary ventilation was doubled, calculate how many times greater the rate of diffusion would be.

Correlations and causal relationships

A **correlation** occurs when a change in one of two variables is reflected by a change in the other variable.

The interpretation of the data in Figure 5 shows that there is a correlation between drinking alcohol and breast cancer. What we *cannot* do from these data, however, is to conclude that drinking alcohol is the **cause** of breast cancer. The data seem to suggest this is the case but there is no actual evidence here to prove it. There needs to be a clear causal connection between drinking alcohol and breast cancer before you we can say that the case is proven. These data alone show only a correlation and not a cause. It could be that women who are stressed drink more alcohol and that it is the stress, rather than the alcohol, that causes breast cancer. To prove that drinking alcohol is the cause of breast cancer we would need experimental evidence to show that some component of the alcoholic drink led directly to women getting breast cancer. Recognising the distinction between a correlation and a causal relationship is a necessary and important skill.

Figure 6 shows how the incidence of lung cancer changes with the number of cigarettes smoked a day. What can we conclude from this data? Well, nothing really. We can see that the more cigarettes that are smoked, the greater are the number of deaths from lung cancer. In other words, there is a positive correlation between the two factors. However, we cannot conclude that it is the cigarette smoke that causes lung cancer. It may just be coincidence, or it could be that smokers are more stressed or drink more alcohol and these factors might be the cause of the cancer. Even though this graph does not itself establish a link, scientists have produced compelling experimental evidence to show that smoking tobacco definitely can cause lung cancer.

1   State a correlation shown in figure 6.
2   Explain why the information provided does not show a causal relationship with the correlation you have identified.

Study tip

It is important to be clear that a correlation does *not* mean that there is a causal link.

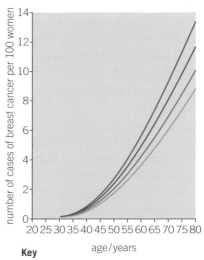

Key
— 6 drinks/day
— 4 drinks/day
— 2 drinks/day
— no alcohol

▲ **Figure 5**

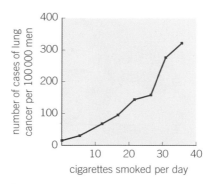

▲ **Figure 6** *Annual incidence of lung cancer per 100 000 men in the USA correlated to the daily consumption of cigarettes*

Risk factors for lung disease

There are a number of specific risk factors that increase the probability of someone suffering from lung disease. In this context 'lung disease' refers to chronic obstructive pulmonary disease (COPD), which includes emphysema and chronic bronchitis. These risk factors include:

- **Smoking**. 90% of people suffering from COPD are, or have been, heavy smokers.

- **Air pollution**. Pollutant particles and gases (e.g., sulfur dioxide) increase the likelihood of COPD, especially in areas of heavy industry.

- **Genetic make-up**. Some people are genetically more likely to get lung disease, others less so; this explains why some lifelong smokers never get lung disease while others die early.

- **Infections**. People who frequently get other chest infections also show a higher incidence of COPD.

- **Occupation**. People working with harmful chemicals, gases and dusts that can be inhaled have an increased risk of lung disease.

Here is an analysis of some data relating to the most significant risk factor — smoking.

The world's longest-running survey of smoking began in the UK in 1951. This survey and other ones elsewhere in the world have revealed a number of general statistical facts about smokers. Look at Figure 7. What does it tell us

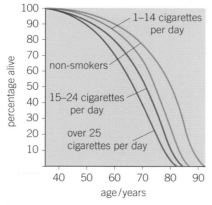

▲ **Figure 7** *Life expectancy related to the number of cigarettes smoked*

- All the lines start at 100%. This shows that the whole of this group of the population were alive at the start of the survey. What else does this tell us? As the scale of the independent variable (age in years) has its origin at 35 it suggests that everyone in this group must have been at least 35 years old at the start of the survey.

Maths link

MS 0.3, see Chapter 11.

- All the lines follow approximately the same pattern: they decline slowly at first and then at an increasing rate until at some point all lines cross the x-axis. This describes the shape but what does it actually show? Namely that only a few people die between the ages of 35 and 60 but that, after age 60, the death rate becomes increasingly rapid until, at some point everyone in the group has died.

- What about the differences between the four separate coloured lines? Each represents a different group distinguished by the number of cigarettes smoked each day. At every age beyond 35 years, the more cigarettes smoked, the fewer people remain alive. This difference is more marked the greater the age.

- At what age did the members of each group die? Well the line representing the group who smoked more than 25 cigarettes a day crosses the x-axis at 82 years, showing that no one in the group lived beyond that age. By contrast, some of the non-smokers lived beyond 90 years.

- What is the overall interpretation? Namely that the more cigarettes smoked per day, the earlier, on average, a smoker dies.

The interpretation of the data in Figure 7 shows there is a **correlation** between smoking and premature death. This does not, however, prove that smoking is the cause of an early death. The data seem to suggest this is the case but there is no evidence here to prove that it is so. There needs to be a clear causal connection between smoking and death before you can say that the case is proven. These data alone show only a correlation and not a cause. To prove that smoking is the cause of early death in smokers the correct scientific process needs to be followed. There are three main stages:

1. Establish a hypothesis to try to explain the correlation; this should be based on current knowledge.
2. Design and perform experiments to test the hypothesis.
3. Establish the causal link and formulate theories to explain it.

This is precisely what happened in establishing the causal link between smoking and lung cancer.

1   List **four** risk factors associated with lung disease.
2   Use Figure 7 to determine what percentage of non-smokers are likely to survive to age 80.
3   Calculate how many times greater is the likelihood of a non-smoker living to age 70 than someone who smokes over 25 cigarettes a day.
4   About 10 to 15 years after giving up smoking the risk of death approaches that of non-smokers. Use this information to explain to a 40-year-old who smokes 30 cigarettes a day the likely impact on her life expectancy of giving up smoking immediately.
5   Data showing a causal link between smoking and lung disease has led to statutory restrictions on the sources of risk factors. Suggest some restrictions that have been introduced and how these might reduce the incidence of lung disease.
6   Pulmonary fibrosis is a lung disease that causes the epithelium of the lungs to become irreversibly thickened. It also leads to reduced elasticity of the lungs.
    One symptom of the disease is shortness of breath, especially when exercising. Suggest why this symptom arises.
7   One measure of lung function is Forced Expiratory Volume (FEV). This is the volume of air that can forcibly be blown out in one second, after full inspiration. Suggest how pulmonary fibrosis might effect FEV and explain why.

Smoking and lung cancer

Life insurance companies have calculated that, on average, smoking a single cigarette lowers an individual's life expectancy by 10.7 minutes – longer than it takes to smoke the cigarette! While this is a statistical deduction rather than a scientific one, there is now clear scientific evidence to support the view that smoking cigarettes damages your health and reduces life expectancy. One type of evidence comes from correlations between cigarette smoking and certain diseases.

Figure 8 shows deaths from lung cancer in the UK correlated to the number of cigarettes smoked per year during a period in the last century. Study it carefully and then answer the questions.

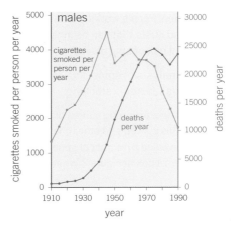

1   Determine in which decade smoking reached its peak for the following:
    a males   **b** females
2   Explain how the graphs show that there is a correlation between the number of cigarettes smoked and deaths from lung cancer in both sexes.
3   In both sexes, the number of deaths per year from lung cancer increased over the period 1910 to 1970. Suggest **three** possible reasons for this.
4   Suggest a reason why there is a time lag between the number of cigarettes smoked and a corresponding change in the number of deaths from lung cancer.

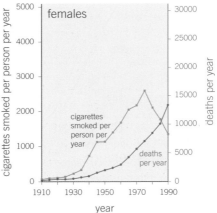

▲ **Figure 8** *Incidence of deaths from lung cancer in the UK correlated to cigarettes smoked per year (1910–90)*

The human digestive system is made up of a long muscular tube and its associated glands. The glands produce **enzymes** that hydrolyse large molecules into small ones ready for absorption. The digestive system (Figure 1) is therefore an exchange surface through which food substances are absorbed.

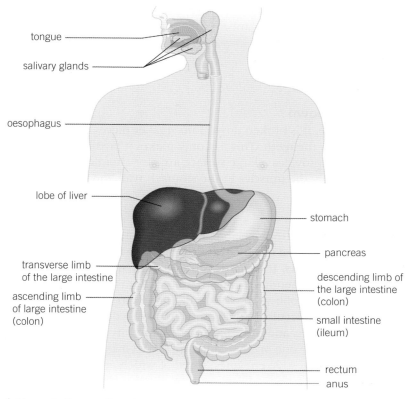

tongue

salivary glands

oesophagus

lobe of liver

transverse limb of the large intestine

ascending limb of large intestine (colon)

stomach

pancreas

descending limb of the large intestine (colon)

small intestine (ileum)

rectum

anus

▲ **Figure 1** *Human digestive system*

Learning objectives
→ Describe the structure and function of the major parts of the digestive system.
→ Explain how the digestive system breaks down food both physically and chemically.
→ Explain the role of enzymes in digestion of carbohydrates, lipids and proteins.

Specification reference: 3.3.3

Study tip
Digestion is the process in which *large* molecules are hydrolysed by enzymes into *small* molecules, which can be absorbed and assimilated.

Major parts of the digestive system

- The **oesophagus** carries food from the mouth to the stomach.
- The **stomach** is a muscular sac with an inner layer that produces enzymes. Its role is to store and digest food, especially proteins. It has glands that produce enzymes which digest protein.
- The **ileum** is a long muscular tube. Food is further digested in the ileum by enzymes that are produced by its walls and by glands that pour their secretions into it. The inner walls of the ileum are folded into villi, which gives them a large surface area. The surface area of these villi is further increased by millions of tiny projections, called microvilli, on the epithelial cells of each villus. This adapts the ileum for its purpose of absorbing the products of digestion into the bloodstream.
- The **large intestine** absorbs water. Most of the water that is absorbed is water from the secretions of the many digestive glands.
- The **rectum** is the final section of the intestines. The faeces are stored here before periodically being removed via the anus in a process called **egestion**.

Hint

The contents of the intestines are *not* inside the body. Molecules and ions only truly enter the body when they cross the cells and cell-surface membranes of the epithelial lining of the intestines.

Hint

All organisms are made up of the same biological molecules and therefore your food consists almost entirely of other organisms, or parts of them. You must first hydrolyse them into molecules that are small enough to pass across cell-surface membranes.

Synoptic link

It will help you understand this Topic if you revisit Topics 1.3, 1.5, 1.6 and 1.7.

- The **salivary glands** are situated near the mouth. They pass their secretions via a duct into the mouth. These secretions contain the enzyme amylase, which hydrolyses starch into maltose.
- The **pancreas** is a large gland situated below the stomach. It produces a secretion called pancreatic juice. This secretion contains proteases to hydrolyse proteins, lipase to hydrolyse lipids and amylase to hydrolyse starch.

What is digestion?

In humans, as with many organisms, digestion takes place in two stages:

1 physical breakdown,
2 chemical digestion.

Physical breakdown

If the food is large, it is broken down into smaller pieces by means of structures such as the teeth. This not only makes it possible to ingest the food but also provides a large surface area for chemical digestion. Food is churned by the muscles in the stomach wall and this also physically breaks it up.

Chemical digestion

Chemical digestion hydrolyses large, insoluble molecules into smaller, soluble ones. It is carried out by enzymes. All digestive enzymes function by **hydrolysis**. Hydrolysis is the splitting up of molecules by adding water to the chemical bonds that hold them together. Enzymes are specific and so it follows that more than one enzyme is needed to hydrolyse a large molecule. Usually one enzyme hydrolyses a large molecule into sections and these sections are then hydrolysed into smaller molecules by one or more additional enzymes. There are different types of digestive enzymes, three of which are particularly important:

- **Carbohydrases** hydrolyse carbohydrates, ultimately to monosaccharides.
- **Lipases** hydrolyse lipids (fats and oils) into glycerol and fatty acids.
- **Proteases** hydrolyse proteins, ultimately to amino acids.

You can now look at these three groups of digestive enzymes in more detail.

Carbohydrate digestion

It usually takes more than one enzyme to completely hydrolyse a large molecule. Typically one enzyme hydrolyses the molecule into smaller sections and then other enzymes hydrolyse these sections further into their monomers. These enzymes are usually produced in different parts of the digestive system. It is obviously important that enzymes are added to the food in the correct sequence. This is true of starch digestion.

Firstly the enzyme **amylase** is produced in the mouth and the pancreas. Amylase hydrolyses the alternate glycosidic bonds of the starch molecule to produce the disaccharide maltose. The maltose is in turn hydrolysed into the monosaccharide α-glucose by a second enzyme, a disaccharidase called **maltase**. Maltase is produced by the lining of ileum.

In humans the process takes place as follows:

- Saliva enters the mouth from the salivary glands and is thoroughly mixed with the food during chewing.
- Saliva contains **salivary amylase**. This starts hydrolysing any starch in the food to maltose. It also contains mineral salts that help to maintain the pH at around neutral. This is the optimum pH for salivary amylase to work.
- The food is swallowed and enters the stomach, where the conditions are acidic. This acid denatures the amylase and prevents further hydrolysis of the starch.
- After a time the food is passed into the small intestine, where it mixes with the secretion from the pancreas called pancreatic juice.
- The pancreatic juice contains **pancreatic amylase**. This continues the hydrolysis of any remaining starch to maltose. Alkaline salts are produced by both the pancreas and the intestinal wall to maintain the pH at around neutral so that the amylase can function.
- Muscles in the intestine wall push the food along the ileum. Its epithelial lining produces the disaccharidase **maltase**. Maltase is not released into the lumen of the ileum but is part of to the cell-surface membranes of the epithelial cells that line the ileum. It is therefore referred to as a **membrane-bound disaccharidase**. The maltase hydrolyses the maltose from starch breakdown into α-glucose.

In addition to the digestion of maltose described above, there are two other common disaccharides in the diet that are hydrolysed – sucrose and lactose.

Sucrose is found in many natural foods, especially fruits. Lactose is found in milk, and hence in milk products, such as yoghurt and cheese. Each disaccharide is hydrolysed by a membrane-bound disaccharidase as follows:

- **Sucrase** hydrolyses the single glycosidic bond in the sucrose molecule. This hydrolysis produces the two monosaccharides glucose and fructose.
- **Lactase** hydrolyses the single glycosidic bond in the lactose molecule. This hydrolysis produces the two monosaccharides glucose and galactose.

Lipid digestion

Lipids are hydrolysed by enzymes called **lipases**. Lipases are enzymes produced in the pancreas that hydrolyse the ester bond found in triglycerides to form fatty acids and monoglycerides. A monoglyceride is a glycerol molecule with a single fatty acid molecule attached. Lipids (fats and oils) are firstly split up into tiny droplets called **micelles** (Topic 6.10) by **bile salts**, which are produced by the liver. This process is called **emulsification** and increases the surface area of the lipids so that the action of lipases is speeded up.

Protein digestion

Proteins are large, complex molecules that are hydrolysed by a group of enzymes called **peptidases** (proteases). There are a number of different peptidases:

- **Endopeptidases** hydrolyse the peptide bonds between amino acids in the central region of a protein molecule forming a series of peptide molecules.

- **Exopeptidases** hydrolyse the peptide bonds on the terminal amino acids of the peptide molecules formed by endopeptidases. In this way they progressively release dipeptides and single amino acids.

- **Dipeptidases** hydrolyse the bond between the two amino acids of a dipeptide. Dipeptidases are membrane-bound, being part of the cell-surface membrane of the epithelial cells lining the ileum.

Summary questions

1 Define hydrolysis.
2 List **two** structures that produce amylase.
3 Suggest why the stomach does not have villi or microvilli.
4 Name the final product of starch digestion in the gut.
5 List **three** enzymes produced by the epithelium of the ileum.

Lactose intolerance

Milk is the only food of human babies and so they produce a relatively large amount of lactase, the enzyme that hydrolyses lactose, the sugar in milk. As milk forms a less significant part of the diet in adults, the production of lactase diminishes as children get older. This reduction can be so great in some adults that they produce little, or no, lactase at all.

This was not a problem to our ancestors but can be to humans of today. Humans that produce no lactase cannot hydrolyse the lactose they consume. When the undigested lactose reaches the large intestines, microorganisms hydrolyse it. This gives rise to small soluble molecules and a large volume of gas. This can result in diarrhoea because the soluble molecules lower the water potential of the material in the colon. The condition is known as lactose intolerance. Some people with the condition cannot consume milk or milk products at all while others can consume them only in small amounts.

▲ **Figure 3** *Milk and milk products*

1 a Suggest the process by which micoorganisms produce 'a large volume of gas' in lactose intolerant individuals.

b Suggest a reason why this gas is unlikely to be carbon dioxide.

2 Suggest an explanation why lactose intolerance is a problem for modern day humans but wasn't for our ancestors.

3 Explain how the lowering of water potential in the colon can cause diarrhoea.

We have seen in Topic 6.9 how enzymes hydrolyse carbohydrates, fats and proteins. The products of this hydrolysis are monosaccharides, amino acids, monoglycerides and fatty acids. We will now see how these products are absorbed by the ileum.

Structure of the ileum

The ileum is adapted to the function of absorbing the products of digestion. The wall of the ileum is folded and possesses finger-like projections, about 1 mm long, called **villi** (Figure 2). They have thin walls, lined with epithelial cells on the other side of which is a rich network of blood capillaries. The villi considerably increase the surface area of the ileum and therefore accelerate the rate of absorption.

Villi are situated at the interface between the lumen (cavity) of the intestines (in effect outside the body) and the blood and other tissues inside the body. They are part of a specialised exchange surface adapted for the absorption of the products of digestion. Their properties increase the efficiency of absorption in the following ways:

- They increase the surface area for diffusion.
- They are very thin walled, thus reducing the distance over which diffusion takes place.
- They contain muscle and so are able to move. This helps to maintain diffusion gradients because their movement mixes the contents of the ileum. This ensures that, as the products of digestion are absorbed from the food adjacent to the villi, new material rich in the products of digestion replaces it.
- They are well supplied with blood vessels so that blood can carry away absorbed molecules and hence maintain a diffusion gradient.
- The epithelial cells lining the villi possess **microvilli** (Figure 1). These are finger-like projections of the cell-surface membrane that further increase the surface area for absorption.

Learning objectives

→ Describe the structure of the ileum.

→ Explain how the ileum is adapted for the function of absorption.

→ Explain how monosaccharides and amino acids are absorbed.

→ Explain how triglycerides are absorbed.

Specification reference: 3.3.3

Synoptic link

You will better understand the contents of this Topic if you first read through Topics 4.2, 4.5 and 7.6.

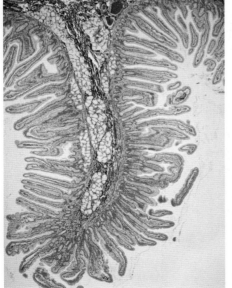

◀ **Figure 1** *Light micrograph of a section through a villus in the small intestine. Villi are projections that increase the surface area for the absorption of food. They are covered in microvilli (smaller, finger-like projections) that further increase this surface area*

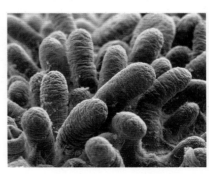

▲ **Figure 2** *False-colour SEM of villi (brown) in the lining of the ileum*

Absorption of amino acids and monosaccharides

The digestion of proteins produces amino acids, while that of carbohydrates produces monosaccharides such as glucose, fructose and galactose. The methods of absorbing these products are the same, namely diffusion and co-transport. We saw how glucose and amino acids are absorbed in the ileum by these processes in Topic 4.2 and 4.5.

Absorption of triglycerides

Once formed during digestion, monoglycerides and fatty acids remain in association with the bile salts that initially emulsified the lipid droplets (see Topic 6.9). The structures formed are called **micelles**. They are tiny, being around 4–7 nm in diameter. Through the movement of material within the lumen of the ileum, the micelles come into contact with the epithelial cells lining the villi of the ileum. Here the micelles break down, releasing the monoglycerides and fatty acids. As these are non-polar molecules, they easily diffuse across the cell-surface membrane into the epithelial cells.

Once inside the epithelial cells, monoglycerides and fatty acids are transported to the endoplasmic reticulum where they are recombined to form triglycerides. Starting in the endoplasmic reticulum and continuing in the Golgi apparatus, the triglycerides associate with cholesterol and lipoproteins to form structures called **chylomicrons**. Chylomicrons are special particles adapted for the transport of lipids.

Chylomicrons move out of the epithelial cells by exocytosis. They enter lymphatic capillaries called **lacteals** that are found at the centre of each villus. The process is illustrated in Figure 2.

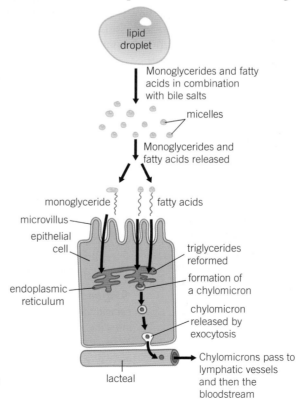

▶ **Figure 2** *The absorption of triglycerides*

From here, the chylomicrons pass, via lymphatic vessels, into the blood system. The triglycerides in the chylomicrons are hydrolysed by an enzyme in the endothelial cells of blood capillaries (see Topic 7.6) from where they diffuse into cells.

Absorption of fatty acids

Bile salts play a role in the digestion and absorption of fatty acids. One end of the bile salt molecule is soluble in fat (lipophilic) but not in water (hydrophobic). The other end is soluble in water (hydrophilic) but not in fat (lipophobic). Bile salt molecules therefore arrange themselves with their lipophilic ends in fat droplets, leaving their lipophobic ends sticking out. In this way they prevent fat droplets from sticking to each other to form large droplets, leaving only tiny ones (micelles). It is in this form that fatty acids reach the epithelial cells of the ileum where they break down, releasing the fatty acids for absorption.

An experiment was carried out to investigate the absorption of fatty acids. Six sections of intestine were filled with a fatty acid called oleic acid. To each section were added different mixtures of other contents as shown in Table 1.

Iodoacetate inhibits an enzyme involved in glycolysis – a stage of the respiratory process in cells that involves phosphorylation.

▼ **Table 1**

Contents of section of intestine					Relative amounts of oleic acid absorbed in 10 hours
Bile salts	Glycerol	Phosphate	Glycerol phosphate	Iodoacetate	
✓	✗	✗	✗	✗	2.9
✓	✗	✓	✗	✗	1.1
✓	✓	✗	✗	✗	2.6
✓	✓	✓	✗	✗	5.8
✓	✗	✗	✓	✗	8.5
✓	✗	✗	✓	✓	0.0

✓ = substance present ✗ = substance absent

From the information in Table 1:

1 ⎷x̄ List **three** pieces of evidence that support the idea that the absorption of fatty acids in the intestine is increased if they are combined with a compound of glycerol and phosphate.

2 ⎷x̄ Recognise the evidence supporting the view that the absorption of fatty acids involves phosphorylation.

Summary questions

1 List **three** organelles that you would expect to be numerous and/or well developed in an epithelial cell of the ileum, giving a reason for your choice in each case.

2 Name the other chemical that moves across epithelial cells with glucose molecules during co-transport.

3 In addition to having microvilli, state **one** other feature of the epithelial cells of the ileum that would increase the rate of absorption of amino acids.

Maths link ⎷x̄

MS 1.3, see Chapter 11.

1 **(a)** Flatworms are small animals that live in water. They have no specialised gas exchange or circulatory systems. The drawing shows one type of flatworm.

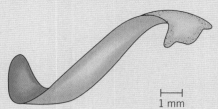

1 mm

      (i) Name the process by which oxygen reaches the cells inside the body of this flatworm. *(1 mark)*

      (ii) The body of a flatworm is adapted for efficient gas exchange between the water and the cells inside the body. Using the diagram, explain how two features of the flatworm's body allow efficient gas exchange. *(2 marks)*

  (b) **(i)** A leaf is an organ. What is an organ? *(1 mark)*

      (ii) Describe how carbon dioxide in the air outside a leaf reaches mesophyll cells inside the leaf. *(3 marks)*

                                           AQA June 2012

2 **(a)** The diagram shows the structure of the human gas exchange system.

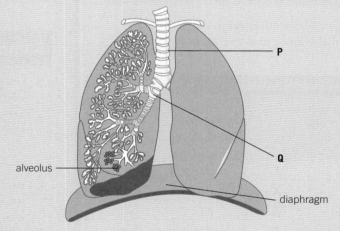

P

Q

alveolus

diaphragm

Name organs
P
Q *(1 mark)*

  (b) Explain how downward movement of the diaphragm leads to air entering the lungs. *(2 marks)*

                                           AQA Jan 2013

3 The diagram shows the position of the diaphragm at times **P** and **Q**.

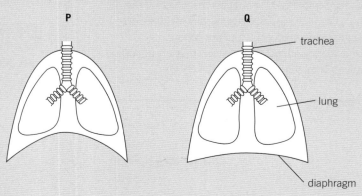

P

Q

trachea

lung

diaphragm

(a) Describe what happens to the diaphragm between times **P** and **Q** to bring
     about the change in its shape.   (*2 marks*)
(b) Air moves into the lungs between times P and Q. Explain how the
     diaphragm causes this.   (*3 marks*)
(c) Describe how oxygen in air in the alveoli enters the blood in capillaries.   (*2 marks*)

                                                                    AQA June 2012

4   Insects such as beetles obtain oxygen by drawing air into their tracheae through spiracles.
    Diving beetles live in ponds. They carry a bubble of air under their wing cases when they
    swim underwater. The bubble supplies air to the spiracles. When the bubble has been used
    up, the beetle comes to the surface to collect a new bubble.

    An investigation was carried out into the effect of temperature on diving beetles. Three
    beetles, **A**, **B** and **C**, of the same species, were observed in thermostatically-controlled
    water baths. The number of times each beetle surfaced to renew its air bubble was
    counted at three different temperatures.

    The results are shown in Table 1 below.

temperature / °C	number of times air bubble was renewed per hour		
	beetle A	beetle B	beetle C
10	10	12	8
20	18	22	18
30	44	48	38

(a) Calculate the mean number of times the air bubble was renewed per hour
     at each temperature.   (*1 mark*)
(b) Sketch a graph to show the relationship between temperature and the
     mean number of times the air bubble was renewed per hour and name
     the shape of the line obtained.   (*2 marks*)
(c) The number of times the air bubble is renewed per hour is related to a
     beetle's need for oxygen to carry out aerobic respiration, which is catalysed
     by enzymes. Explain what the data reveal about the size of the effect
     of each 10°C rise in temperature on the rate of respiration.   (*2 marks*)

5   Forced expiratory volume (FEV) is the greatest volume of air a person can
    breathe out in 1 second.

    Forced vital capacity (FVC) is the greatest volume of air a person can breathe out in a
    single breath. Figure 2 shows results for the volume of air breathed out by three groups
    of people, **A**, **B** and **C**. Group **A** had healthy lungs. Groups **B** and **C** had different lung
    conditions that affect breathing.

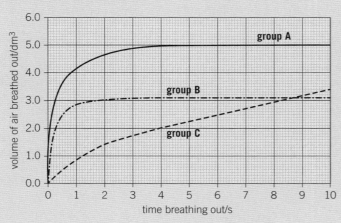

(a) Calculate the percentage drop in FEV for group **C** compared with the healthy people. *(1 mark)*

(b) Asthma affects bronchioles and reduces flow of air in and out of the lungs. Fibrosis does not affect bronchioles; it reduces the volume of the lungs. Which group, **B** or **C**, was the one containing people with fibrosis of their lungs? Use the information provided and evidence from **Figure 2** to explain your answer. *(3 marks)*

AQA SAMS A LEVEL PAPER 1

6 An animal cell takes in oxygen over its surface area but uses oxygen in proportion to its volume. Size and shape affect the ratio of surface area to volume of a cell, and therefore affect the efficiency of oxygen uptake.

(a) Complete the table to compare the surface area to volume ratios of the four model cells described. *(4 marks)*

model cell description	surface area / μm^2	volume / μm^3	ratio of surface area to volume
cube, side length 4 μm	96		
sphere, diameter 4 μm	50.3	33.5	
cube, side length 6 μm		216	
sphere, diameter 6 μm			1:1

(b) Summarise what the results show about the effect of size and shape on the ability of a cell to obtain enough oxygen for its needs. *(2 marks)*

7 (a) Describe how you would use a simple respirometer to measure the oxygen uptake of 5g of maggots. *(5 marks)*

(b) A student takes respirometer readings by measuring the distance moved by the marker fluid along a capillary tube in ten minutes. Explain what calculations need to be performed to obtain an hourly oxygen uptake rate per gram of maggots. *(3 marks)*

8 Breathing out as hard as you can is called forced expiration.

(a) Describe and explain the mechanism that causes forced expiration. *(4 marks)*

Two groups of people volunteered to take part in an experiment.
- People in group **A** were healthy.
- People in group **B** were recovering from an asthma attack.

Each person breathed in as deeply as they could. They then breathed out by forced expiration.
A scientist measured the volume of air breathed out during forced expiration by each person.
Forced expiration volume (FEV) is the volume of air a person can breathe out in 1 second.

(b) Using data from the first second of forced expiration, calculate the percentage decrease in the FEV for group **B** compared with group **A**. *(1 mark)*

(c) The people in group **B** were recovering from an asthma attack. Explain how an asthma attack caused the drop in the mean FEV shown in **Figure 4**. *(4 marks)*

AQA SAMS AS PAPER

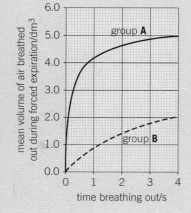

In the last chapter we looked at how substances are exchanged between the internal and external environments. This chapter looks at how these substances are distributed throughout an organism. Before considering mass transport systems, it begins by looking at an important group of molecules that are highly adapted for transporting oxygen – the haemoglobins.

Haemoglobin molecules

The haemoglobins are a group of chemically similar molecules found in a wide variety of organisms. In Topic 1.6 we investigated the structure of proteins and how their shape is important to their functions. Haemoglobins are protein molecules with a quaternary structure that has evolved to make it efficient at loading oxygen under one set of conditions but unloading it under a different set of conditions. The structure of a haemoglobin molecule is shown in Figure 1. It is made up as follows:

- **primary structure**, sequence of amino acids in the four polypeptide chains
- **secondary structure**, in which each of these polypeptide chains is coiled into a helix
- **tertiary structure**, in which each polypeptide chain is folded into a precise shape – an important factor in its ability to carry oxygen
- **quaternary structure**, in which all four polypeptides are linked together to form an almost spherical molecule. Each polypeptide is associated with a haem group – which contains a ferrous (Fe^{2+}) ion. Each Fe^{2+} ion can combine with a single oxygen molecule (O_2), making a total of four O_2 molecules that can be carried by a single haemoglobin molecule in humans.

Loading and unloading oxygen

The process by which haemoglobin binds with oxygen is called **loading**, or **associating**. In humans this takes place in the lungs.

The process by which haemoglobin releases its oxygen is called **unloading**, or **dissociating**. In humans this takes place in the tissues.

Haemoglobins with a high affinity for oxygen take up oxygen more easily, but release it less easily. Haemoglobins with a low affinity for oxygen take up oxygen less easily, but release it more easily.

The role of haemoglobin

The role of haemoglobin is to transport oxygen. To be efficient at transporting oxygen, haemoglobin must:

- readily associate with oxygen at the surface where gas exchange takes place
- readily dissociate from oxygen at those tissues requiring it.

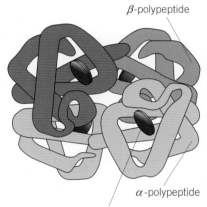

each chain is attached to a haem group that can combine with oxygen

▲ **Figure 1** *Quaternary structure of a haemoglobin molecule*

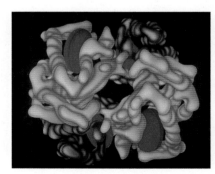

▲ **Figure 2** *Computer graphic representation of a haemoglobin molecule showing two pairs of polypeptide chains (orange and blue) associated with a haem group (red)*

Study tip

A change in the environment of any protein changes its tertiary structure and therefore affects the way it functions. This explains why haemoglobin binds with oxygen in the lungs and releases it in the tissues.

These two requirements may appear to contradict each other, but they are achieved by a remarkable property of haemoglobin. It changes its affinity (chemical attraction) for oxygen under different conditions (Table 1). It achieves this because its shape changes in the presence of certain substances, such as carbon dioxide. In the presence of carbon dioxide, the new shape of the haemoglobin molecule binds more loosely to oxygen. As a result haemoglobin releases its oxygen.

▼ **Table 1** *Affinity of haemoglobin for oxygen under different conditions*

Region of body	Oxygen concentration	Carbon dioxide concentration	Affinity of haemoglobin for oxygen	Result
gas exchange surface	high	low	high	oxygen is associated
respiring tissues	low	high	low	oxygen is dissociated

Why are there different haemoglobins?

Scientists long ago observed that many organisms possessed haemoglobin. They proposed that it carried oxygen from the gas-exchange surface to the tissues that required it for respiration. If so, this meant that it must readily combine with oxygen. Consequently they investigated the ability of haemoglobin from different organisms to combine with oxygen. Results showed that there were different types of haemoglobins. These exhibited different properties relating to the way they took up and released oxygen.

Why do different haemoglobins have different affinities for oxygen?

The answer, scientists discovered, lies in the shape of the molecule. Each species produces a haemoglobin with a slightly different amino acid sequence. The haemoglobin of each species therefore has a slightly different tertiary and quaternary structure and hence different oxygen binding properties. Depending on its structure haemoglobin molecules range from those that have a high affinity for oxygen to those that have a low affinity for oxygen.

Summary questions

1 Describe the quaternary structure of haemoglobin.

2 Explain how DNA leads to different haemoglobin molecules having different affinities for oxygen.

3 When the body is at rest, only one of the four oxygen molecules carried by haemoglobin is normally released into the tissues. Suggest why this could be an advantage when the organism becomes more active.

4 Carbon monoxide occurs in car exhaust fumes. It binds permanently to haemoglobin in preference to oxygen. Suggest a reason why a person breathing in car-exhaust fumes might lose consciousness.

Having looked at haemoglobin in topic 7.1, let us now consider its properties. How does it load and unload oxygen and what effect does carbon dioxide have on this process?

Oxygen dissociation curves

When haemoglobin is exposed to different partial pressures of oxygen, it does not bind the oxygen evenly. The graph of the relationship between the saturation of haemoglobin with oxygen and the partial pressure of oxygen is known as the **oxygen dissociation curve** (see Figure 1). The explanation for the shape of the oxygen dissociation curve is as follows:

- The shape of the haemoglobin molecule makes it difficult for the first oxygen molecule to bind to one of the sites on its four polypeptide subunits because they are closely united. Therefore at low oxygen concentrations, little oxygen binds to haemoglobin. The gradient of the curve is shallow initially.

- However, the binding of this first oxygen molecule changes the quaternary structure of the haemoglobin molecule, causing it to change shape. This change makes it easier for the other subunits to bind to an oxygen molecule. In other words, the binding of the first oxygen molecule induces the other subunits to bind to an oxygen molecule.

- It therefore takes a smaller increase in the partial pressure of oxygen to bind the second oxygen molecule than it did to bind the first one. This is known as **positive cooperativity** because binding of the first molecule makes binding of the second easier and so on. The gradient of the curve steepens.

- The situation changes, however, after the binding of the third molecule. While in theory it is easier for haemoglobin to bind the fourth oxygen molecule, in practice it is harder. This is simply due to probability. With the majority of the binding sites occupied, it is less likely that a single oxygen molecule will find an empty site to bind to. The gradient of the curve reduces and the graph flattens off.

We saw in Topic 7.1 that there are different types of haemoglobin molecules in different species, each with a different shape and hence a different affinity for oxygen. In addition, the shape of any one type of haemoglobin molecule can change under different conditions. These facts both mean that there are a large number of different oxygen dissociation curves. They all have a roughly similar shape but differ in their position on the axes.

The many different oxygen dissociation curves are better understood if two facts are always kept in mind:

- The further to the left the curve, the greater is the affinity of haemoglobin for oxygen (so it loads oxygen readily but unloads it less easily).

- The further to the right the curve, the lower is the affinity of haemoglobin for oxygen (so it loads oxygen less readily but unloads it more easily).

Learning objectives

→ Describe the nature of an oxygen dissociation curve.

→ Explain the effect of carbon dioxide concentration on the curve and the reasons why.

→ Explain how the properties of the haemoglobins in different organisms relate to the environment and way of life of the organism concerned.

Specification reference: 3.3.4.1

Hint

Haemoglobin that is 100% saturated with oxygen has the maximum number of oxygen molecules it can bind to. If 50% saturated, it has half the maximum number it can bind to.

Maths link $\sqrt{x}$

MS 3.1, see Chapter 11.

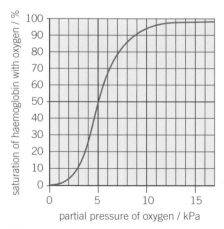

▲ **Figure 1** *Oxygen dissociation curve for adult human haemoglobin*

Hint

Measuring oxygen concentration
The amount of a gas that is present in a mixture of gases is measured by the pressure it contributes to the total pressure of the gas mixture. This is known as the **partial pressure** of the gas and, in the case of oxygen, is written as pO_2. It is measured in kiloPascals *kPa*. Normal atmospheric pressure is 100 kPa. As oxygen makes up 21% of the atmosphere, its partial pressure is normally 21 kPa.

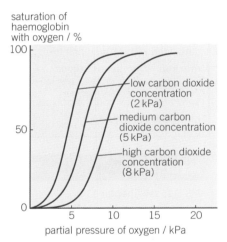

▲ **Figure 2** *The effect of carbon dioxide concentration on the oxygen dissociation curve*

Effects of carbon dioxide concentration

Haemoglobin has a reduced affinity for oxygen in the presence of carbon dioxide. The greater the concentration of carbon dioxide, the more readily the haemoglobin releases its oxygen (the Bohr effect). This explains why the behaviour of haemoglobin changes in different regions of the body.

- At the gas-exchange surface (e.g., lungs), the concentration of carbon dioxide is low because it diffuses across the exchange surface and is excreted from the organism. The affinity of haemoglobin for oxygen is increased, which, coupled with the high concentration of oxygen in the lungs, means that oxygen is readily loaded by haemoglobin. The reduced carbon dioxide concentration has shifted the oxygen dissociation curve to the left (Figure 2).

- In rapidly respiring tissues (e.g., muscles), the concentration of carbon dioxide is high. The affinity of haemoglobin for oxygen is reduced, which, coupled with the low concentration of oxygen in the muscles, means that oxygen is readily unloaded from the haemoglobin into the muscle cells. The increased carbon dioxide concentration has shifted the oxygen dissociation curve to the right (Figure 2).

We have seen that the greater the concentration of carbon dioxide, the more readily haemoglobin releases its oxygen. This is because dissolved carbon dioxide is acidic and the low pH causes haemoglobin to change shape. Let us see how this works in the transport of oxygen by haemoglobin.

Loading, transport and unloading of oxygen

- At the gas-exchange surface carbon dioxide is constantly being removed.
- The pH is slightly raised due to the low concentration of carbon dioxide.
- The higher pH changes the shape of haemoglobin into one that enables it to load oxygen readily.
- This shape also increases the affinity of haemoglobin for oxygen, so it is not released while being transported in the blood to the tissues.
- In the tissues, carbon dioxide is produced by respiring cells.
- Carbon dioxide is acidic in solution, so the pH of the blood within the the tissues is lowered.
- The lower pH changes the shape of haemoglobin into one with a lower affinity for oxygen.
- Haemoglobin releases its oxygen into the respiring tissues.

The above process is a flexible way of ensuring that there is always sufficient oxygen for respiring tissues. The more active a tissue, the more oxygen is unloaded. This works as follows:

The higher the rate of respiration → the more carbon dioxide the tissues produce → the lower the pH → the greater the haemoglobin shape change → the more readily oxygen is unloaded → the more oxygen is available for respiration.

In humans, haemoglobin normally becomes saturated with oxygen as it passes through the lungs. In practice not all haemoglobin molecules are loaded with their maximum four oxygen molecules. As a consequence, the overall saturation of haemoglobin at atmospheric pressure is normally around 97%. When this haemoglobin reaches a tissue with a low respiratory rate, only one of these molecules will normally be released. The blood returning to the lungs will therefore contain haemoglobin that is still 75 per cent saturated with oxygen. If a tissue is very active, for example, an exercising muscle, then three oxygen molecules will usually be unloaded from each haemoglobin molecule. These events are shown in Figure 3.

Different species have different types of haemoglobin, each with its own different oxygen dissociation curve. These different types have evolved within species as adaptations to different environments and conditions. For example, species of animals that live in an environment with a lower partial pressure of oxygen have evolved haemoglobin that has a higher affinity for oxygen than the haemoglobin of animals that live where the partial pressure of oxygen is higher.

Take for example the lugworm, an animal that lives on the seashore.

The lugworm is not very active, spending almost all its life in a U-shaped burrow. Most of the time the lugworm is covered by sea water, which it circulates through its burrow. Oxygen diffuses into the lugworm's blood from the water and it uses haemoglobin to transport oxygen to its tissues.

When the tide goes out, the lugworm can no longer circulate a fresh supply of oxygenated water through its burrow. As a result, the water in the burrow contains progressively less oxygen as the lugworm uses it up. The lugworm has to extract as much oxygen as possible from the water in the burrow if it is to survive until the tide covers it again. Figure 4 shows the oxygen dissociation curve of lugworm haemoglobin compared to that of adult human haemoglobin.

The dissociation curve is shifted far to the left of that of a human. This means that the haemoglobin of the lugworm is fully loaded with oxygen even when there is little available in its environment.

Another example is the llama. It is an animal that lives at high altitudes. At these altitudes the atmospheric pressure is lower and so the partial pressure of oxygen is also lower. It is therefore difficult to load haemoglobin with oxygen. Llamas also have a type of haemoglobin that has a higher affinity for oxygen than human haemoglobin. In other words it is shifted to the left of that of human haemoglobin.

Key
— Haemoglobin molecule that is fully loaded with oxygen in the lungs

— Haemoglobin molecule in a resting tissue unloads only 25% of its oxygen

— Haemoglobin molecule in an active tissue unloads 75% of its oxygen

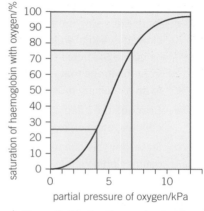

▲ **Figure 3** *The loading and unloading of a haemoglobin with oxygen*

Maths link √x̄

MS 3.1 and 3.4, see Chapter 11.

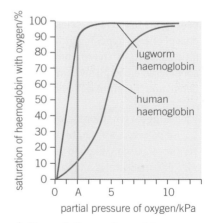

▲ **Figure 4** *Comparison of the oxygen dissociation curves of lugworm and human haemoglobin*

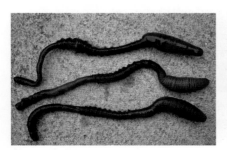 

▲ **Figure 5** *Three lugworms lying on sand (left); lugworm casts at the entrances to their burrows (right)*

Summary questions

1 Study Figure 3 on the previous page and answer the following questions:

   a State at what partial pressure of oxygen the haemoglobin is 50% saturated with oxygen.

   b Determine the percentage saturation of haemoglobin with oxygen when the partial pressure of oxygen is 9 kPa.

   c In an exercising muscle the partial pressure of oxygen is 4 kPa while in the lungs it is 12 kPa. Calculate the percentage of the oxyhaemoglobin from the lungs that will have released its oxygen to an exercising muscle.

2 a Describe the effect of increased carbon dioxide concentration on oxygen dissociation.

   b State how this changes the saturation of haemoglobin with oxygen.

3 A rise in temperature shifts the oxygen dissociation curve to the right. Suggest how this enables an exercising muscle to work more efficiently.

4 In Figure 4, line A is drawn at a partial pressure of oxygen of 2 kPa. This is the partial pressure of oxygen found in lugworm burrows after the sea no longer covers them. Use figures from the graph to explain why a lugworm can survive at these concentrations of oxygen while a human could not.

5 Haemoglobin usually loads oxygen less readily when the concentration of carbon dioxide is high (the Bohr effect). The haemoglobin of lugworms does not exhibit this effect. Explain why to do so could be harmful.

6 In terms of obtaining oxygen, suggest a reason why lugworms are not found higher up the seashore.

Activity counts

Flight in birds and swimming in fish are both energy-demanding processes. The muscles that move a bird's wings are powerful and require a lot of oxygen to enable them to respire at a sufficient rate to keep the body airborne. Flight muscles have a very high metabolic rate and, during flight, much of the blood pumped by the heart goes to these muscles. While birds use a great deal of energy opposing gravity in a medium that gives little support, fish have a different problem. They expend considerable energy swimming in a medium that is very dense and therefore difficult to move through.

1 Suggest whether the oxygen dissociation curve of a pigeon is shifted to the right or left of the curve for a human. Explain your answer.

2 The mackerel is a type of fish that swims freely in the surface waters of the sea. These fish rely on their ability to swim very fast in order to escape from predators. The plaice is a marine fish that uses a different strategy. These fish spend much of their lives stationary or moving very slowly on the sea bed, where they are camouflaged by their skin colour. The two fish are of relatively similar mass. Sketch a graph to show what you would expect to be the relative positions of the oxygen dissociation curves of these two fish.

▲ **Figure 6** *Mackerel (top) live in surface waters and swim rapidly. Plaice (bottom) live on the sea bed and move very slowly*

Size matters √x̄

Mice are small mammals and therefore have a large surface area to volume ratio. As a result they tend to lose heat rapidly when the environmental temperature is lower than their body temperature. Figure 7 shows the oxygen dissociation curve for the haemoglobin of a mouse compared to that of adult human haemoglobin.

1 The partial pressure of oxygen at which haemoglobin is 50 per cent saturated is known as the unloading pressure. Calculate the difference between the unloading pressure of human haemoglobin and that of mouse haemoglobin.

2 The oxygen dissociation curve of the mouse is shifted to the right of that for a human.

   a Explain what difference this makes to the way oxygen is unloaded from mouse haemoglobin compared to human haemoglobin.

   b Suggest an advantage this has for the maintenance of body temperature in mice.

   c The position of the oxygen dissociation curve for a mouse means that its haemoglobin loads oxygen less readily than human haemoglobin. Given that the partial pressure of oxygen in air is normally 21 kPa, use the graph to explain why this is of no disadvantage to the mouse.

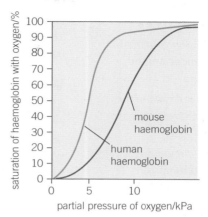

▲ **Figure 7** *Oxygen dissociation curves of mouse and human haemoglobin*

3 Sketch a graph to show the shapes and relative positions of the oxygen dissociation curves of the following mammals:

   a a human

   b an elephant

   c a shrew.

4 Ice fish live in the Antarctic and are the only vertebrates to completely lack haemoglobin. Suggest **one** reason why they can survive in the seas around Antarctica without haemoglobin in their blood.

▲ **Figure 1** *Large organisms require a transport system to take materials from exchange surfaces to the cells that need them*

Diffusion is fast enough for transport over short distances (see Topic 4.2). The efficient supply of materials over larger distances requires a mass transport system.

Why large organisms have a transport system

All organisms exchange materials between themselves and their environment. We have seen that in small organisms this exchange takes place over the surface of the body (see Topic 6.2). However, with increasing size, the surface area to volume ratio decreases to a point where the needs of the organism cannot be met by the body surface alone (see Topic 6.1). Specialist exchange surfaces are therefore required to absorb nutrients and respiratory gases, and remove excretory products. These exchange surfaces are located in specific regions of the organism. A transport system is required to take materials from cells to exchange surfaces and from exchange surfaces to cells. Materials have to be transported between exchange surfaces and the environment. They also need to be transported between different parts of the organism. As organisms have evolved into larger and more complex structures, the tissues and organs of which they are made have become more specialised and dependent upon one another. This makes a transport system all the more essential.

Whether or not there is a specialised transport medium, and whether or not it is circulated by a pump, depends on two factors:

- the surface area to volume ratio,
- how active the organism is.

The lower the surface area to volume ratio, and the more active the organism, the greater is the need for a specialised transport system with a pump.

Features of transport systems

Any large organism encounters the same problems in transporting materials within itself. Not surprisingly, the transport systems of many organisms have many common features:

- A suitable medium in which to carry materials, for example blood. This is normally a liquid based on water because water readily dissolves substances and can be moved around easily, but can be a gas such as air breathed in and out of the lungs.
- A form of mass transport in which the transport medium is moved around in bulk over large distances – more rapid than diffusion.
- A closed system of tubular vessels that contains the transport medium and forms a branching network to distribute it to all parts of the organism.
- A mechanism for moving the transport medium within vessels. This requires a pressure difference between one part of the system and another.

It is achieved in two main ways:

a Animals use muscular contraction either of the body muscles or of a specialised pumping organ, such as the heart (see Topic 7.4).

b Plants rely on natural, passive processes such as the evaporation of water (see Topic 7.8).

- A mechanism to maintain the mass flow movement in one direction, for example, valves.

- A means of controlling the flow of the transport medium to suit the changing needs of different parts of the organism.

- A mechanism for the mass flow of water or gases, for example, intercostal muscles and diaphragm during breathing in mammals.

Circulatory systems in mammals

Mammals have **a closed, double circulatory system** in which blood is confined to vessels and passes twice through the heart for each complete circuit of the body (Figure 2). This is because, when blood is passed through the lungs, its pressure is reduced. If it were to pass immediately to the rest of the body its low pressure would make circulation very slow. Blood is therefore returned to the heart to boost its pressure before being circulated to the rest of the tissues. As a result, substances are delivered to the rest of the body quickly, which is necessary as mammals have a high body temperature and hence a high rate of **metabolism**. The vessels that make up the circulatory system of a mammal are divided into three types: arteries, veins and capillaries. We will look in more detail at these in Topic 7.6.

The arrangement of the main arteries and veins that make up the circulatory system of a mammal is shown in Figure 3.

Although a transport system is used to move substances longer distances, the final part of the journey to cells is by diffusion. The final exchange from blood vessels into cells is rapid because it takes place over a large surface area, across short distances and there is a steep diffusion gradient.

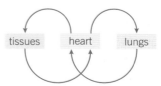

▲ **Figure 2** *Double circulation of a mammal*

Study tip

Almost all cells in the body are within 1 mm of a capillary – a short diffusion path.

Summary questions

1 Name the blood vessel in each of the following descriptions:

  a joins the right ventricle of the heart to the capillaries of the lungs

  b carries oxygenated blood away from the heart

  c carries deoxygenated blood away from the kidney

  d the first main blood vessel that an oxygen molecule reaches after being absorbed from an alveolus

  e has the highest blood pressure.

2 State **two** factors that make it more likely that an organism will have a circulatory pump such as the heart.

3 State the main advantage of the double circulation found in mammals.

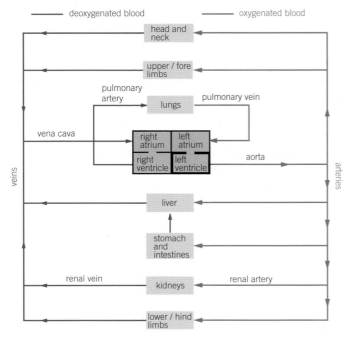

▲ **Figure 3** *Plan of the mammalian circulatory system*

Learning objectives

→ Describe the appearance of the heart and its associated blood vessels.

→ Explain why the heart is made up of two adjacent pumps.

→ Explain how the structure of the heart is related to its functions.

Specification reference: 3.3.4.1

The heart is a muscular organ that lies in the thoracic cavity behind the sternum (breastbone). It operates continuously and tirelessly throughout the life of an organism.

Structure of the human heart

The human heart is really two separate pumps lying side by side. The pump on the left deals with oxygenated blood from the lungs, while the one on the right deals with deoxygenated blood from the body. Each pump has two chambers:

- The **atrium** is thin-walled and elastic and stretches as it collects blood.
- The **ventricle** has a much thicker muscular wall as it has to contract strongly to pump blood some distance, either to the lungs or to the rest of the body.

Why have two separate pumps? Why not just pump the blood through the lungs to collect oxygen and then straight to the rest of the body before returning it to the heart? The problem with such a system is that the blood has to pass through tiny capillaries in the lungs in order to present a large surface area for the exchange of gases (see Topic 6.8). In doing so, there is a very large drop in pressure and so the blood flow to the rest of the body would be very slow. This drop in pressure is illustrated in Figure 1. Mammals therefore have a system in which the blood is returned to the heart to increase its pressure before it is distributed to the rest of the body. It is essential to keep the oxygenated blood in the pump on the left side separate from the deoxygenated blood in the pump on the right.

The right ventricle pumps blood only to the lungs, and it has a thinner muscular wall than the left ventricle. The left ventricle, in contrast, has a thick muscular wall, enabling it to contract to create enough pressure to pump blood to the rest of the body. Although the two sides of the heart are separate pumps and, after birth, there is no mixing of the blood in each of them, they nevertheless pump in time with each other. Both atria contract together and then both ventricles contract together, pumping the same volume of blood.

Between each atrium and ventricle are valves that prevent the backflow of blood into the atria when the ventricles contract. There are two valves:

- the **left atrioventricular (bicuspid) valve**
- the **right atrioventricular (tricuspid) valve**

Each of the four chambers of the heart is connected to large blood vessels that carry blood towards or away from the heart. The ventricles pump blood away from the heart and into the arteries. The atria receive blood from the veins.

> **Study tip**
>
> Although the left ventricle has a thicker wall than the right ventricle, their internal volumes are the same. They have to be, otherwise more blood would be pumped out of one side of the heart than the other.

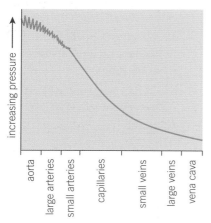

▲ **Figure 1** *Pressure changes in blood vessels*

> **Study tip**
>
> The left and right sides of the heart both contract together.

Vessels connecting the heart to the lungs are called **pulmonary** vessels. The vessels connected to the four chambers are therefore as follows:

- The **aorta** is connected to the left ventricle and carries oxygenated blood to all parts of the body except the lungs.

- The **vena cava** is connected to the right atrium and brings deoxygenated blood back from the tissues of the body (except the lungs).

- The **pulmonary artery** is connected to the right ventricle and carries deoxygenated blood to the lungs, where its oxygen is replenished and its carbon dioxide is removed. Unusually for an artery, it carries deoxygenated blood.

- The **pulmonary vein** is connected to the left atrium and brings oxygenated blood back from the lungs. Unusually for a vein, it carries oxygenated blood.

The structure of the heart and its associated blood vessels is shown in Figure 2.

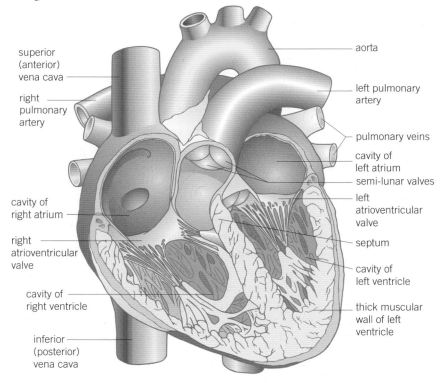

▲ **Figure 2** *Section through the human heart*

Supplying the heart muscle with oxygen

Although oxygenated blood passes through the left side of the heart, the heart does not use this oxygen to meet its own great respiratory needs. Instead, the heart muscle is supplied by its own blood vessels, called the **coronary arteries**, which branch off the aorta shortly after it leaves the heart. Blockage of these arteries, for example by a blood clot, leads to myocardial infarction, or heart attack, because an area of the heart muscle is deprived of blood and therefore oxygen also. The muscle cells in this region are unable to respire (aerobically) and so die.

> ### Hint
>
> An easy way to recall which heart chambers are attached to which type of blood vessel is to remember that A and V always go together. Hence: **A**tria link to **V**eins and **A**rteries link to **V**entricles.

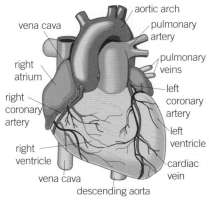

▲ **Figure 3** *External appearance of the human heart showing the blood supply to the heart muscle*

Summary questions

1 Name the blood vessel that supplies the heart muscle with oxygenated blood.

2 State whether the blood in each of the following structures is oxygenated or deoxygenated:

   a vena cava

   b pulmonary artery

   c left atrium.

3 List the correct sequence of four main blood vessels and four heart chambers that a red blood cell passes through on its journey from the lungs, though the heart and body, and back again to the lungs.

4 Suggest why it is important to prevent mixing of the blood in the two sides of the heart.

Risk factors associated with cardiovascular disease

There are a number of factors that separately increase the risk of an individual suffering from cardiovascular disease. When combined together, four or five of these factors produce a disproportionately greater risk (Figure 4). These risk factors include the following.

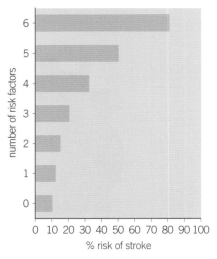

▲ **Figure 4** *The combined impact of six risk factors on the likelihood of a 70-year-old man experiencing a stroke in the next ten years*

Smoking

Smokers are between two and six times more likely to suffer from heart disease than non-smokers. Giving up smoking is the single most effective way of increasing life expectancy. There are two main constituents of tobacco smoke that increase the likelihood of heart disease:

- **Carbon monoxide** combines easily, but irreversibly, with the haemoglobin in red blood cells to form carboxyhaemoglobin. It thereby reduces the oxygen-carrying capacity of the blood. To supply the equivalent quantity of oxygen to the tissues, the heart works harder. This can lead to raised blood pressure that increases the risk of coronary heart disease and strokes. In addition, the reduction in the oxygen-carrying capacity of the blood means that it may be insufficient to supply the heart muscle during exercise. This leads to chest pain (angina) or, in severe cases, a myocardial infarction (heart attack).

- **Nicotine** stimulates the production of the hormone adrenaline, which increases heart rate and raises blood pressure. As a consequence there is a greater risk of smokers suffering coronary heart disease or a stroke. Nicotine also makes the platelets in the blood more

'sticky', and this leads to a higher risk of thrombosis and hence of strokes or myocardial infarction.

High blood pressure

If your genes cause you to have a high blood pressure, altering your lifestyle will not change this fact. Lifestyle factors such as excessive prolonged stress, certain diets and lack of exercise, increase the risk of high blood pressure. These are factors over which the individual can exert control. High blood pressure increases the risk of heart disease for the following reasons:

- As there is already a higher pressure in the arteries, the heart must work harder to pump blood into them and is therefore more prone to failure.

- Higher blood pressure within the arteries means that they are more likely to develop an aneurysm (weakening of the wall) and burst, causing haemorrhage.

- To resist the higher pressure within them, the walls of the arteries tend to become thickened and may harden, restricting the flow of blood.

> **Hint**
>
> Always remember that risk factors increase the *probability* of getting heart disease, but they do not mean that someone will certainly get it. Heavy smokers, with high blood pressure and high blood cholesterol, may never develop heart disease, they are just more likely to (see Figure 5).

Blood cholesterol

Cholesterol is an essential component of membranes. As such, it is an essential biological molecule which must be transported in the blood. It is carried in the plasma as tiny spheres of lipoproteins (lipid and protein). There are two main types:

- **high-density lipoproteins (HDLs)**, which remove cholesterol from tissues and transport it to the liver for excretion. They help protect arteries against heart disease

- **low-density lipoproteins (LDLs)**, which transport cholesterol from the liver to the tissues, including the artery walls, which they infiltrate, leading to the development of atheroma, which may lead to heart disease.

Diet

There are a number of aspects of diet that increase the risk of heart disease, both directly and indirectly:

- **High levels of salt** raise blood pressure.
- **High levels of saturated fat** increase low-density lipoprotein levels and hence blood cholesterol concentration.

By contrast, foods that act as antioxidants, for example, vitamin C, reduce the risk of heart disease, and so does non-starch polysaccharide (dietary fibre).

A calculated risk

Figure 5 shows the effect of three of the above risk factors on the chance of heart attack in American men. Study the data and answer the questions:

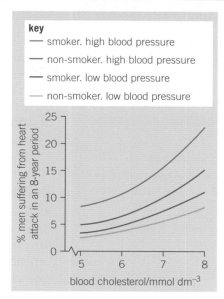

▲ **Figure 5** *Effects of blood pressure, smoking and blood cholesterol on the risk of heart attack in American men*

1 A smoker with high blood pressure wishes to reduce his risk of heart attack. If he could only alter one factor, would he be better giving up smoking or reducing his blood pressure? Explain your answer.

2 A non-smoker with high blood pressure has a blood cholesterol level of 5 mmol dm^{-3}. Over a period of 3 years this concentration increases to 8 mmol dm^{-3}. Calculate how many times greater his risk of heart disease is. Show your working.

3 Two non-smoking men with low blood pressure both have a blood cholesterol level of 5 mmol dm^{-3}. One of them starts to smoke and the blood cholesterol level of the other increases to 7 mmol dm^{-3}. State which man is now at the greater risk from heart disease. Explain your answer.

Learning objectives

→ Describe the stages of the cardiac cycle.

→ Explain how valves control the flow of blood through the heart.

→ Explain the volume and pressure changes which take place in the heart during the cardiac cycle.

Specification reference: 3.3.4.1

The heart undergoes a sequence of events that is repeated in humans around 70 times each minute when at rest. This is known as the **cardiac cycle**. There are two phases to the beating of the heart: contraction (systole) and relaxation (diastole). Contraction occurs separately in the ventricles and the atria and is therefore described in two stages. For some of the time, relaxation takes place simultaneously in all chambers of the heart and is therefore treated as a single phase in the account below, which is illustrated in Figure 1.

Relaxation of the heart (diastole)

Blood returns to the atria of the heart through the pulmonary vein (from the lungs) and the vena cava (from the body). As the atria fill, the pressure in them rises. When this pressure exceeds that in the ventricles, the atrioventricular valves open allowing the blood to pass into the ventricles. The passage of blood is aided by gravity. The muscular walls of both the atria and ventricles are relaxed at this stage. The relaxation of the ventricle walls causes them to recoil and reduces the pressure within the ventricle. This causes the pressure to be lower than that in the aorta and the pulmonary artery, and so the semi-lunar valves in the aorta and the pulmonary artery close, accompanied by the characteristic 'dub' sound of the heart beat.

1. Blood enters atria and ventricles from pulmonary veins and vena cava.

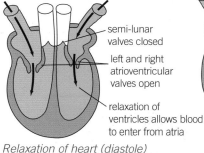

semi-lunar valves closed

left and right atrioventricular valves open

relaxation of ventricles allows blood to enter from atria

Relaxation of heart (diastole)

Atria are relaxed and fill with blood. Ventricles are also relaxed.

▲ **Figure 1** *The cardiac cycle*

2.

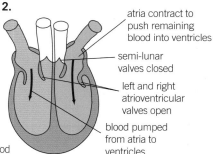

atria contract to push remaining blood into ventricles

semi-lunar valves closed

left and right atrioventricular valves open

blood pumped from atria to ventricles

Contraction of atria (atrial systole)

Atria contract, pushing blood into the ventricles. Ventricles remain relaxed.

3. Blood pumped into pulmonary arteries and the aorta.

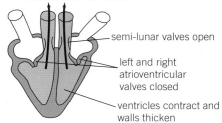

semi-lunar valves open

left and right atrioventricular valves closed

ventricles contract and walls thicken

Contraction of ventricles (ventricular systole)

Atria relax. Ventricles contract, pushing blood away from heart through pulmonary arteries and the aorta.

Contraction of the atria (atrial systole)

The contraction of the atrial walls, along with the recoil of the relaxed ventricle walls, forces the remaining blood into the ventricles from the atria. Throughout this stage the muscle of the ventricle walls remains relaxed.

Contraction of the ventricles (ventricular systole)

After a short delay to allow the ventricles to fill with blood, their walls contract simultaneously. This increases the blood pressure within them, forcing shut the atrioventricular valves and preventing backflow of blood into the atria. The 'lub' sound of these valves closing is a characteristic of the heart beat. With the atrioventricular valves closed, the pressure in the ventricles rises further. Once it exceeds that in the aorta and pulmonary artery, blood is forced from the ventricles into these vessels. The ventricles have thick muscular walls which mean they contract forcefully. This creates the high pressure necessary to pump blood around the body. The thick wall of the left ventricle has to pump blood to the extremities of the body while the relatively thinner wall of the right ventricle, has to pump blood to the lungs.

Valves in the control of blood flow

Blood is kept flowing one direction through the heart and around the body by the pressure created by the heart muscle. Blood will always move from a region of higher pressure to one of lower pressure. There are, however, situations within the circulatory system when pressure differences would result in blood flowing in the opposite direction from that which is desirable. In these circumstances, valves are used to prevent any unwanted backflow of blood.

Valves in the cardiovascular system are designed so that they open whenever the difference in blood pressure either side of them favours the movement of blood in the required direction. When pressure differences are reversed, that is, when blood would tend to flow in the opposite direction to that which is desirable, the valves are designed to close. Examples of such valves include:

- **Atrioventricular valves** between the left atrium and ventricle and the right atrium and ventricle. These prevent backflow of blood when contraction of the ventricles means that ventricular pressure exceeds atrial pressure. Closure of these valves ensures that, when the ventricles contract, blood within them moves to the aorta and pulmonary artery rather than back to the atria.
- **Semi-lunar valves** in the aorta and pulmonary artery. These prevent backflow of blood into the ventricles when the pressure in these vessels exceeds that in the ventricles. This arises when the elastic walls of the vessels recoil increasing the pressure within them and when the ventricle walls relax reducing the pressure within the ventricles.
- **Pocket valves** in veins (see Topic 7.6) that occur throughout the venous system. These ensure that when the veins are squeezed, e.g. when skeletal muscles contract, blood flows back towards the heart rather than away from it.

The design of these valves is basically the same. They are made up of a number of flaps of tough, but flexible, fibrous tissue, which are cusp-shaped, in other words like deep bowls. When pressure is greater on the convex side of these cusps, rather than on the concave side, they move apart to let blood pass between the cusps. When pressure is greater on the concave side than on the convex side, blood collects within the 'bowl' of the cusps. This pushes them together to form a tight fit that prevents the passage of blood (Figure 2).

Pressure and volume changes of the heart

Mammals have a closed circulatory system, in other words the blood is confined to vessels, and this allows the pressure within them to be maintained and regulated. Figure 4 illustrates the pressure and volume changes, and associated valve movements, that take place in the heart during a typical cardiac cycle.

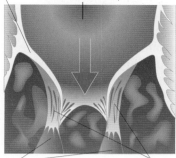

a *Valve open*

cusp of valve

higher blood pressure above valve forces it open

pillar muscles   Lower blood pressure beneath valve   string-like tendons

b *Valve closed*

lower blood pressure cannot open valve

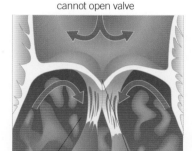

higher blood pressure beneath valve forces it closed

cusps of valves fit closely together

▲ **Figure 2** *Action of the valves*

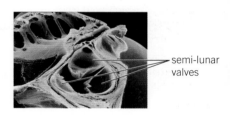

semi-lunar valves

▲ **Figure 3** *False-colour SEM of the semi-lunar valve of the aorta*

Maths link $\sqrt{x}$

MS 2.4, see Chapter 11.

Cardiac output

Cardiac output is the volume of blood pumped by one ventricle of the heart in one minute. It is usually measured in $dm^3 \, min^{-1}$ and depends upon two factors:

- the heart rate (the rate at which the heart beats)
- the stroke volume (volume of blood pumped out at each beat).

Cardiac output = heart rate × stroke volume

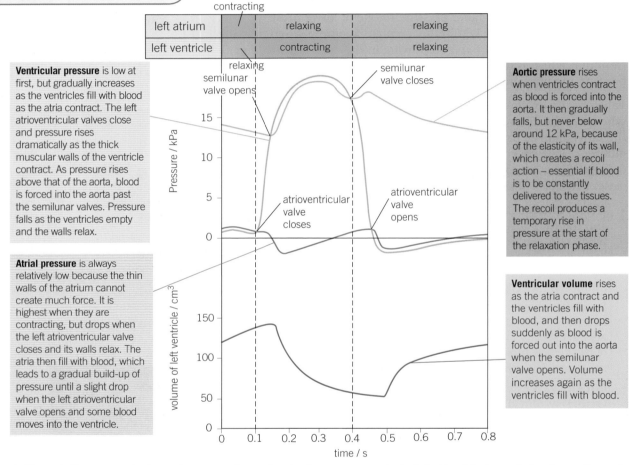

Ventricular pressure is low at first, but gradually increases as the ventricles fill with blood as the atria contract. The left atrioventricular valves close and pressure rises dramatically as the thick muscular walls of the ventricle contract. As pressure rises above that of the aorta, blood is forced into the aorta past the semilunar valves. Pressure falls as the ventricles empty and the walls relax.

Atrial pressure is always relatively low because the thin walls of the atrium cannot create much force. It is highest when they are contracting, but drops when the left atrioventricular valve closes and its walls relax. The atria then fill with blood, which leads to a gradual build-up of pressure until a slight drop when the left atrioventricular valve opens and some blood moves into the ventricle.

Aortic pressure rises when ventricles contract as blood is forced into the aorta. It then gradually falls, but never below around 12 kPa, because of the elasticity of its wall, which creates a recoil action – essential if blood is to be constantly delivered to the tissues. The recoil produces a temporary rise in pressure at the start of the relaxation phase.

Ventricular volume rises as the atria contract and the ventricles fill with blood, and then drops suddenly as blood is forced out into the aorta when the semilunar valve opens. Volume increases again as the ventricles fill with blood.

▲ **Figure 4** *Pressure and volume changes, and associated valve movements, in the left side of the heart during the cardiac cycle*

Summary questions

1. Name the chamber of the heart that produces the greatest pressure.

2. State whether each of the following statements is true or false.

   a. The left and right ventricles contract together.

   b. Veins have pocket valves.

   c. Semi-lunar valves occur between the atria and ventricles.

   d. $\sqrt{x}$ If a person's cardiac output is $4.9 \, dm^3 \, min^{-1}$ and their heart rate is 70 beats a minute, then their stroke volume is $0.7 \, dm^3$.

3 In each case, state what is being described.

   a On contraction it forces blood into the ventricles.

   b The relaxation phase of the heart.

   c Structures that prevent flow of blood from the aorta into the left ventricle.

4 After a period of training, the heart rate is often decreased when at rest although the cardiac output is unchanged. Suggest an explanation for this.

5 √x Use Figure 4 to calculate the heart rate in beats per minute. Show your working.

6 √x If a person has a stroke volume of 0.065 dm^3 and a cardiac output of 5.2 dm^3 min^{-1}, calculate their heart rate.

Hint

Two facts will help you to understand the rather complex graph shown in Figure 4.

- Pressure and volume within a closed container are inversely related. When pressure increases, volume decreases, and vice versa.

- Blood, like all fluids, moves from a region where its pressure is greater to one where it is lower, i.e. it moves down a pressure gradient.

➕ Electrocardiogram

During the cardiac cycle, the heart undergoes a series of electrical current changes. These are related to the waves of electrical activity created by the sinoatrial node and the heart's response to these. If displayed on a cathode ray oscilloscope, these changes can produce a trace known as an **electrocardiogram**. Doctors can use this trace to provide a picture of the heart's electrical activity and hence its health. In a normal electrocardiogram (ECG) there is a pattern of large peaks and small troughs that repeat identically at regular intervals. An ECG produced during a heart attack shows less pronounced peaks and larger troughs that are repeated in a similar, but not identical, way. During a condition called fibrillation, the heart muscle contracts in a disorganised way that is reflected in an irregular ECG.

a
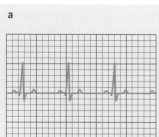

b
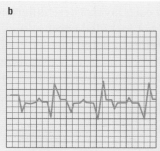

c
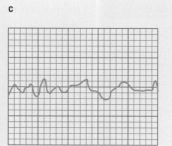

▲ **Figure 6** *Three different electrocardiogram (ECG) traces*

1 The three ECG traces shown in Figure 6 represent an ECG trace for:
- a normal heart
- a heart in fibrillation
- during a heart attack

Using the letters a, b and c, suggest which trace corresponds to which heart condition. Give reasons for your answers.

Maths link √x

Maths skill 1.3, see Chapter 11.

Learning objectives

→ Describe the structures of arteries, arterioles and veins.

→ Explain how the structure of each of the above vessels is related to its function.

→ Explain the structure of capillaries and how it is related to their function.

Specification reference: 3.3.4.1

In Topic 7.3 we saw that, in larger organisms, materials are transported around the body by the blood that is confined to blood vessels. This topic looks in more detail at these vessels.

Structure of blood vessels

There are different types of blood vessels:

- **Arteries** carry blood away from the heart and into arterioles.
- **Arterioles** are smaller arteries that control blood flow from arteries to capillaries.
- **Capillaries** are tiny vessels that link arterioles to veins.
- **Veins** carry blood from capillaries back to the heart.

Arteries, arterioles and veins all have the same basic layered structure. From the outside inwards, these layers are:

- **tough fibrous outer layer** that resists pressure changes from both within and outside
- **muscle layer** that can contract and so control the flow of blood
- **elastic layer** that helps to maintain blood pressure by stretching and springing back (recoiling)
- **thin inner lining (endothelium)** that is smooth to reduce friction and thin to allow diffusion
- **lumen** that is not actually a layer but the central cavity of the blood vessel through which the blood flows.

> **Study tip**
>
> Arteries, arterioles and veins carry out transport not exchange; only capillaries carry out exchange.

> **Study tip**
>
> The elastic tissue of arteries will stretch and recoil. It is not muscle and will not contract and relax.

What differs between each type of blood vessel is the relative proportions of each layer. These differences are shown in Figure 1. Arterioles are not included because they are similar to arteries. They differ from arteries in being smaller in diameter and having a relatively larger muscle layer and lumen. The differences in structure are related to the differences in the function that each type of vessel performs.

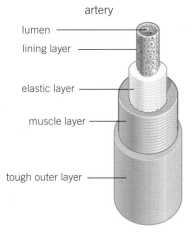
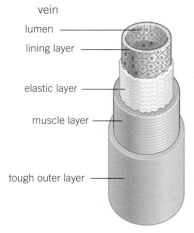
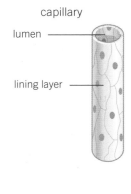

artery — lumen, lining layer, elastic layer, muscle layer, tough outer layer

vein — lumen, lining layer, elastic layer, muscle layer, tough outer layer

capillary — lumen, lining layer

▲ **Figure 1** *Comparison of arteries, veins and capillaries*

Artery structure related to function

The function of arteries is to transport blood rapidly under high pressure from the heart to the tissues. Their structure is adapted to this function as follows:

- **The muscle layer is thick compared to veins.** This means smaller arteries can be constricted and dilated in order to control the volume of blood passing through them.

- **The elastic layer is relatively thick compared to veins** because it is important that blood pressure in arteries is kept high if blood is to reach the extremities of the body. The elastic wall is stretched at each beat of the heart (systole). It then springs back when the heart relaxes (diastole) in the same way as a stretched elastic band. This stretching and recoil action helps to maintain high pressure and smooth pressure surges created by the beating of the heart.

- **The overall thickness of the wall is great.** This also resists the vessel bursting under pressure.

- **There are no valves** (except in the arteries leaving the heart) because blood is under constant high pressure due to the heart pumping blood into the arteries. It therefore tends not to flow backwards.

Arteriole structure related to function

Arterioles carry blood, under lower pressure than arteries, from arteries to capillaries. They also control the flow of blood between the two. Their structure is related to these functions as follows:

- **The muscle layer is relatively thicker than in arteries.** The contraction of this muscle layer allows constriction of the lumen of the arteriole. This restricts the flow of blood and so controls its movement into the capillaries that supply the tissues with blood.

- **The elastic layer is relatively thinner than in arteries** because blood pressure is lower.

Vein structure related to function

Veins transport blood slowly, under low pressure, from the capillaries in tissues to the heart. Their structure is related to this function as follows:

- **The muscle layer is relatively thin** compared to arteries because veins carry blood away from tissues and therefore their constriction and dilation cannot control the flow of blood to the tissues.

- **The elastic layer is relatively thin** compared to arteries because the low pressure of blood within the veins will not cause them to burst and pressure is too low to create a recoil action.

- **The overall thickness of the wall is small** because there is no need for a thick wall as the pressure within the veins is too low to create any risk of bursting. It also allows them to be flattened easily, aiding the flow of blood within them (see below).

- **There are valves at intervals throughout** to ensure that blood does not flow backwards, which it might otherwise do because the pressure is so low. When body muscles contract, veins are compressed, pressurising the blood within them. The valves ensure that this pressure directs the blood in one direction only: towards the heart (Figure 3).

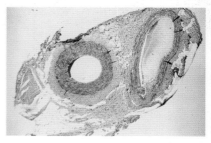

▲ **Figure 2** *Artery (left) and vein (right)*

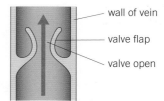

Blood flowing towards the heart passes easily through the valves.

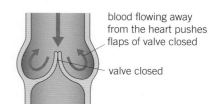

Blood flowing away from the heart pushes valves closed and so blood is prevented from flowing any further in this direction.

▲ **Figure 3** *Action of valves in veins in ensuring one-way flow of blood*

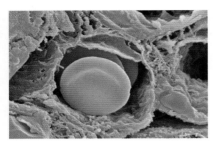

▲ **Figure 4** *False-colour SEM of a section through a capillary with red blood cells passing through it*

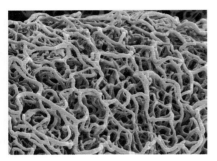

▲ **Figure 5** *Resin cast of a capillary network from the large intestine*

Capillary structure related to function

The function of capillaries (Figures 4 and 5) is to exchange metabolic materials such as oxygen, carbon dioxide and glucose between the blood and the cells of the body. The flow of blood in capillaries is much slower. This allows more time for the exchange of materials.

The structure of capillaries is related to their function as follows:

- **Their walls consist mostly of the lining layer**, making them extremely thin, so the distance over which diffusion takes place is short. This allows for rapid diffusion of materials between the blood and the cells.
- **They are numerous and highly branched**, thus providing a large surface area for exchange.
- **They have a narrow diameter** and so permeate tissues, which means that no cell is far from a capillary and there is a short diffusion pathway.
- **Their lumen is so narrow** that red blood cells are squeezed flat against the side of a capillary. This brings them even closer to the cells to which they supply oxygen. This again reduces the diffusion distance.
- **There are spaces between the lining (endothelial) cells** that allow white blood cells to escape in order to deal with infections within tissues.

Although capillaries are small, they cannot serve every single cell directly. Therefore the final journey of metabolic materials is made in a liquid solution that bathes the tissues. This liquid is called **tissue fluid**.

Tissue fluid and its formation

Tissue fluid is a watery liquid that contains glucose, amino acids, fatty acids, ions in solution and oxygen. Tissue fluid supplies all of these substances to the tissues. In return, it receives carbon dioxide and other waste materials from the tissues. Tissue fluid is therefore the means by which materials are exchanged between blood and cells and, as such, it bathes all the cells of the body. It is the immediate environment of cells and is, in effect, where they live. Tissue fluid is formed from blood plasma, and the composition of blood plasma is controlled by various homeostatic systems. As a result tissue fluid provides a mostly constant environment for the cells it surrounds.

Formation of tissue fluid

Blood pumped by the heart passes along arteries, then the narrower arterioles and, finally, the even narrower capillaries. Pumping by the heart creates a pressure, called **hydrostatic pressure**, at the arterial end of the capillaries. This hydrostatic pressure causes tissue fluid to move out of the blood plasma. The outward pressure is, however, opposed by two other forces:

- hydrostatic pressure of the tissue fluid outside the capillaries, which resists outward movement of liquid
- the lower water potential of the blood, due to the plasma proteins, that causes water to move back into the blood within the capillaries.

However, the combined effect of all these forces is to create an overall pressure that pushes tissue fluid out of the capillaries at the arterial end. This pressure is only enough to force small molecules out of the capillaries, leaving all cells and proteins in the blood because these are too large to cross the membranes. This type of filtration under pressure is called **ultrafiltration**.

Hint

To help prevent cells and proteins from leaking out, capillaries have a little fibrous tissue around them.

Return of tissue fluid to the circulatory system

Once tissue fluid has exchanged metabolic materials with the cells it bathes, it is returned to the circulatory system. Most tissue fluid returns to the blood plasma directly via the capillaries. This return occurs as follows:

- The loss of the tissue fluid from the capillaries reduces the hydrostatic pressure inside them.
- As a result, by the time the blood has reached the venous end of the capillary network its hydrostatic pressure is usually lower than that of the tissue fluid outside it.
- Therefore tissue fluid is forced back into the capillaries by the higher hydrostatic pressure outside them.
- In addition, the plasma has lost water and still contains proteins. It therefore has a lower water potential than the tissue fluid.
- As a result, water leaves the tissue by osmosis down a water potential gradient.

The tissue fluid has lost much of its oxygen and nutrients by diffusion into the cells that it bathed, but it has gained carbon dioxide and waste materials in return. These events are summarised in Figure 6.

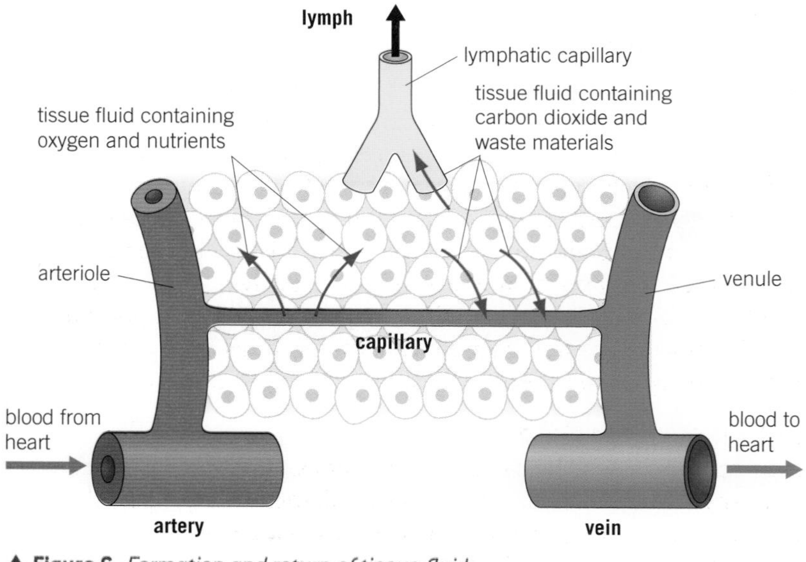

▲ **Figure 6** *Formation and return of tissue fluid*

Not all the tissue fluid can return to the capillaries; the remainder is carried back via the lymphatic system. This is a system of vessels that begin in the tissues. Initially they resemble capillaries (except that they have dead ends), but they gradually merge into larger vessels that form a network throughout the body. These larger vessels drain their contents back into the bloodstream via two ducts that join veins close to the heart.

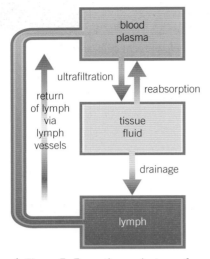

▲ **Figure 7** *Formation and return of tissue fluid to the bloodstream*

The contents of the lymphatic system (lymph) are not moved by the pumping of the heart. Instead they are moved by:

- **hydrostatic pressure** of the tissue fluid that has left the capillaries
- **contraction of body muscles** that squeeze the lymph vessels – valves in the lymph vessels ensure that the fluid inside them moves away from the tissues in the direction of the heart.

A summary of the methods of tissue fluid formation and its return to the bloodstream is shown in Figure 7.

Summary questions

1 State **one** advantage of having:

 a thick elastic tissue in the walls of arteries

 b relatively thick muscle walls in arterioles

 c valves in veins

 d only a lining layer in capillaries.

2 Table 1 shows the mean wall thickness of different blood vessels in a mammal. Suggest the letter that is most likely to refer to **a** the aorta, **b** a capillary, **c** a vein, **d** an arteriole and **e** the renal artery.

3 State what forces tissue fluid out of the blood plasma in capillaries and into the surrounding tissues.

4 List the **two** routes by which tissue fluid returns to the bloodstream.

▼ **Table 1**

Blood vessel	Mean wall thickness/mm
A	1.000
B	0.001
C	2.000
D	0.500
E	0.030

Maths link

MS 1.3, see Chapter 11.

 Blood flow in various blood vessels

The graph in Figure 8 shows certain features of the flow of blood from and to the heart through a variety of blood vessels.

1 Describe the changes in the rate of blood flow as blood passes from the aorta to the vena cava.

2 Explain why blood pressure in region A fluctuates up and down.

3 Explain why the rate of blood flow decreases between the aorta and capillaries.

4 Explain how the rate of blood flow in the capillaries increases the rate of exchange of metabolic materials.

5 Explain why the structure of capillaries increases the efficiency of the exchange of metabolic substances.

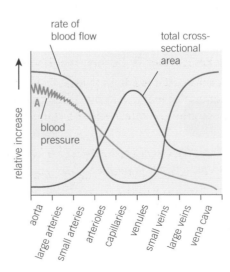

▲ **Figure 8** *Flow of blood to and from the heart*

In plants water is absorbed by the roots through extensions called root hairs. In flowering plants the vast majority of water is transported through hollow, thick-walled tubes called **xylem vessels**. The main force that pulls water through the xylem vessels in the stem of a plant is the evaporation of water from leaves – a process called **transpiration**. The energy for this is supplied by the sun and the process is therefore passive. It is therefore logical to begin an explanation of how water moves through the xylem from the point where water evaporates from the surfaces of cells surrounding the stomatal air space and water vapour diffuses out of the stomatal pore.

Movement of water out through stomata

The humidity of the atmosphere is usually less than that of the air spaces next to the stomata (see Figure 2). As a result there is a water potential gradient from the air spaces through the stomata to the air. Provided the stomata are open, water vapour molecules diffuse out of the air spaces into the surrounding air. Water lost by diffusion from the air spaces is replaced by water evaporating from the cell walls of the surrounding mesophyll cells. By changing the size of the stomatal pores, plants can control their rate of transpiration.

Movement of water across the cells of a leaf

Water is lost from mesophyll cells by evaporation from their cell walls to the air spaces of the leaf. This is replaced by water reaching the mesophyll cells from the xylem either via cell walls or via the cytoplasm. In the case of the cytoplasmic route the water movement occurs because:

- mesophyll cells lose water to the air spaces by evaporation due to heat supplied by the sun
- these cells now have a lower **water potential** and so water enters by **osmosis** from neighbouring cells
- the loss of water from these neighbouring cells lowers their water potential
- they, in turn, take in water from their neighbours by osmosis.

In this way, a water potential gradient is established that pulls water from the xylem, across the leaf mesophyll, and finally out into the atmosphere. These events are summarised in Figure 2 on the next page.

Movement of water up the stem in the xylem

The main factor that is responsible for the movement of water up the xylem, from the roots to the leaves, is cohesion–tension. The movement of water up the stem occurs as follows:

- Water evaporates from mesophyll cells due to heat from the sun leading to transpiration.
- Water molecules form hydrogen bonds between one another and hence tend to stick together. This is known as **cohesion**.
- Water forms a continuous, unbroken column across the mesophyll cells and down the xylem.

Learning objectives

→ Define what transpiration is.

→ Explain how water moves through the leaf.

→ Explain how water moves up the xylem.

→ Explain the cohesion–tension theory of water transport.

Specification reference: 3.3.4.2

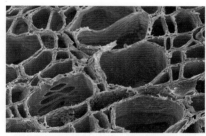

▲ **Figure 1** *False-colour SEM showing hollow, tubular xylem vessels adapted to transport water*

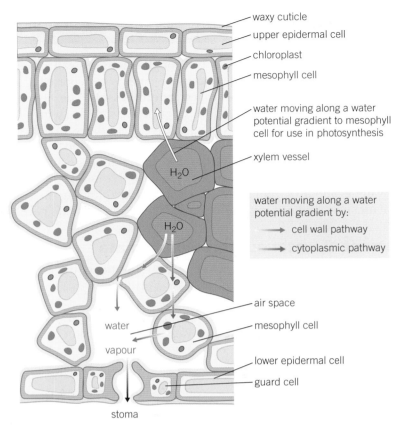

▲ **Figure 2** *Movement of water across leaf*

▲ **Figure 3** *Section through a leaf showing the tissues involved in the movement of water*

- As water evaporates from the mesophyll cells in the leaf into the air spaces beneath the stomata, more molecules of water are drawn up behind it as a result of this cohesion.
- A column of water is therefore pulled up the xylem as a result of transpiration. This is called the **transpiration pull**.
- Transpiration pull puts the xylem under tension, that is, there is a negative pressure within the xylem, hence the name **cohesion–tension theory**.

Such is the force of the transpiration pull that it can easily raise water up the 100 m or more of the tallest trees. There are several pieces of evidence to support the cohesion–tension theory. These include:

- Change in the diameter of tree trunks according to the rate of transpiration. During the day, when transpiration is at its greatest, there is more tension (more negative pressure) in the xylem. This pulls the walls of the xylem vessels inwards and causes the trunk to shrink in diameter. At night, when transpiration is at its lowest, there is less tension in the xylem and so the diameter of the trunk increases.
- If a xylem vessel is broken and air enters it, the tree can no longer draw up water. This is because the continuous column of water is broken and so the water molecules can no longer stick together.
- When a xylem vessel is broken, water does not leak out, as would be the case if it were under pressure. Instead air is drawn in, which is consistent with it being under tension.

Transpiration pull is a passive process and therefore does not require metabolic energy to take place. Indeed, the xylem vessels through which the water passes are dead and so cannot actively move the water. Xylem vessels have no end walls which means that xylem forms a series of continuous, unbroken tubes from root to leaves, which is essential to the cohesion–tension theory of water flow up the stem. Energy is nevertheless needed to drive the process of transpiration. This energy is in the form of heat that evaporates water from the leaves and it ultimately comes from the sun.

Figure 4 summarises the movement of water from the soil, through the plant, and into the atmosphere.

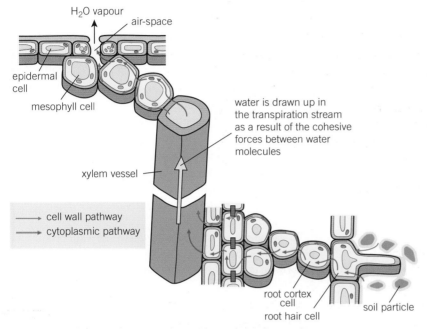

▲ **Figure 4** *Summary of water transport through a plant*

 ## Hug a tree $\sqrt{x}$

If you put your arms around a suitably sized tree trunk in the middle of the day your fingers will just touch on the far side of the tree. Try to hug the same tree at night and your fingers will probably no longer meet. The graph in Figure 5 shows why. It shows the rate of water flow up a tree and the diameter of the tree trunk over a 24-hour period.

1 At what time of day is transpiration rate greatest? Explain your answer using information in Figure 5.
2 Describe the changes in the rate of flow of water during the 24-hour period.
3 Explain in terms of the cohesion–tension theory the changes in the rate of flow of water during the 24-hour period.
4 Explain the changes in the diameter of the tree trunk over the 24-hour period.
5 If the tree was sprayed with ammonium sulfamate, a herbicide that kills living cells, the rate of water flow would be unchanged. Explain why.

Summary question

1 State the most suitable word, or words, represented by **a**–**f** in the passage below.

Water leaves from the air spaces in a plant by a process called **a**. This takes place mainly through pores called **b** in the epidermis of the leaf. Water evaporates into the air spaces from mesophyll cells. As a result these cells have a **c** water potential and so draw water by **d** from neighbouring cells. In this way, a water potential gradient is set up that draws water from the xylem. Water is pulled up the xylem because water molecules stick together – a phenomenon called **e**. During the night the diameter of a tree trunk **f**.

Maths link $\sqrt{x}$

MS 1.3, see Chapter 11.

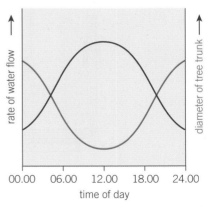

▲ **Figure 5** *Variation of rate of water flow and diameter of a tree trunk*

Measurement of water uptake using a potometer

It is almost impossible to measure transpiration because it is extremely difficult to condense and collect all the water vapour that leaves all the parts of a plant. What we can easily measure, however, is the amount of water that is taken up in a given time by a part of the plant such as a leafy shoot. About 99% of the water taken up by a plant is lost during transpiration, which means that the rate of uptake is almost the same as the rate at which transpiration is occurring. We can then measure water uptake by the same shoot under different conditions, e.g. various humidities, wind speeds or temperatures. In this way we get a reasonably accurate measure of the effects of these conditions on the rate of transpiration.

The rate of water loss in a plant can be measured using a potometer (Figure 6). The experiment is carried out in the following stages:

- A leafy shoot is cut under water. Care is taken not to get water on the leaves.
- The potometer is filled completely with water, making sure there are no air bubbles.
- Using a rubber tube, the leafy shoot is fitted to the potometer under water.
- The potometer is removed from under the water and all joints are sealed with waterproof jelly.
- An air bubble is introduced into the capillary tube.
- The distance moved by the air bubble in a given time is measured a number of times and the mean is calculated.

- Using this mean value, the volume of water lost is calculated.
- The volume of water lost against the time in minutes can be plotted on a graph.
- Once the air bubble nears the junction of the reservoir tube and the capillary tube, the tap on the reservoir is opened and the syringe is pushed down until the bubble is pushed back to the start of the scale on the capillary tube. Measurements then continue as before.
- The experiment can be repeated to compare the rates of water uptake under different conditions, for example at different temperatures, humidity, light intensity, or the differences in water uptake between different species under the same conditions.

1 From your knowledge of how water moves up the stem, suggest a reason why each of the following procedures is carried out:
   a The leafy shoot is cut under water rather than in the air.
   b All joints are sealed with waterproof jelly.
2 State what assumption must be made if a potometer is used to measure the rate of transpiration.
3 The volume of water taken up in a given time can be calculated using the formula $\pi r^2 l$ (where $\pi = 3.142$, r = radius of the capillary tube, and l = the distance moved by the air bubble). In an experiment the mean distance moved by the air bubble in a capillary tube of radius 0.5 mm during 1 min was 15.28 mm. Calculate the rate of water uptake in $cm^3\,h^{-1}$. Show your working.
4 If a potometer is used to compare the transpiration rates of two different species of plant, suggest one feature of both plant shoots that should, as far as possible, be kept the same.
5 Suggest reasons why the results obtained from a laboratory potometer experiment may not be representative of the transpiration rate of the same plant in the wild.

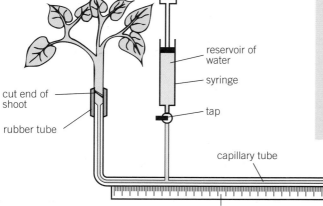

▲ **Figure 6** *A potometer*

Specialised plant cells

The root hair cell

Figure 7 shows the structure of a root hair cell. Each root hair is an extension of a root epidermal cell. Root hairs are the exchange surfaces in plants that are responsible for the absorption of water by osmosis and mineral ions by active transport.

1 State **two** features shown in Figure 7 that suit a root hair cell to its function of absorbing water and mineral ions.
2 Define osmosis.
3 Explain in terms of water potential how water might be absorbed into a root hair cell.
4 Suggest the name of an organelle that you might expect to occur in large numbers in a root hair cell, giving a reason for your answer.

Xylem vessels

Xylem vessels vary in appearance, depending on the type and amount of thickening of their cell walls. As they mature, their walls incorporate a substance called lignin and the cells die. The lignin often forms rings or spirals around the vessel. The structure of xylem vessels is shown in Figure 8.

5 The process of transporting water in plants in the transpiration stream involves water being pulled up the plant, which causes a negative pressure in the xylem vessels. Explain how xylem vessels are adapted to cope with this.
6 Name **two** other features shown in Figure 8 that suit xylem vessels to their function of transporting water up a plant.
7 Suggest one advantage of xylem vessels being dead cells in order to carry out their function effectively.
8 Suggest another possible feature of lignin, other than its mechanical strength, that would be useful in ensuring that water was carried up the plant.
9 The thickening of the cell wall in xylem vessels is often spiral. Suggest **three** advantages to the plant of having this arrangement rather than continuous thickening.

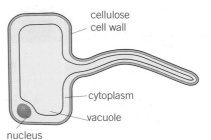

cellulose cell wall

cytoplasm

vacuole

nucleus

▲ **Figure 7** *A root hair cell (top); root hairs on radish seedlings are for the absorption of water and mineral ions (bottom)*

longitudinal section

transverse section

lignified wall of xylem vessel

spiral thickening in xylem vessel

hollow centre of xylem vessel

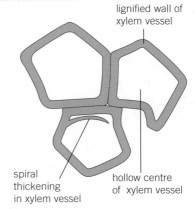

▲ **Figure 8** *Xylem vessels seen in longitudinal and transverse section*

Transport of organic substances in the phloem

Learning objectives

→ Describe the mass flow mechanism for the transport of organic substances in the phloem.

→ Summarise the evidence for and against the mass flow mechanism.

Specification reference: 3.3.4.2

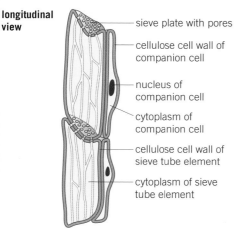

longitudinal view

- sieve plate with pores
- cellulose cell wall of companion cell
- nucleus of companion cell
- cytoplasm of companion cell
- cellulose cell wall of sieve tube element
- cytoplasm of sieve tube element

transverse view

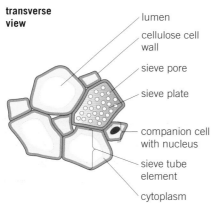

- lumen
- cellulose cell wall
- sieve pore
- sieve plate
- companion cell with nucleus
- sieve tube element
- cytoplasm

▲ **Figure 1** *Phloem as seen under a light microscope*

The process by which organic molecules and some mineral ions are transported from one part of a plant to another is called **translocation**. In flowering plants, the tissue that transports biological molecules is called phloem. Phloem is made up of sieve tube elements, long thin structures arranged end to end. Their end walls are perforated to form sieve plates. Associated with the sieve tube elements are cells called companion cells. The structure of phloem is shown in Figure 1.

Having produced sugars during photosynthesis, the plant transports them from the sites of production, known as **sources**, to the places where they will be used directly or stored for future use – known as **sinks**. As sinks can be anywhere in a plant – sometimes above and sometimes below the source – it follows that the translocation of molecules in phloem can be in either direction. Organic molecules to be transported include sucrose and amino acids. The phloem also transports inorganic ions such as potassium, chloride, phosphate and magnesium ions.

Mechanism of translocation

It is accepted that materials are transported in the phloem and that the rate of movement is too fast to be explained by diffusion. What is in doubt is the precise mechanism by which translocation is achieved. Current thinking favours the **mass flow theory**, a theory that can be divided into three phases:

1. Transfer of sucrose into sieve elements from photosynthesising tissue

- Sucrose is manufactured from the products of photosynthesis in cells with chloroplasts.
- The sucrose diffuses down a concentration gradient by facilitated diffusion from the photosynthesising cells into companion cells.
- Hydrogen ions are actively transported from companion cells into the spaces within cell walls using ATP.
- These hydrogen ions then diffuse down a concentration gradient through carrier proteins into the sieve tube elements.
- Sucrose molecules are transported along with the hydrogen ions in a process known as co-transport (Topic 4.5). The protein carriers are therefore also known as **co-transport proteins**.

2. Mass flow of sucrose through sieve tube elements

Mass flow is the bulk movement of a substance through a given channel or area in a specified time. Mass flow of sucrose through sieve tube elements takes place as follows:

- The sucrose produced by photosynthesising cells (source) is actively transported into the sieve tubes as described above.
- This causes the sieve tubes to have a lower (more negative) water potential.
- As the xylem has a much higher (less negative) water potential (see Topic 7.7), water moves from the xylem into the sieve tubes by osmosis, creating a high hydrostatic pressure within them.

- At the respiring cells (sink), sucrose is either used up during respiration or converted to starch for storage.
- These cells therefore have a low sucrose content and so sucrose is actively transported into them from the sieve tubes lowering their water potential.
- Due to this lowered water potential, water also moves into these respiring cells, from the sieve tubes, by osmosis.
- The hydrostatic pressure of the sieve tubes in this region is therefore lowered.
- As a result of water entering the sieve tube elements at the source and leaving at the sink, there is a high hydrostatic pressure at the source and a low one at the sink.
- There is therefore a mass flow of sucrose solution down this hydrostatic gradient in the sieve tubes.

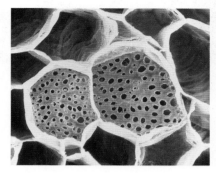

▲ **Figure 2** *Colourised SEM of sieve plates*

While mass flow is a passive process, it occurs as a result of the active transport of sugars. Therefore the process as a whole is active which is why it is affected by, for example, temperature and metabolic poisons. A model of this theory is shown in Figure 3 and the evidence for and against the mass flow theory is listed in Table 1.

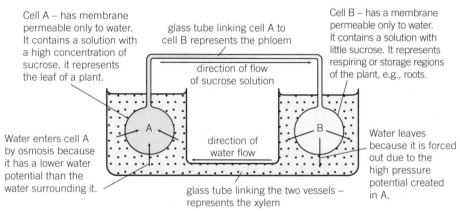

Cell A – has membrane permeable only to water. It contains a solution with a high concentration of sucrose. it represents the leaf of a plant.

glass tube linking cell A to cell B represents the phloem

direction of flow of sucrose solution

Cell B – has a membrane permeable only to water. It contains a solution with little sucrose. It represents respiring or storage regions of the plant, e.g., roots.

Water enters cell A by osmosis because it has a lower water potential than the water surrounding it.

direction of water flow

Water leaves because it is forced out due to the high pressure potential created in A.

glass tube linking the two vessels – represents the xylem

Provided sucrose is continually produced in A (leaf) and continually removed at B (e.g., root), the mass flow of sucrose from A to B continues.

▲ **Figure 3** *Model illustrating the movement of sucrose by mass flow in phloem*

▼ **Table 1** *Evidence for and against the mass flow theory*

Evidence supporting the mass flow hypothesis	Evidence questioning the mass flow hypothesis
• there is a pressure within sieve tubes, as shown by sap being released when they are cut. • the concentration of sucrose is higher in leaves (source) than in roots (sink). • downward flow in the phloem occurs in daylight, but ceases when leaves are shaded, or at night. • increases in sucrose levels in the leaf are followed by similar increases in sucrose levels in the phloem a little later. • metabolic poisons and/or lack of oxygen inhibit translocation of sucrose in the phloem. • companion cells possess many mitochondria and readily produce ATP.	• the function of the sieve plates is unclear, as they would seem to hinder mass flow (it has been suggested that they may have a structural function, helping to prevent the tubes from bursting under pressure). • not all solutes move at the same speed – they should do so if movement is by mass flow. • sucrose is delivered at more or less the same rate to all regions, rather than going more quickly to the ones with the lowest sucrose concentration, which the mass flow theory would suggest.

3. Transfer of sucrose from the sieve tube elements into storage or other sink cells

The sucrose is actively transported by companion cells, out of the sieve tubes and into the sink cells.

The process of translocation of sucrose in phloem is illustrated in Figure 4.

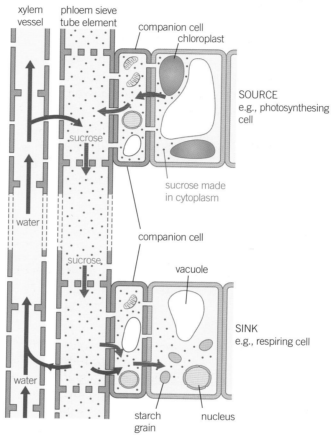

▲ **Figure 4** *Movement of sucrose from source to sink through the phloem of a plant*

Summary questions

State the most suitable word or words represented by the letters **a–k** in the passage below.

Transport of sucrose in plants occurs in the tissue called **a**, from places where it is produced, known as **b**, to places where it is used up or stored, called **c**. One theory of how it is translocated is called the **d** theory. Initially the sucrose is transferred into **e** elements by the process of **f**. The sucrose is produced by **g** cells that therefore have a **h** water potential due to this sucrose. Water therefore moves into them from the nearby **i** tissue that has a **j** water potential. The opposite occurs in those cells (sinks) using up sucrose, and water therefore leaves them by the process of **k**.

We have seen that water is carried in xylem while sugars and amino acids are carried in phloem. How can we be sure that this is the case? This topic looks at some of the evidence and how it is obtained.

Ringing experiments

Woody stems have an outer protective layer of bark on the inside of which is a layer of phloem that extends all round the stem. Inside the phloem layer is xylem (Figure 1).

At the start of a ringing experiment, a section of the outer layers (protective layer and phloem) is removed around the complete circumference of a woody stem while it is still attached to the rest of the plant. After a period of time, the region of the stem immediately above the missing ring of tissue is seen to swell (Figure 1). Samples of the liquid that has accumulated in this swollen region are found to be rich in sugars and other dissolved organic substances. Some non-photosynthetic tissues in the region below the ring (towards the roots) are found to wither and die, while those above the ring continue to grow.

These observations suggest that removing the phloem around the stem has led to:

- the sugars of the phloem accumulating above the ring, leading to swelling in this region
- the interruption of flow of sugars to the region below the ring and the death of tissues in this region.

The conclusion drawn from this type of ringing experiment is that phloem, rather than xylem, is the tissue responsible for translocating sugars in plants. As the ring of tissue removed had not extended into the xylem, its continuity had not been broken. If it were the tissue responsible for translocating sugars you would not have expected sugars to accumulate above the ring nor tissues below it to die.

Tracer experiments

Radioactive **isotopes** are useful for tracing the movement of substances in plants. For example the isotope ^{14}C can be used to make radioactively labelled carbon dioxide ($^{14}CO_2$). If a plant is then grown in an atmosphere containing $^{14}CO_2$, the ^{14}C isotope will be incorporated into the sugars produced during photosynthesis. These radioactive sugars can then be traced as they move within the plant using autoradiography. In our example, this involves taking thin cross-sections of the plant stem and placing them on a piece of X-ray film. The film becomes blackened where it has been exposed to the radiation produced by the ^{14}C in the sugars. The blackened regions are found to correspond to where phloem tissue is in the stem. As the other tissues do not blacken the film, it follows that they do not carry sugars and that phloem alone is responsible for their translocation.

Learning outcomes

→ Describe the use of ringing experiments to investigate transport in plants.

→ Describe the use of tracer experiments to investigate transport in plants.

→ Explain the evidence that translocation of organic molecules occurs in the phloem.

Specification reference: 3.3.4.2

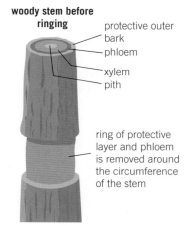

woody stem before ringing

protective outer bark
phloem
xylem
pith

ring of protective layer and phloem is removed around the circumference of the stem

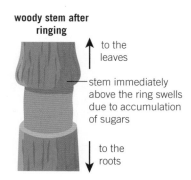

woody stem after ringing

to the leaves

stem immediately above the ring swells due to accumulation of sugars

to the roots

▲ **Figure 1** *Ringing of a woody stem and its results*

Evidence that translocation of organic molecules occurs in phloem

The techniques described are only two of the pieces of evidence that support the view that translocation of organic molecules such as sugars takes place in phloem. A more complete list of evidence is given below.

- When phloem is cut, a solution of organic molecules flow out.
- Plants provided with radioactive carbon dioxide can be shown to have radioactively labelled carbon in phloem after a short time.
- Aphids are a type of insect that feed on plants. They have needle-like mouthparts which penetrate the phloem. They can therefore be used to extract the contents of the sieve tubes. These contents show daily variations in the sucrose content of leaves that are mirrored a little later by identical changes in the sucrose content of the phloem Figure 2.
- The removal of a ring of phloem from around the whole circumference of a stem leads to the accumulation of sugars above the ring and their disappearance from below it.

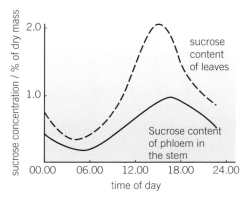

▲ **Figure 2** *Diurnal variation in sucrose content of leaves and phloem*

Summary questions

1. **a** Suggest what difference there would be between the results of a ringing experiment carried out in the summer and one carried out in the winter.

   b Explain the reason for the difference you have suggested.

2. Squirrels sometimes strip sections of bark from around branches. Explain why these branches might die.

3. Suggest how a branch with a complete ring of phloem stripped from it by squirrels might still survive.

4. Explain why squirrels are unlikely to cause the death of a large mature tree by stripping some bark from its trunk.

5. Study Figure 2 and suggest an explanation for:

   a Why there is a time lag between the maximum sucrose content in the leaves and the phloem in the stem and roots.

   b Why the sucrose concentration in the phloem in the stem is lower than that in the leaves.

Using radioactive tracers to find which tissue transports minerals

In an experiment to determine whether minerals are transported in xylem or phloem, a plant was grown in a pot. One branch (Y) on each plant had a 225 mm section of its phloem and xylem separated by inserting strips of impervious wax paper between them as shown in Figure 3. A 225 mm section of another branch (X) of the same plant that had *not* had its xylem and phloem separated by wax paper was used as a control.

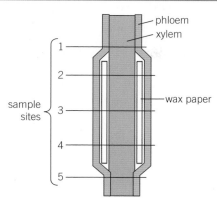

▲ **Figure 3** *Portion of branch of plant showing how xylem and phloem are separated by wax paper and where samples were taken*

The plant was watered with a solution that contained radioactive potassium (^{42}K). After 5 hours absorbing radioactive ^{42}K, sections of the experimental branch were tested for the quantity of ^{42}K in the xylem and phloem. The sections tested are labelled on Figure 3.

The equivalent positions on the control branch were also tested for ^{42}K.

The results are shown in Table 1.

▼ **Table 1**

	Percentage of total ^{42}K			
Section of stem	Branch X (phloem and xylem together)		Branch Y (phloem and xylem separated)	
	Phloem	Xylem	Phloem	Xylem
1	53	47	53	47
2			09	91
3	56	44	01	99
4			15	85
5	52	48	59	41

1 Draw a conclusion from the data in the table.
2 Justify your conclusion with supporting evidence.
3 Explain the fact that the levels of ^{42}K are similar in the xylem and phloem of branch Y in sections 1 and 5.
4 The control (branch X) was an identical length of a different branch that had not had wax paper placed between the xylem and phloem. Suggest a way in which this control could have been improved. Explain why the change you suggest is an improvement.

1   Lugworms live in mud where the partial pressure of oxygen is low. The graph shows oxygen dissociation curves for a lugworm and for a human.

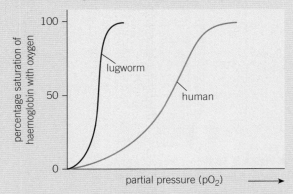

(a) Explain the advantage to the lugworm of having haemoglobin with a dissociation curve in the position shown.   (*2 marks*)

(b) In humans, substances move out of the capillaries to form tissue fluid. Describe how this tissue fluid is returned to the circulatory system.   (*3 marks*)

AQA June 2011

2   The table shows pressure changes in the left side of the heart during one cardiac cycle.

Time / s	Blood pressure / kPa	
	Left atrium	Left ventricle
0.1	0.7	0.3
0.2	1.0	2.0
0.3	0.1	12.5
0.4	0.2	15.3
0.5	1.0	4.5
0.6	0.5	1.0
0.7	0.6	0.3
0.8	0.7	0.3

(a) Between which times is the valve between the atrium and the ventricle closed? Explain your answer.   (*2 marks*)

(b) The maximum pressure in the ventricle is much higher than that in the atrium. Explain what causes this.   (*2 marks*)

(c) Use the information in the table to calculate the heart rate in beats per minute.   (*1 mark*)

AQA June 2011

3   (a) (i)   An arteriole is described as an organ. Explain why.   (*1 mark*)

(ii)   An arteriole contains muscle fibres. Explain how these muscle fibres reduce blood flow to capillaries.   (*2 marks*)

(b) (i)   A capillary has a thin wall. This leads to rapid exchange of substances between the blood and tissue fluid. Explain why.   (*1 mark*)

(ii)   Blood flow in capillaries is slow. Give the advantage of this.   (*1 mark*)

(c) Kwashiorkor is a disease caused by a lack of protein in the blood. This leads to a swollen abdomen due to a build up of tissue fluid. Explain why a lack of protein in the blood causes a build up of tissue fluid.   (*3 marks*)

AQA Jan 2013

4   (a) Scientists measured the rate of water flow and the pressure in the xylem in a small branch. Their results are shown in the graph.

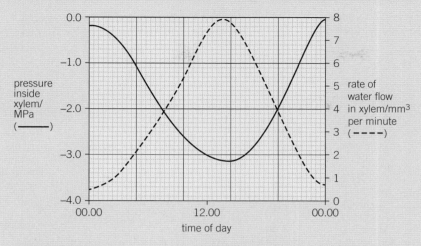

(i) Use your knowledge of transpiration to explain the changes in the rate of flow in the xylem shown in the graph. *(3 marks)*

(ii) Explain why the values for the pressure in the xylem are negative. *(1 mark)*

(b) Doctors measured the thickness of the walls of three blood vessels in a large group of people. Their results are given in the table.

Name of vessel	Mean wall thickness / mm (± standard deviation)
Aorta	5.7 ± 1.2
Pulmonary artery	1.0 ± 0.2
Pulmonary vein	0.5 ± 0.2

(i) Explain the difference in thickness between the pulmonary artery and the pulmonary vein. *(1 mark)*

(ii) The thickness of the aorta wall changes all the time during each cardiac cycle. Explain why. *(3 marks)*

(iii) Which of the three blood vessels shows the greatest variation in wall thickness? Explain your answer. *(1 mark)*

(c) Describe how tissue fluid is formed **and** how it is returned to the circulatory system. *(6 marks)*

AQA June 2012

5 Explain how water enters xylem from the endodermis in the root and is then transported to the leaves. *(6 marks)*

AQA June 2013

6 The table shows measurements of pulse rate, systolic blood pressure and diastolic blood pressure of an individual after sitting in a chair or walking fast or running.

level of activity	pulse rate / beats min⁻¹	systolic pressure / kPa	diastolic pressure / kPa	stroke volume / ml
sitting	62	15.5	10.4	55
after walking	58	19.2	10.7	74
after running	106	23.8	11.1	88

(a) Calculate the changes in systolic pressure as the level of activity increases. *(2 marks)*

(b) Explain the difference in the effect of exercise on systolic and diastolic pressure. *(3 marks)*

(c) Calculate the cardiac output of this individual after running. Give you answer in $dm^3\ min^{-1}$. *(1 mark)*

(d) Predict and explain the effect on potential cardiac output of daily exercise sessions over a six month period. *(2 marks)*

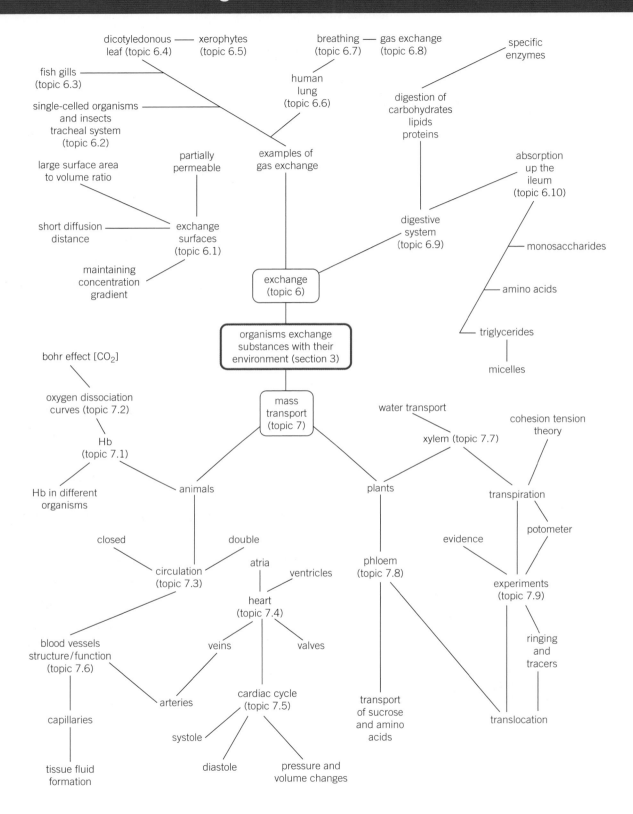

Practical skills

In this section you have met the following practical skills:

- Evaluating the results of scientific experiments
- Using appropriate apparatus, such as a potometer, to obtain quantitative measurements
- Commenting on experimental design and suggesting improvements.

Maths skills

In this section you have met the following maths skills:

- Calculating surface area to volume ratios
- Changing the subject in pulmonary ventilation and cardiac output equations
- Using appropriate units in calculations
- Substituting values in, and solving, algebraic equations
- Interpreting graphs and translating information between graphical and numerical forms
- Recognising expressions in decimal and standard forms
- Understanding simple probability
- Interpreting bar charts.

Extension tasks

Using the knowledge that you have gained from this section make a comprehensive list of each of the following:

a   The general features of all transport systems.

b   The differences between transport systems in plants and transport systems in animals.

c   An explanation for each of the differences you have listed under b).

Figure 1 shows you how to take a person's pulse at the wrist. Each pulse is equivalent to a single heartbeat.

You should take each person's pulse for 30 seconds and double the reading to give the number of heart beats per minute (heart rate).

Find out what is meant by the 'recovery heart rate' and then design and carry out an experiment to determine any difference between the recovery heart rate of people who exercise frequently and those who do not. Draw conclusions from your results and suggest an explanation for them.

Section 3 practice questions

1   Large insects contract muscles associated with the abdomen to force air in and out of the spiracles. This is known as 'abdominal pumping'. The table shows the mean rate of abdominal pumping of an insect before and during flight.

Stage of flight	Mean rate of abdominal pumping / dm^3 of air kg^{-1} hour^{-1}
Before	42
During	186

(a) Calculate the percentage increase in the rate of abdominal pumping before and during flight.
Show your working.   (2 marks)

(b) Abdominal pumping increases the efficiency of gas exchange between the tracheoles and muscle tissue of the insect. Explain why.   (2 marks)

(c) Abdominal pumping is an adaptation not found in many small insects.
These small insects obtain sufficient oxygen by diffusion.
Explain how their small size enables gas exchange to be efficient without the need for abdominal pumping.   (1 mark)
The graph shows the concentration of oxygen inside the tracheoles of an insect when at rest. It also shows when the spiracles are fully open.

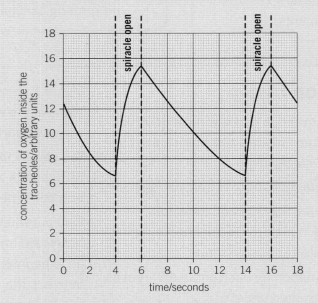

(d) Use the graph to calculate the frequency of spiracle opening. Show your working.   (2 marks)

(e) The insect opens its spiracles at a lower frequency in very dry conditions.
Suggest **one** advantage of this.   (1 mark)

(f) The ends of tracheoles connect directly with the insect's muscle tissue and are filled with water. When flying, water is absorbed into the muscle tissue.
Removal of water from the tracheoles increases the rate of diffusion of oxygen between the tracheoles and muscle tissue. Suggest **one** reason why.   (1 mark)
AQA June 2013

2   (a) Describe how proteins are digested in the human gut.   (4 marks)
AQA SAMS A LEVEL PAPER 1

(b) The enzyme lipase catalyses the digestion of lipids. The optimum pH of one version of this enzyme is 4.7. Calculate the concentration of hydrogen ions in a solution of pH 4.7.   (2 marks)

3 Newborn babies can be fed with breast milk or with formula milk. Both types of milk contain carbohydrates, lipids and proteins.
   * Human breast milk also contains a bile-activated lipase. This enzyme is thought to be inactive in milk but activated by bile in the small intestine of the newborn baby.
   * Formula milk does not contain a bile-activated lipase.

   Scientists investigated the benefits of breast milk compared with formula milk.
   The scientists used kittens (newborn cats) as model organisms in their laboratory investigation.
   (a) Other than ethical reasons, suggest **two** reasons why they chose to use cats as model organisms.   (*2 marks*)
   (b) Before starting their experiments, the scientists confirmed that, like human breast milk, cat's milk also contained bile-activated lipase. To do this, they added bile to cat's milk and monitored the pH of the mixture.
        Explain why monitoring the pH of the mixture could show whether the cat's milk contained lipase.   (*2 marks*)

   The scientists then took 18 kittens. Each kitten had been breastfed by its mother for the previous 48 hours.
   The scientists divided the kittens randomly into three groups of six.
   * The kittens in group **1** were fed formula milk.
   * The kittens in group **2** were fed formula milk plus a supplement containing bile-activated lipase.
   * The kittens in group **3** were fed breast milk taken from their mothers.

   Each kitten was fed 2 cm³ of milk each hour for 5 days.
   The scientists weighed the kittens at the start of the investigation and on each day for 5 days.

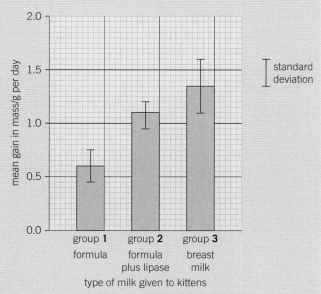

   ▲ **Figure 5** *shows the scientists' results.*

   (c) What can you conclude from **Figure 5** about the importance of bile-activated lipase in breast milk?   (*3 marks*)

                                                                                 AQA SAMS AS PAPER 2

HAMPTON SCHOOL
BIOLOGY DEPARTMENT

Section 4
Genetic information, variation and relationships between organisms

Introduction

A look at the living world around us is enough to demonstrate the striking variety of life. This biodiversity is reflected not just in the multiplicity of different species, but also in the range of different individuals within a single species. Then there is the variety of different tissues, organs and cells that make up one individual. All this diversity is primarily the result of the genes that each organism possesses. It is not entirely genetic however, the environment plays its part by modifying the characteristics determined by an individual's genes.

A gene is found on a specific region of a DNA molecule known as the locus. DNA is a remarkable molecule, carrying vast amounts of information in the form of its sequence of nucleotide bases. The base sequence of each gene carries coded genetic information that determines, in turn, the sequence of amino acids in an organism's proteins. The genetic code is universal, being the same for all living organisms and therefore provides indirect evidence that organisms have evolved from one another.

DNA is a very stable molecule. It has to be if it is to reliably transfer information from one generation to the next in a way that faithfully reproduces the characteristics of the parents in the offspring. How then has the genetic diversity that is so evident arisen if DNA is not easily altered? There are a number of mechanisms. The process of sexual reproduction introduces variety in a number of ways, not least the combining of two different sets of genes – one from each parent. In addition, meiosis at some point in the life cycle leads to a random shuffling of the chromosomes. Then there are spontaneous random changes to DNA called mutations. Despite being rare, they are the basis of genetic change and evolution.

From the diverse range of individuals in a population of a species there will be some that are better adapted to the particular conditions that exist at the time they are alive. These individuals will be more likely to survive and breed and therefore pass on their alleles to the next generation. In this way populations may evolve by natural selection into new species.

Measuring diversity within a species can be achieved by comparing differences in the sequence of nucleotide bases in an individual's DNA or in the sequence of amino acids in the proteins that this DNA codes for. Measuring biodiversity within a community can be achieved by using an index of diversity or species richness.

Working scientifically

The study of genetic information, variation and relationships between organisms provides scope to perform practical work and to develop practical skills. A required practical activity is the use of aseptic techniques to investigate the effect of antimicrobial substances on bacterial growth.

A range of mathematical skills will be needed in your work, in particular the ability to use power and logarithmic functions of a calculator, use a logarithmic scale, find arithmetic means, understand measures of dispersion including standard deviation and substitute values in algebraic equations.

What you already know

The material in this unit is intended to be self-explanatory, but there are certain facts from GCSE that will help your understanding of this section. These facts include:

○ Petri dishes and culture media must be sterilised before use to kill unwanted microorganisms.

○ Inoculating loops are used to transfer microorganisms to culture media and they must be sterilised by passing them through a flame.

○ The lid of the Petri dish should be secured with adhesive tape to prevent microorganisms from the air contaminating the culture.

○ The type of cell division in which a cell divides to form gametes is called meiosis.

○ Sexual reproduction gives rise to variation because, when gametes fuse, one of each pair of alleles comes from each parent.

○ Chromosomes are made up of large molecules of DNA (deoxyribonucleic acid), which has a double helix structure.

○ A gene is a small section of DNA.

○ Each gene codes for a particular combination of amino acids that make a specific protein.

○ Genetic variation is due to a population having a wide range of alleles that control their characteristics.

○ In each population, the alleles that control the characteristics which help the organism to survive are selected.

○ Quantitative data on the distribution of organisms can be obtained by random sampling with quadrats and sampling along a transect.

Once it had been established that DNA was the means by which genetic information was passed from generation to generation, scientists puzzled as to exactly how DNA determined the features of organisms. Before we look at this problem, we need first to be clear about what is meant by a gene.

What is a gene?

A gene is a section of DNA that contains the coded information for making polypeptides and functional RNA. The coded information is in the form of a specific sequence of bases along the DNA molecule. Polypeptides make up proteins and so genes determine the proteins of an organism. Enzymes are proteins. As enzymes control chemical reactions they are responsible for an organism's development and activities. In other words genes, along with environmental factors, determine the nature and development of all organisms. A gene is a section of DNA located at a particular position, called a **locus**, on a DNA molecule. The gene is a base sequence of DNA that codes for:

- the amino acid sequence of a polypeptide
- or a functional RNA, including ribosomal RNA and transfer RNAs (Topic 8.3).

One DNA molecule carries many genes.

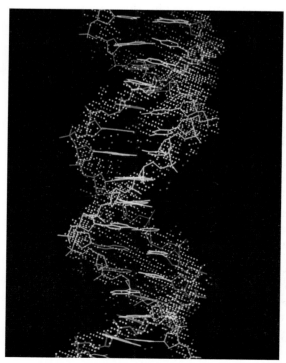

▲ **Figure 1** *Computer graphics representation of a short section of DNA*

The genetic code

In trying to discover how DNA bases coded for amino acids, scientists suggested that there must be a minimum of three bases that coded for each amino acid. Their reasoning was as follows:

- Only 20 different amino acids regularly occur in proteins.
- Each amino acid must have its own code of bases on the DNA.
- Only four different bases (adenine, guanine, cytosine and thymine) are present in DNA.
- If each base coded for a different amino acid, only four different amino acids could be coded for.
- Using a pair of bases, 16 (4^2) different codes are possible, which is still inadequate.
- Three bases produce 64 (4^3) different codes, more than enough to satisfy the requirements of 20 amino acids.

As the code has three bases for each amino acid, each one is called a triplet. As there are 64 possible triplets and only 20 amino acids, it follows that some amino acids are coded for by than one triplet.

Features of the genetic code

Further experiments have revealed the following features of the genetic code.

- A few amino acids are coded for by only a single triplet.
- The remaining amino acids are coded for by between two and six triplets each.
- The code is known as a '**degenerate code**' because most amino acids are coded for by more than one triplet.
- A triplet is always read in one particular direction along the DNA strand.
- The start of a DNA sequence that codes for a polypeptide is always the same triplet. This codes for the amino acid methionine. If this first methionine molecule does not form part of the final polypeptide, it is later removed.
- Three triplets do not code for any amino acid. These are called 'stop codes' and mark the end of a polypeptide chain. They act in much the same way as a full stop at the end of a sentence.
- The code is **non-overlapping**, in other words each base in the sequence is read only once. Thus six bases numbered 123456 are read as triplets 123 and 456, rather than as triplets 123, 234, 345, 456.
- The code is **universal**, with a few minor exceptions each triplet codes for the same amino acid in all organisms. This is indirect evidence for evolution.

Much of the DNA in eukaryotes does not code for polypeptides. For example, between genes there are non-coding sequences made up of multiple repeats of base sequences. Even within genes, only certain sequences code for amino acids. These coding sequences are called **exons**. Within the gene these exons are separated by further non-coding sequences called **introns**. Some genes code for ribosomal RNA and transfer RNAs.

Summary questions

1 Describe what a gene is.

2 Calculate how many bases are required to code for a chain of six consecutive amino acids.

3 Explain how a change in one base along a DNA molecule may result in an enzyme becoming non-functional.

4 A section of DNA has the following sequence of bases along it:
TAC GCT CCG CTG TAC. All of the bases are part of the code for amino acids. The first base in the sequence is the start of the code.

   a Calculate the number of amino acids that the section of DNA codes for.

   b Determine which two sequences code for the same amino acid.

   c It is possible that this sequence codes for many different amino acids or many copies of the same amino acid. From your knowledge of the genetic code explain how this can happen.

Interpreting the genetic code

Table 1 is a genetic code table showing the amino acids that each codon (set of three nucleotides in mRNA) is translated into during protein synthesis. An amino acid is indicated by three letters of its name, for example arg = **arg**inine, ile = **iso**l**e**ucine. To find the code for any amino acid you find the relevant three letters (usually the first three) of its name in Table 1 and then read:

▼ **Table 1** *The genetic code. The base sequences shown are those on mRNA*

First position	Second position				Third position
	U	C	A	G	
U	Phe	Ser	Tyr	Cys	U
	Phe	Ser	Tyr	Cys	C
	Leu	Ser	Stop	Stop	A
	Leu	Ser	Stop	Trp	G
C	Leu	Pro	His	Arg	U
	Leu	Pro	His	Arg	C
	Leu	Pro	Gln	Arg	A
	Leu	Pro	Gln	Arg	G
A	Ile	Thr	Asn	Ser	U
	Ile	Thr	Asn	Ser	C
	Ile	Thr	Lys	Arg	A
	Met	Thr	Lys	Arg	G
G	Val	Ala	Asp	Gly	U
	Val	Ala	Asp	Gly	C
	Val	Ala	Glu	Gly	A
	Val	Ala	Glu	Gly	G

- the first base in the sequence from the column on the left
- the second base in the sequence from the row at the top
- the third base in the sequence from the column to the right.

You can also use the table to find an amino acid that is coded for by a particular codon. For example, UGC codes for the amino acid Cys (cysteine):

- the first letter (U) is in the column on the left
- the second letter (G) is in the row at the top
- the third letter (C) is in the column to the right.

You will notice that most amino acids have more than one codon(s), for example alanine (ala) has four codon(s) GCU, GCC, GCA and GCG.

Using Table 1, answer the following questions. In each case identify amino acids by their three-letter codon(s).

1 List the two amino acids that have only one codon and state what it is in each case.
2 Name the amino acids that have each of the following codon.
   a CUC
   b AAA
   c GAU
3 For each of the following base sequences on a DNA molecule, deduce the sequence of amino acids in the order in which they would occur in the resultant polypeptide.
   a ATGCGTTAAGGCAGT
   b GCTAAGTTTCCAGAT

In Topic 3.6 we saw that, according to their organisation, there are two types of cell: **prokaryotic cells** and **eukaryotic cells**. We looked at some of the differences between the two. These differences extend to their DNA:

- In prokaryotic cells, such as bacteria, the DNA molecules are shorter, form a circle and are not associated with protein molecules. Prokaryotic cells therefore do not have chromosomes.

- In eukaryotic cells, the DNA molecules are longer, form a line (are linear) rather than a circle and occur in association with proteins called **histones** to form structures called **chromosomes**. The mitochondria and chloroplasts of eukaryotic cells also contain DNA which, like the DNA of prokaryotic cells, is short, circular and not associated with proteins.

Chromosome structure

Chromosomes are only visible as distinct structures when a cell is dividing. For the rest of the time they are widely dispersed throughout the nucleus. When they first become visible at the start of cell division chromosomes appear as two threads, joined at a single point (Figure 1). Each thread is called a **chromatid** because DNA has already replicated to give two identical DNA molecules. The DNA in chromosomes is held by histones. The considerable length of DNA found in each cell (around 2 m in every human cell) is highly coiled and folded as illustrated in Figure 2. Let us look carefully at this diagram.

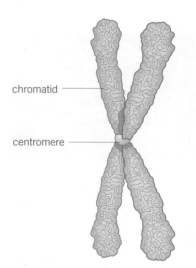

▲ **Figure 1** *Structure of a chromosome*

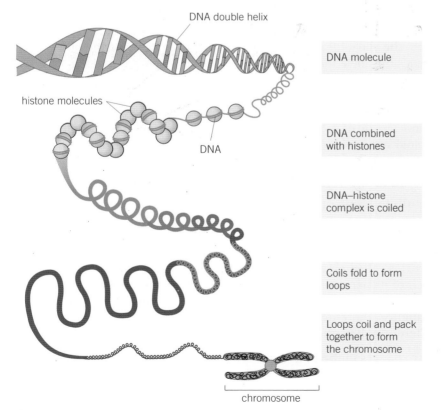

▲ **Figure 2** *How DNA is packed into a chromosome*

Hint

It is often mistakenly thought that humans have just 46 chromosomes throughout the body, rather than 46 in every single cell (except sperm and eggs).

We already know that DNA is a double helix. From the diagram we will see that this helix is wound around histones to fix it in position. This DNA–histone complex is then coiled. The coil, in turn, is looped and further coiled before being packed into the chromosome. In this way a lot of DNA is condensed into a single chromosome. If we follow the diagram carefully we will see that a chromosome contains just a single molecule of DNA, although this is very long. This single DNA molecule has many genes along its length (see Topic 8.1). Each gene occupies a specific position (locus) along the DNA molecule.

Although the number of chromosomes is always the same for normal individuals of a species, it varies from one species to another. For example, while humans have 46 chromosomes, potato plants have 48 and dogs have 78. In most species, there is an even number of chromosomes in the cells of adults.

Homologous chromosomes

Sexually produced organisms, such as humans, are the result of the fusion of a sperm and an egg, each of which contributes one complete set of chromosomes to the offspring. Therefore, one of each pair is derived from the chromosomes provided by the mother in the egg (maternal chromosomes) and the other is derived from the chromosomes provided by the father in the sperm (paternal chromosomes). These are known as **homologous pairs** and the total number is referred to as the **diploid** number. In humans this is 46.

A homologous pair is always two chromosomes that carry the same genes but not necessarily the same alleles of the genes.

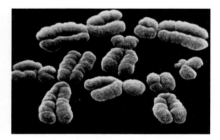

▲ **Figure 3** *False-colour SEM of a group of human chromosomes*

For instance, a homologous pair of chromosomes may each possess genes for tongue rolling and blood group, but one chromosome may carry the allele for non-roller and blood group A, while the other carries the allele for roller and blood group B. During meiosis, the halving of the number of chromosomes is done in a manner which ensures that each daughter cell receives one chromosome from each homologous pair. In this way each cell receives one gene for each characteristic of the organism. When these haploid cells combine, the diploid state, with paired homologous chromosomes, is restored.

What is an allele?

An **allele** is one of a number of alternative forms of a gene. We have seen that genes are sections of DNA that contain coded information in the form of specific sequences of bases. Each gene exists in two, occasionally more, different forms. Each of these forms is called an allele. Each individual inherits one allele from each of its parents. These two alleles may be the same or they may be different. When they are different, each allele has a different base sequence, therefore a different amino acid sequence, so produces a different polypeptide.

Any changes in the base sequence of a gene produces a new allele of that gene (=mutation) and results in a different sequence of amino acids being coded for. This different amino acid sequence will lead

Hint

Do not confuse genes and alleles. A *gene* refers to a particular characteristic such as blood groups. Genes can exist in two or more different forms called **alleles**.

to the production of a different polypeptide, and hence a different protein. Sometimes this different protein may not function properly or may not function at all. When the protein produced is an enzyme, it may have a different shape. The new shape may not fit the enzyme's substrate. As a result the enzyme may not function and this can have serious consequences for the organism.

Synoptic link

To remind you how the shape of an enzyme is important to the way it works look back at Topic 1.7. We shall also learn more about the importance of meiosis in producing genetic variation in Topic 9.2

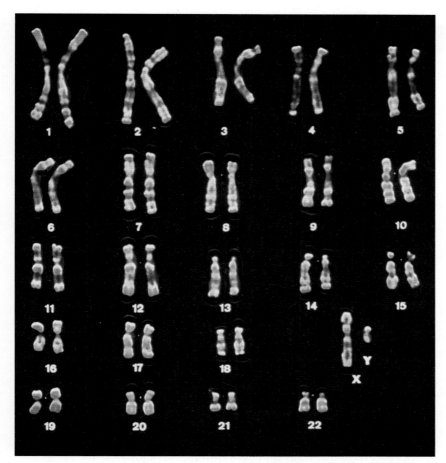

▲ **Figure 4** *The 46 chromosomes of a human showing them in their 22 pairs, as well as the X and Y sex chromosomes*

Summary questions

1 Contrast the DNA of a prokaryotic cell with that of a eukaryotic cell.

2 State the function of the protein found in chromosomes.

3 Explain how the considerable length of a DNA molecule is compacted into a chromosome.

4 ⓥ Suppose the total length of all the DNA in a single human muscle cell is 2.3 m.

   a If all the DNA were distributed equally between the chromosomes, calculate the mean length of DNA in each one.

   b Calculate in mm the length of DNA in a human brain cell.

Learning objectives

→ Describe what the genetic code is and its main features

→ Describe the structure of ribonucleic acid (RNA).

→ Describe the structure and the role of messenger RNA (mRNA).

→ Describe the structure and the role of transfer RNA (tRNA).

Specification reference: 3.4.1

In Topic 8.1 we learned about the importance of DNA and how it carries coded information for the sequence of amino acids that make up a protein. In this topic, we will turn our attention to exactly how DNA triplets are used to make the proteins that they code for.

Transferring the coded information

We know that the sequence of nucleotide bases in DNA determines the sequence of amino acids in the proteins of an organism. In eukaryotic cells DNA is largely confined to the nucleus. However, the synthesis of proteins takes place in the cytoplasm. So how is the coded information on the DNA in the nucleus transferred to the cytoplasm where it is translated into proteins? The answer is that sections of the DNA code are transcribed onto a single-stranded molecule called ribonucleic acid (RNA).

There are a number of types of RNA. The one that transfers the DNA code from the nucleus to the cytoplasm acts as a type of messenger and is hence given the name **messenger RNA**, or **mRNA** for short. This mRNA is small enough to leave the nucleus through the nuclear pores and to enter the cytoplasm, where the coded information that it contains is used to determine the sequence of amino acids in the proteins which are synthesised there.

The term **codon** refers to the sequence of three bases on mRNA that codes for a single amino acid.

There are two other terms that are relevant:

- **Genome** – the complete set of genes in a cell, including those in mitochondria and/or chloroplasts.
- **Proteome** – the full range of proteins produced by the genome. This is sometimes called the **complete proteome**, in which case the term proteome refers to the proteins produced by a given type of cell under a certain set of conditions.

We saw in Topic 2.1 that DNA is composed of two nucleotide chains wound around each other (double helix). We will now look at the structure of a related molecule that is usually made up of a single nucleotide chain: **ribonucleic acid (RNA)**.

Ribonucleic acid (RNA) structure

Ribonucleic acid (RNA) is a polymer made up of repeating mononucleotide sub-units (Figure 1). It forms a single strand in which each nucleotide is made up of:

- the pentose sugar ribose
- one of the organic bases adenine (A), guanine (G), cytosine (C) and uracil (U)
- a phosphate group.

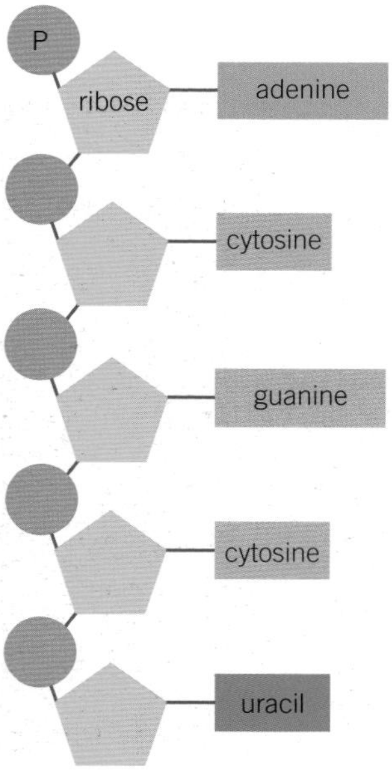

▲ **Figure 1** *Section of ribonucleic acid (RNA) molecule*

The two types of RNA that are important in protein synthesis are:

* messenger RNA (mRNA)
* transfer RNA (tRNA).

Messenger RNA (mRNA)

Consisting of thousands of mononucleotides, mRNA is a long strand that is arranged in a single helix. The base sequence of mRNA is determined by the sequence of bases on a length of DNA in a process called transcription. There is a great variety of different types of mRNA. Once formed, mRNA leaves the nucleus via pores in the nuclear envelope and enters the cytoplasm, where it associates with the ribosomes. There it acts as a template for protein synthesis. Its structure is suited to this function because it possesses information in the form of codons (three bases that are complementary to a triplet in DNA). The sequence of codons determines the amino acid sequence of a specific polypeptide that will be made.

Transfer RNA (tRNA)

Transfer RNA (tRNA) is a relatively small molecule that is made up of around 80 nucleotides. It is a single-stranded chain folded into a clover-leaf shape, with one end of the chain extending beyond the other. This is the part of the tRNA molecule to which an amino acid can easily attach. There are many types of tRNA, each of which binds to a specific amino acid. At the opposite end of the tRNA molecule is a sequence of three other organic bases, known as the **anticodon**. Given that the genetic code is degenerate there must be as many tRNA molecules as there are coding triplets. However, each tRNA is specific to one amino acid and has an anticodon that is specific to that amino acid.

> ### Synoptic link
>
> A third type of RNA, ribosomal RNA, was covered in Topic 3.4.

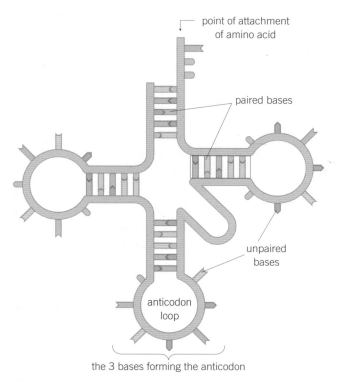

▲ **Figure 3** *Clover-leaf structure of tRNA*

(labels: point of attachment of amino acid; paired bases; unpaired bases; anticodon loop; the 3 bases forming the anticodon)

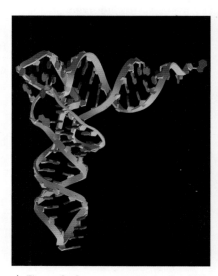

▲ **Figure 2** *Computer artwork of a tRNA molecule*

Summary questions

1 Distinguish between the structure of mRNA and the structure of tRNA.

2 State **three** ways in which the molecular structure of RNA differs from DNA.

3 Distinguish between a codon and an anticodon.

You will recall from Topic 8.1 that the organic bases in DNA pair up in a precise way, for example, guanine pairs with cytosine, and adenine pairs with thymine. These are known as complementary base pairs. In RNA, however, the base thymine is always replaced by a similar base called uracil. RNA can join with both DNA and other RNA molecules by complementary base pairing. The complementary base pairings that RNA forms are therefore:

- guanine with cytosine
- adenine with uracil (in RNA) or thymine (in DNA).

During protein synthesis, an anticodon pairs with the three complementary organic bases that make up the codon on mRNA. The tRNA structure (Figure 3), with its end chain for attaching amino acids and its anticodon for complementary base pairing with the codon of the mRNA, is structurally suited to its role of lining up amino acids on the mRNA template during protein synthesis.

Comparison of DNA, messenger RNA and transfer RNA

Table 1 compares the structure, function and composition of DNA, mRNA and tRNA.

▼ **Table 1** *Comparison of DNA, mRNA and tRNA*

DNA	Messenger RNA	Transfer RNA
double polynucleotide chain	single polynucleotide chain	single polynucleotide chain
largest molecule of the three	molecule is smaller than DNA but larger than tRNA	smallest molecule of the three
double-helix molecule	single-helix molecule (except in a few viruses)	clover-shaped molecule
pentose sugar is deoxyribose	pentose sugar is ribose	pentose sugar is ribose
organic bases are adenine, guanine, cytosine and thymine	organic bases are adenine, guanine, cytosine and uracil	organic bases are adenine, guanine, cytosine and uracil
found mostly in the nucleus	manufactured in the nucleus but found throughout the cell	manufactured in the nucleus but found throughout the cell
quantity is constant for all cells of a species (except gametes)	quantity varies from cell to cell and with level of metabolic activity	quantity varies from cell to cell and with level of metabolic activity
chemically very stable	Less stable than DNA or tRNA, individual molecules are usually broken down in cells within a few days.	chemically more stable than mRNA but less stable than DNA

1 Table 1 states that, for DNA, the 'quantity is constant for all cells of a species (except gametes)'.
   a State how the quantity in a gamete differs from that in a body cell.
   b Explain the significance of the difference you have described.
2 Explain an advantage of:
   a DNA being a chemically stable molecule
   b mRNA being broken down relatively quickly.

8.4 Polypeptide synthesis – transcription and splicing

We saw in Topic 1.6 that proteins are made up of one or more polypeptides. Proteins, especially enzymes, are essential to all aspects of life. Every organism needs to make its own, unique, proteins. The biochemical machinery in the cytoplasm of each cell has the capacity to make every protein from just 20 amino acids. Exactly which proteins it manufactures depends upon the instructions that are provided, at any given time, by the DNA in the cell's nucleus. The basic process is as follows.

- DNA provides the instructions in the form of a long sequence of bases.
- A complementary section of part of this sequence is made in the form of a molecule called pre-mRNA – a process called **transcription**.
- The pre-mRNA is spliced to form mRNA.
- The mRNA is used as a template to which complementary tRNA molecules attach and the amino acids they carry are linked to form a polypeptide – a process called **translation**.

The process can be likened to a bakery, where the basic equipment and ovens (cell organelles) can manufacture any variety of cake (protein) from relatively few basic ingredients (amino acids). Which particular variety of cake is made depends on the recipe (genetic code) that the baker uses on any particular day. By choosing different recipes at different times, rather than making everything all the time, the baker can meet seasonal demands, adapt to changing customer needs and avoid waste.

DNA replication can be likened to the publication of many copies of a recipe book (genome); making a photocopy of a recipe to use in the bakery is therefore transcription. Making the cakes, using the photocopied recipe, is translation. If the book is not removed from the library, many copies of the recipe can be made, and the same cakes can be produced in many places at the same time or over many years.

Transcription

Transcription is the process of making pre-mRNA using part of the DNA as a template. The process, which is illustrated in Figure 1, is as follows.

- An enzyme acts on a specific region of the DNA causing the two strands to separate and expose the **nucleotide** bases in that region.
- The nucleotide bases on one of the two DNA strands, known as the **template strand**, pair with their complementary nucleotides from the pool which is present in the nucleus. The enzyme RNA polymerase then moves along the strand and joins the nucleotides together to form a pre mRNA molecule.

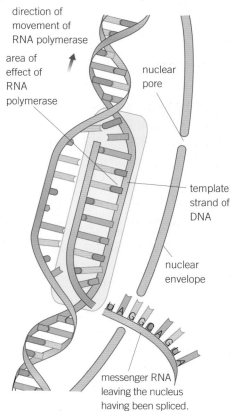

direction of movement of RNA polymerase

area of effect of RNA polymerase

nuclear pore

template strand of DNA

nuclear envelope

messenger RNA leaving the nucleus having been spliced.

▲ **Figure 1** *Summary of transcription*

- In this way an exposed guanine base on the DNA binds to the cytosine base of a free nucleotide. Similarly, cytosine links to guanine, and thymine joins to adenine. The exception is adenine, which links to uracil rather than thymine.
- As the RNA polymerase adds the nucleotides one at a time to build a strand of pre-mRNA, the DNA strands rejoin behind it. As a result, only about 12 base pairs on the DNA are exposed at any one time.
- When the RNA polymerase reaches a particular sequence of bases on the DNA that it recognises as a 'stop' triplet code, it detaches, and the production of pre-mRNA is then complete.

Splicing of pre-mRNA

In prokaryotic cells, transcription results directly in the production of mRNA from DNA. In **eukaryotic cells** transcription results in the production of pre-mRNA, which is then spliced to form mRNA. The DNA of a gene eukaryotic cells is made up of sections called exons that code for proteins and sections called introns that do not. These intervening introns would prevent the synthesis of a polypeptide. In the pre-mRNA of eukaryotic cells. The base sequences corresponding to the introns are removed and the functional exons are joined together during a process called **splicing**. As most prokaryotic cells do not have introns, splicing of their DNA is unnecessary. The process of splicing is shown in Figure 2.

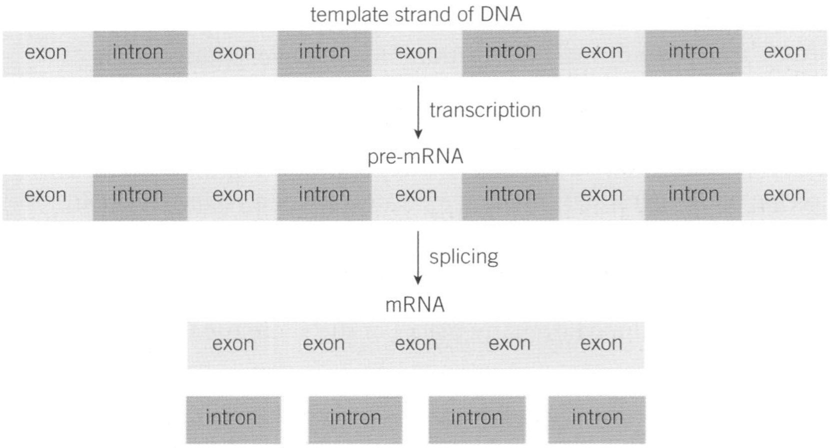

▲ **Figure 2** *Splicing of pre-mRNA to form mRNA*

The mRNA molecules are too large to diffuse out of the nucleus and so, once they have been spliced, they leave via a nuclear pore. Outside the nucleus, the mRNA is attracted to the ribosomes to which it becomes attached, ready for the next stage of the process: translation.

Summary questions

1. Describe the role of RNA polymerase in transcription.
2. State which other enzyme is involved in transcription and describe its role.
3. Explain why splicing of pre-mRNA is necessary.
4. A sequence of bases along the template strand of DNA is ATGCAAGTCCAG.
   - **a** Deduce the sequence of bases on a pre-messenger RNA molecule that has been transcribed from this part of the DNA molecule.
   - **b** Calculate how many amino acids the sequence codes for.
5. A gene is made up of 756 base pairs. The mRNA that is transcribed from this gene is only 524 nucleotides long. Explain why there is this difference.

8.5 Polypeptide synthesis – translation

In Topic 8.4 we looked at how the triplet code of DNA is transcribed into a sequence of **codons** (genetic code) on messenger RNA (mRNA). The next stage is to translate the codons on the mRNA into a sequence of amino acids that make up a polypeptide.

There are about 60 different tRNAs. A particular tRNA has a specific **anticodon** and attaches to a specific amino acid. Each amino acid therefore has one or more tRNA molecule, with its own anticodon of bases.

Synthesising a polypeptide

Once mRNA has passed out of the nuclear pore it determines the synthesis of a polypeptide. The following explanation of how a polypeptide is made is illustrated in Figures 3 and 4. (The information given in brackets below is only to help you follow the process and does not need to be learned.)

- A ribosome (Figure 4, part 1) becomes attached to the starting codon (AUG) at one end of the mRNA molecule.
- The tRNA molecule with the complementary anticodon sequence (UAC) moves to the ribosome and pairs up with the codon on the mRNA. This tRNA carries a specific amino acid (methionine).
- A tRNA molecule with a complementary anticodon (UGC) pairs with the next codon on the mRNA (ACG). This tRNA molecule carries another amino acid (threonine).
- The ribosome moves along the mRNA, bringing together two tRNA molecules at any one time, each pairing up with the corresponding two codons on the mRNA.
- The two amino acids (methionine and threonine) on the tRNA are joined by a **peptide bond** using an enzyme and ATP which is hydrolysed to provide the required energy.
- The ribosome moves on to the third codon (GAU) in the sequence on the mRNA, thereby linking the amino acids (threonine and aspartic acid) on the second and third tRNA molecules (Figure 4, part 2).
- As this happens, the first tRNA is released from its amino acid (methionine) and is free to collect another amino acid (methionine) from the amino acid pool in the cell.
- The process continues in this way, with up to 15 amino acids being added each second, until a polypeptide chain is built up (Figure 4, part 3).
- Up to 50 ribosomes can pass immediately behind the first, so that many identical polypeptides can be assembled simultaneously (Figure 3).
- The synthesis of a polypeptide continues until a ribosome reaches a stop codon. At this point, the ribosome, mRNA and the last tRNA molecule all separate and the polypeptide chain is complete.

In summary, the DNA sequence of triplets that make up a gene determine the sequence of codons on mRNA. The sequence of codons on mRNA determine the order in which the tRNA molecules line up.

Learning objectives

→ Explain how a polypeptide is synthesised during the process of translation.

→ Describe the roles of messenger RNA and transfer RNA in translation.

Specification reference: 3.4.2

point of attachment of amino acid

Anticodon – this sequence of ACG means that the amino acid cysteine will attach to the other end of this tRNA molecule. This anticodon will combine with the codon UGC on a mRNA molecule during the formation of a polypeptide. The mRNA codon UGC therefore translates into the amino acid cysteine.

▲ **Figure 1** *Simplified structure of one type of tRNA*

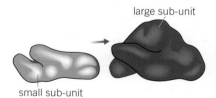

large sub-unit

small sub-unit

▲ **Figure 2** *Structure of a ribosome. The smaller sub-unit fits into a depression on the surface of the larger one*

Hint

Remember that there is no thymine in any RNA molecule. It is uracil in RNA that pairs with adenine.

Study tip

ATP has two roles in translation. It is required to provide energy to attach amino acids to tRNA and also to attach amino acids together.

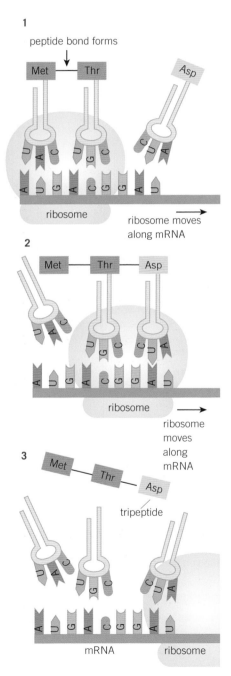

▲ **Figure 4** *Translation*

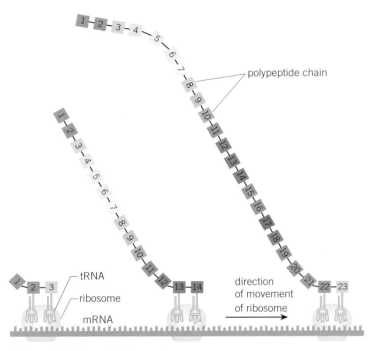

▲ **Figure 3** *Polypeptide formation*

They, in turn, determine the sequence of amino acids in the polypeptide. In this way genes precisely determine which proteins a cell manufactures. As many of these proteins are enzymes, genes effectively control the activities of cells.

Assembling a protein

Sometimes a single polypeptide chain is a functional protein. Often, a number of polypeptides are linked together to give a functional protein (quaternary structure). What happens to the polypeptide next depends upon the protein being made, but usually involves the following:

- The polypeptide is coiled or folded, producing its secondary structure.
- The secondary structure is folded, producing the tertiary structure.
- Different polypeptide chains, along with any non-protein groups, are linked to form the quaternary structure.

Summary questions

1 Name the cell organelle involved in translation.

2 A codon found on a section of mRNA has the sequence of bases AUC. List the sequence of bases found on:

   a the tRNA anticodon that attaches to this codon;

   b the template strand of DNA that formed the mRNA codon.

3 Describe the role of tRNA in the process of translation.

4 A strand of mRNA has 64 codons but the protein produced from it has only 63 amino acids. Suggest a reason for this difference.

Protein synthesis

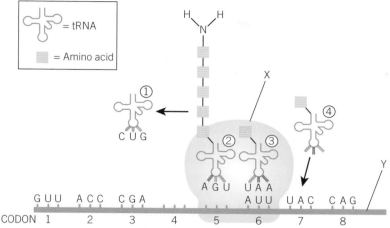

▲ Figure 5

Figure 5 shows the formation of part of a polypeptide along a section of eight codons. Codons 4 and 5 have been left blank. Using Figure 5 and Table 1 answer the following questions.

1 Name the structures X and Y.
2 Recall the chemical group shown on the end of the polypeptide chain.
3 Determine the anticodon sequence on tRNA molecule 4.
4 Deduce the sequence of the first five amino acids in the polypeptide.
5 Determine the sequence of bases on that portion of DNA from which codons 1 – 3 are transcribed.
6 A DNA mutation results in the base cytosine being replaced by uracil in codon 8. Explain the significance of this change.
7 Another mutant form of a gene causes the inversion (reversal) of the code for the amino acid glutamine (Glu).
   a Consider all possible outcomes from this change and explain the effect on the polypeptide in each case.
   b If the polypeptide formed from this mutant gene forms part of an enzyme, suggest **two** reasons why it might fail to function. Explain your answer.

▼ **Table 1** *The base sequences shown are those on mRNA*

First position	Second position				Third position
	U	C	A	G	
U	Phe	Ser	Tyr	Cys	U
	Phe	Ser	Tyr	Cys	C
	Leu	Ser	Stop	Stop	A
	Leu	Ser	Stop	Trp	G
C	Leu	Pro	His	Arg	U
	Leu	Pro	His	Arg	C
	Leu	Pro	Gln	Arg	A
	Leu	Pro	Gln	Arg	G
A	Ile	Thr	Asn	Ser	U
	Ile	Thr	Asn	Ser	C
	Ile	Thr	Lys	Arg	A
	Met	Thr	Lys	Arg	G
G	Val	Ala	Asp	Gly	U
	Val	Ala	Asp	Gly	C
	Val	Ala	Glu	Gly	A
	Val	Ala	Glu	Gly	G

Cracking the code

How exactly did scientists decipher which amino acid was coded for by which codon? Nirenberg and others did so by making synthetic mRNA and using this to make polypeptides.

The basic stages of the experiments were as follows:

- Cell extracts with the necessary components to make polypeptides were obtained and treated with DNase.
- Synthetic mRNA was added to the extract and all 20 amino acids attached to their appropriate tRNA.
- One amino acid was radioactively labelled with carbon 14 (^{14}C) while the remaining 19 had normal, non-radioactive carbon 12 (^{12}C).
- The extracts were incubated and the polypeptide produced was later extracted.
- The radioactivity level of the polypeptide produced in each case was measured.

1   Suggest a reason why DNase was added to the cell extract.

- In one experiment the radioactive amino acid was phenylalanine and four mixtures, differing only in their mRNA, were set up as follows:
  - mRNA made up of a chain of nucleotides containing only the base adenine = poly A
  - mRNA made up of a chain of nucleotides containing only the base uracil = poly U
  - mRNA made up of a chain of nucleotides containing only the base cytosine = poly C
  - no mRNA was present.

The results are shown in Table 2.

▼ **Table 2** *Results of experiment using radioactively labelled phenylalanine*

Type of synthetic mRNA	Radioactivity / counts min^{-1}
poly A	50
poly U	39 800
poly C	38
none	44

2   State **one** codon for the amino acid phenylalanine that is suggested by the results of this experiment. Explain your answer.
3   Explain why a mixture without any synthetic RNA was used.

Using this method Nirenberg deciphered 47 of the 64 possible codons in the genetic code. The remaining 17 codons, however, gave ambiguous results. This led another scientist, called Khorana, to devise a different technique. He formed very long mRNA molecules that had a repeating sequence of nucleotide bases, such as GUGUGUGUGUGUG. The polypeptide produced by this mRNA was made up of alternating cysteine and valine amino acids. The question was, what was the codon for each amino acid?

4   Suggest why it is not possible to say what the codon is for each amino acid.

From Nirenberg's earlier experiments, Khorana knew that UGU was a codon for cysteine. This meant that GUG was a codon for valine. By analysing the results of similar experiments using specific sequences of mRNA he was able to decipher the complete genetic code and to show that the code was degenerate. Further experiments by Nirenberg verified these findings.

5   The genetic code can be described as degenerate but not ambiguous. Discuss this statement.
6   Despite all these experiments, it was still not possible to find the amino acids coded for by certain codons. Explain why not.

1   The diagram shows a short sequence of DNA bases.

T T T G T A T A C T A G T C T A C T T C G T T A A T A

(a) (i)   What is the maximum number of amino acids for which this
            sequence of DNA bases could code?   *(1 mark)*
      (ii) The number of amino acids coded for could be fewer than your
            answer to part **(a)(i)**.
            Give **one** reason why.   *(1 mark)*
(b) Explain how a change in the DNA base sequence for a protein may result
      in a change in the structure of the protein.   *(3 marks)*
(c) A piece of DNA consisted of 74 base pairs. The two strands of the DNA,
      strands **A** and **B**, were analysed to find the **number** of bases of each type
      that were present. Some of the results are shown in the table.

	Number of bases			
	C	G	A	T
Strand **A**	26			
Strand **B**	19		9	

Complete the table by writing in the missing values.   *(2 marks)*

AQA June 2011

2   **Figure 1** shows a short section of a DNA molecule.

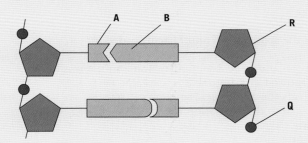

(a) Name parts **R** and **Q**.
      (i)   **R**
      (ii) **Q**   *(2 marks)*
(b) Name the bonds that join **A** and **B**.   *(1 mark)*
(c) Ribonuclease is an enzyme. It is 127 amino acids long.
      What is the minimum number of DNA bases needed to code for ribonuclease?   *(1 mark)*
(d) **Figure 2** shows the sequence of DNA bases coding for seven amino acids in
      the enzyme ribonuclease.

GTTTACTACTCTTCTTCTTTA

▲ Figure 2

The number of each type of amino acid coded for by this sequence of DNA bases is shown
in the table.

Amino acid	Number present
Arg	3
Met	2
Gln	1
Asn	1

Use the table and **Figure 2** to work out the sequence of amino acids in this part of the
enzyme. Write your answer in the boxes below.

Gln						

(1 mark)

(e) Explain how a change in a sequence of DNA bases could result in a non-functional enzyme. *(3 marks)*

AQA Jan 2010

3 (a) What name is used for the non-coding sections of a gene? *(1 mark)*

Figure 1 shows a DNA base sequence. It also shows the effect of two mutations on this base sequence. **Figure 2** shows DNA triplets that code for different amino acids.

Original DNA base sequence	A	T	T	G	G	C	G	T	G	T	C	T
Amino acid sequence												
Mutation **1** DNA base sequence	A	T	T	G	G	A	G	T	G	T	C	T
Mutation **2** DNA base sequence	A	T	T	G	G	C	C	T	G	T	C	T

▲ **Figure 1**

DNA triplets	Amino acid
GGT, GGC, GGA, GGG	Gly
GTT, GTA, GTG, GTC	Val
ATC, ATT, ATA	Ile
TCC, TCT, TCA, TCG	Ser
CTC, CTT, CTA, CTG	Leu

▲ **Figure 2**

(b) Complete **Figure 1** to show the sequence of amino acids coded for by the original DNA base sequence. *(1 mark)*

(c) Some gene mutations affect the amino acid sequence. Some mutations do not. Use the information from **Figure 1** and **Figure 2** to explain

    (i) whether mutation 1 affects the amino acid sequence *(2 marks)*

    (ii) how mutation 2 could lead to the formation of a non-functional enzyme. *(3 marks)*

(d) Gene mutations occur spontaneously.

    (i) During which part of the cell cycle are gene mutations most likely to occur? *(1 mark)*

    (ii) Suggest an explanation for your answer. *(1 mark)*

AQA June 2010

4 The diagram shows part of a pre-mRNA molecule.

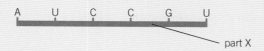

part X

(a) (i) Name the two substances that make up part **X**. *(1 mark)*

    (ii) Give the sequence of bases on the DNA strand from which this pre-mRNA has been transcribed. *(1 mark)*

(b) (i) Give **one** way in which the structure of an mRNA molecule is different from the structure of a tRNA molecule. *(1 mark)*

    (ii) Explain the difference between pre-mRNA and mRNA. *(1 mark)*

(c) The table shows the percentage of different bases in two pre-mRNA molecules.

The molecules were transcribed from the DNA in different parts of a chromosome.

Part of chromosome	Percentage of base			
	A	G	C	U
Middle	38	20	24	
End	31	22	26	

(i) Complete the table by writing the percentage of uracil (U) in the appropriate boxes. *(1 mark)*

(ii) Explain why the percentages of bases from the middle part of the chromosome and the end part are different. *(2 marks)*

AQA June 2011

5 **Figure 3** represents one process that occurs during protein synthesis.

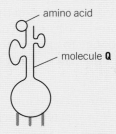

amino acid

molecule **Q**

A U G C C G U A C C G A C U

(a) Name the process shown. *(1 mark)*

(b) Identify the molecule labelled **Q**. *(1 mark)*

(c) In **Figure 3**, the first codon is AUG. Give the base sequence of the complementary DNA base sequence and the missing anticodon. *(2 marks)*

Table 1 shows the base triplets that code for two amino acids.

▼ Table 1

Amino acid	Encoding base triplet
Aspartic acid	GAC, GAU
Proline	CCA, CCG, CCC, CCU

Aspartic acid and proline are both amino acids.

(d) Describe how two amino acids differ from one another. You may use a diagram to help your description. *(1 mark)*

(e) Deletion of the sixth base (G) in the sequence shown in **Figure 3** would change the nature of the protein produced but substitution of the same base would not. Use the information in **Table 1** and your own knowledge to explain why. *(3 marks)*

AQA SAMS PAPER 2

6 **Table 6** lists the chromosome numbers and genome sizes of four plant species. One Mbp (mega base pair) is equal to 1 000 000 base pairs of DNA.

▼ Table 6

name	chromosome number(s)	genome size (Mbp)
Amborella	2n = 26	870
sweet rush	2n = 18	392
monkey flower	2n = 28	430

(a) Calculate the mean number of base pairs per chromosome in sweet rush, expressing your answer in standard form. *(2 marks)*

(b) Modern techniques allow the genomes to be sequenced, but only 750 base pairs can be read at a time.

Calculate the smallest number of DNA fragments that would need to be made to sequence the DNA of monkey flower. *(2 marks)*

9 Genetic diversity and adaptation

9.1 Gene mutation

Learning objectives

→ Describe gene mutations.

→ Explain how deletion and substitution of bases result in different amino acid sequences in polypeptides.

→ Explain why some mutations do not result in a changed amino acid sequence.

→ Describe what chromosome mutations are.

Specification reference: 3.4.3

Synoptic link

You will find it easier to follow this topic if you first revise DNA structure and replication (Topics 2.1 and 2.2), genes and the triplet code (Topic 8.1) as well as chromosome structure (Topic 8.2).

Hint

The various gene mutations are illustrated by specific examples that name bases and amino acids. These are only to illustrate the points being made and do not need to be remembered.

Any change to the quantity or the base sequence of the DNA of an organism is known as a **mutation**. Mutations occurring during the formation of gametes may be inherited, often producing sudden and distinct differences between individuals. Any change to one or more nucleotide bases, or a change in the sequence of the bases, in DNA is known as a **gene mutation**.

We have seen that a sequence of triplets on DNA is transcribed into mRNA and is then translated into a sequence of amino acids that make up a polypeptide. It follows that any changes to one or more bases in the DNA triplets could result in a change in the amino acid sequence of the polypeptide. Gene mutations can arise spontaneously during DNA replication and include base substitution and base deletion.

Substitution of bases

The type of gene mutation in which a nucleotide in a DNA molecule is replaced by another nucleotide that has a different base is known as a substitution. As an example, consider the DNA triplet of bases, guanine–thymine–cytosine (GTC) that codes for the amino acid glutamine. If the final base, cytosine, is replaced by guanine, then GTC becomes GTG. GTG is one of the DNA triplet that codes for the amino acid histidine and this then replaces the original amino acid: glutamine. The polypeptide produced will differ in a single amino acid. The significance of this difference will depend upon the precise role of the original amino acid. If it is important in forming bonds that determine the tertiary structure of the final protein, then the replacement amino acid may not form the same bonds. The protein may then be a different shape and therefore not function properly. For example, if the protein is an enzyme, its active site may no longer fit the substrate and it will no longer catalyse the reaction.

The effect of the mutation is different if the new triplet of bases still codes for the same amino acid as before. This is due to the degenerate nature of the genetic code, in which most amino acids have more than one codon. For instance, if the third base in our example is replaced by thymine, then GTC becomes GTT. However, as both DNA triplets code for glutamine, there is no change in the polypeptide produced and so the mutation will have no effect.

Deletion of bases

A gene mutation by deletion arises when a nucleotide is lost from the normal DNA sequence. The loss of a single nucleotide from the thousands in a typical gene may seem a minor change but the consequences can be considerable. Usually the amino acid sequence of the polypeptide is entirely different and so the polypeptide is unlikely to function correctly. This is because the sequence of bases in DNA is read in units of three bases (triplet). One deleted nucleotide causes all triplets in a sequence to be read differently because each has been shifted to the left by one base as shown in Figure 1.

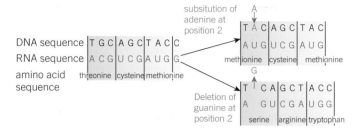

▲ **Figure 1** *Effect of substitution and deletion mutations on amino acid sequence.*

Chromosome mutations

Changes in the structure or number of whole chromosomes are called **chromosome mutations**.

Chromosome mutations can arise spontaneously and take two forms:

- **Changes in whole sets of chromosomes** occur when organisms have three or more sets of chromosomes rather than the usual two. This condition is called **polyploidy** and occurs mostly in plants.
- **Changes in the number of individual chromosomes**. Sometimes individual homologous pairs of chromosomes fail to separate during meiosis (see Topic 9.2). This is known as **non-disjunction** and usually results in a gamete having either one more or one fewer chromosome. On fertilisation with a gamete that has the normal complement of chromosomes, the resultant offspring have more or fewer chromosomes than normal in all their body cells. An example of a non-disjunction in humans is Down's syndrome, where individuals have an additional chromosome 21.

Hybridisation and polyploidy

Around 10 000 years ago, in regions of the Middle East known as the 'fertile crescent', groups of humans are thought to have gathered the grain of the wild **einkorn wheat** and **emmer wheat** using sickles made of antlers. In time, these farmers began to sow their crops. Although they knew nothing of genetics, they will naturally have selected the seed from the varieties that suited their needs rather than types that did not. Continued selection over the intervening years has produced the modern high-yielding varieties of wheat that we cultivate today. These modern varieties are the result of a process known as **hybridisation**. Hybridisation is combining the genes of different varieties or species of organisms to produce a **hybrid**. Sometimes this is followed by organisms that have additional complete sets of chromosomes (polyploidy). How then can polyploidy arise?

Polyploidy can come about in several different ways. One is for the chromosomes not to separate into two distinct sets during meiosis. Gametes could then be produced that have both sets, in other words they

▲ **Figure 2** *Einkorn wheat* (Triticum urartu)

▲ **Figure 3** *Goat grass* (Aegilops speltoides)

▲ **Figure 4** *Emmer wheat* (Triticum turgidum)

are diploid rather than haploid. If these fused with one another the offspring could have four sets of chromosomes – they would be tetraploid. Alternatively, if a diploid gamete fused with a haploid gamete, the offspring would have three sets of chromosomes – they would be triploid.

Sometimes hybrids can be formed by combining sets of chromosomes from two different species. For example, by cross-pollination between two closely related plants leading to successful fertilisation. These hybrids are usually sterile. If, however, the hybrid has a chromosome number that is a multiple of the original chromosome number, a new fertile species can arise because chromosomes have homologous partners and so meiosis is possible. Modern wheat plants have arisen due to a combination of both these forms of polyploidy.

Hybridisation in wheat

Einkorn wheat (*Triticum urartu*) is a wild form of wheat that has 14 chromosomes. It is thought that around 500 000 years ago it hybridised by cross-pollination with *Aegilops speltoides*, a type of goat grass that also had 14 chromosomes. This gave rise to a new hybrid species called emmer wheat (*Triticum turgidum*). This species therefore had four sets of chromosomes (28 chromosomes) and was therefore tetraploid. Emmer and einkorn wheat were the first ones harvested by humans around 10 000 years ago. Sometime shortly after, further hybridisation occurred when the tetraploid emmer wheat with its 28 chromosomes cross-pollinated with another type of goat grass (*Aegilops tauschii*) with 14 chromosomes. The new hybrid species now had 42, or six sets of chromosomes, in other words it was hexaploid. This species is *Triticum aestivum* or bread wheat, which is the type of wheat grown today.

▲ **Figure 5** *Bread wheat* (Triticum aestivum)

The development of modern wheat is summarised in Figure 6.

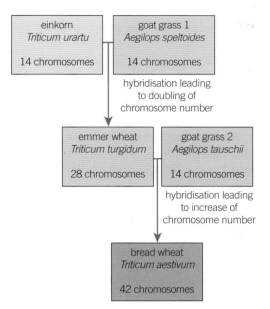

▲ **Figure 6** *The development of modern varieties of wheat by hybridisation*

1 Explain why hybrids formed by combining sets of chromosomes from two different species are often sterile.
2 Selective breeding has led to strains of wheat with shorter stems. Suggest an advantage to farmers of these strains of wheat.
3 Explain why *Triticum urartu* and *Triticum turgidum* are classified as different species.

Summary questions

1 The following is a sequence of 12 nucleotides within a much longer mRNA molecule: AUGCAUGUUACU. Following a gene mutation the same 12-nucleotide portion of the mRNA molecule is AUGCUGUUACUG. Name the type of gene mutation that has occurred. Show your reasoning.

2 Explain why a deletion gene mutation is more likely to result in a change to an organism than a substitution gene mutation.

3 Explain why a mutation that is transcribed onto mRNA may not result in any change to the polypeptide that it codes for.

4 Errors in transcription occur about 100 000 times more often than errors in DNA replication. Explain why errors in DNA replication can be far more damaging than errors in transcription.

HAMPTON SCHOOL
BIOLOGY DEPARTMENT

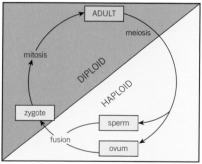

life cycle of most animals

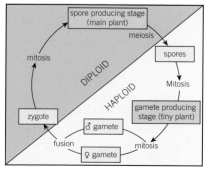

life cycle of fern

▲ **Figure 1** *A comparison of the life cycles of most animals with that of a fern*

Cell division occurs in one of two ways:

- **Mitosis** produces two daughter cells with the same number of chromosomes as the parent cell and as each other.
- **Meiosis** usually produces four daughter cells, each with half the number of chromosomes as the parent cell.

Importance of meiosis

In sexual reproduction two **gametes** fuse to give rise to new offspring. If each gamete had a full set of chromosomes (**diploid** number) then the cell that they produce has double this number. In humans, the diploid number of chromosomes is 46, which means that this cell would have 92 chromosomes. This doubling of the number of chromosomes would continue at each generation. It follows that, in order to maintain a constant number of chromosomes in the adults of a species, the number of chromosomes must be halved at some stage in the life cycle. This halving occurs as a result of meiosis. In most animals meiosis occurs in the formation of gametes. In some plants such as ferns, however, gametes are produced by mitosis. In the fern life cycle meiosis occurs in the formation of spores (Figure 1).

Every diploid cell of an organism has two complete sets of chromosomes: one set provided by each parent. During meiosis, homologous pairs of chromosomes separate, so that only one chromosome from each pair enters a daughter cell. This is known as the **haploid** number of chromosomes which, in humans, is 23. When two haploid gametes fuse at fertilisation, the diploid number of chromosomes is restored.

The process of meiosis

Meiosis involves two nuclear divisions that normally occur immediately one after the other:

1   In the **first division (meiosis 1)** homologous chromosomes pair up and their chromatids wrap around each other. Equivalent portions of these chromatids may be exchanged in a process called **crossing over**. We shall see the significance of this later. By the end of this division the homologous pairs have separated, with one chromosome from each pair going into one of the two daughter cells.

2   In the **second meiotic division (meiosis 2)** the chromatids move apart. At the end of meiosis 2, four cells have usually been formed. In humans, each of these cells contains 23 chromosomes.

This is summarised in Figure 1.

In addition to halving the number of chromosomes, meiosis also produces genetic variation among the offspring, which may lead to adaptations that improve survival chances. Meiosis brings about this genetic variation in the following two ways:

- independent segregation of homologous chromosomes
- new combinations of maternal and paternal alleles by crossing over.

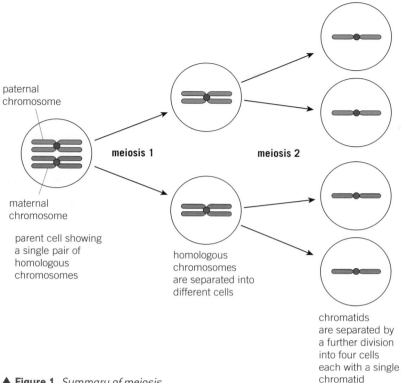

paternal chromosome

maternal chromosome

parent cell showing a single pair of homologous chromosomes

meiosis 1

homologous chromosomes are separated into different cells

meiosis 2

chromatids are separated by a further division into four cells each with a single chromatid

▲ **Figure 1** *Summary of meiosis*

Before we look at these two processes in more detail, let us recall the meaning of three important terms:

- **gene** – a length of DNA that codes for a polypeptide
- **locus** – the position of a gene on a chromosome or DNA molecule
- **allele** – one of the different forms of a particular gene.
- **homologous chromosomes** – a pair of chromosomes, one maternal and one paternal, that have the same gene loci.

Independent segregation of homologous chromosomes

During meiosis 1, each chromosome lines up alongside its homologous partner (see Figure 2). In humans, for example, this means that there will be 23 homologous pairs of chromosomes lying side by side. When these homologous pairs arrange themselves in this line they do so at random. One of each pair will pass to each daughter cell. Which one of the pair goes into the daughter cell, and with which one of any of the other pairs, depends on how the pairs are lined up in the parent cell. Since the pairs are lined up at random, the combination of chromosomes of maternal and paternal origin that go into the daughter cell at meiosis 1 is also a matter of chance. This is called **independent segregation**.

Variety from new genetic combinations

Each member of a homologous pair of chromosomes has exactly the same genes and therefore determines the same characteristics (e.g., tongue rolling and blood group). However, the alleles of these genes may differ (e.g., they may code for rollers or non-rollers, or blood group A or B). The independent assortment, of these chromosomes therefore produces new genetic combinations. An example is shown in Figure 2.

> ### Hint
>
> Imagine your chromosomes as two packs of 23 cards, red and blue, in which the cards are labelled from A to W. You were given the red pack by your mother and the blue pack by your father. Independent segregation is like dealing a card, of each letter in turn, at random from either of these two packs. Your final hand of 23 cards could contain any proportion of red and blue cards. In fact there are 2^{23} (over 8 million) different possible combinations.

In this diagram we look at just two homologous pairs. The stages shown on the figure are:

- **Stage 1.** One of the pair of chromosomes includes the gene for tongue rolling and carries one allele for roller and one for non-roller. The other chromosome includes the gene for blood group and carries the allele for blood group A and the allele for blood group B. There are two possible arrangements, P and Q, of the two chromosomes at the start of meiosis. Both are equally probable, but each produces a different outcome in terms of the characteristics that may be passed on via the gametes.

- **Stage 2.** At the end of meiosis 1, the homologous chromosomes have segregated into two separate cells.

- **Stage 3.** At the end of meiosis 2, the chromosomes have segregated into chromatids producing four gametes for each arrangement. The actual gametes are different, depending on the original arrangement (P or Q) of the chromosomes at stage 1.

Arrangement P produces the following types of gamete with alleles for:

- roller and blood group B
- non-roller and blood group A.

Arrangement Q produces the following types of gamete with alleles for:

- non-roller and blood group B
- roller and blood group A.

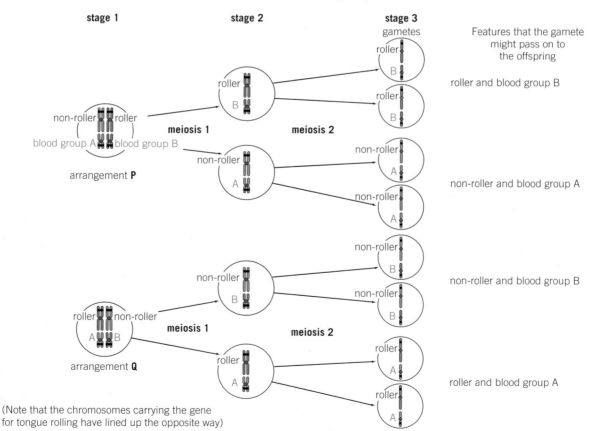

(Note that the chromosomes carrying the gene for tongue rolling have lined up the opposite way)

▲ **Figure 2** *Genetic variation produced as a result of independent segregation of chromosomes during meiosis. This diagram illustrates the independent segregation of two features, tongue rolling and blood group, that are carried on separate chromosomes*

Where the cells produced in meiosis are gametes these will be genetically different as a result of the different combinations of the maternal and paternal chromosomes/alleles they contain. These haploid gametes fuse randomly at fertilisation. The haploid gametes produced by meiosis fuse to restore the diploid state. Each gamete has a different make-up and their random fusion therefore produces variety in the offspring. Where the gametes come from different parents (as is usually the case) two different genetic make-ups are combined and even more variety results.

Genetic recombination by crossing over

We saw above that, during meiosis 1, each chromosome lines up alongside its homologous partner. The following events then take place:

*   The chromatids of each pair become twisted around one another.
*   During this twisting process tensions are created and portions of the chromatids break off.
*   These broken portions might then rejoin with the chromatids of its homologous partner.
*   Usually it is the equivalent portions of homologous chromosomes that are exchanged.
*   In this way new genetic combinations of maternal and paternal alleles are produced (Figure 3).

The chromatids cross over one another many times and so the process is known as **crossing over**. The broken-off portions of chromatid recombine with another chromatid, so this process is called **recombination**.

The effect of this recombination by crossing over on the cells produced at the end of meiosis is illustrated in Figure 4. Compare the four cells

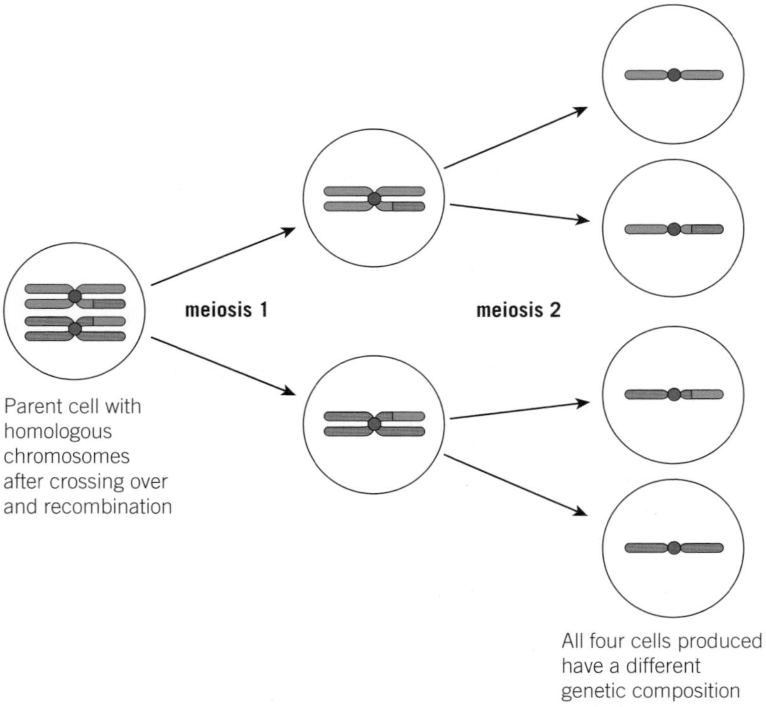

Chromatids of homologous chromosomes twist around one another, crossing over many times

Simplified representation of a single cross over

point of breakage

Result of a single cross over showing equivalent portions of the chromatid having been exchanged

▲ **Figure 3** *Crossing over*

Parent cell with homologous chromosomes after crossing over and recombination

meiosis 1

meiosis 2

All four cells produced have a different genetic composition

▲ **Figure 4** *Genetic variation as a result of recombination by crossing over*

Hint

Imagine your packs of blue and red cards again. Recombination is like taking a red card and a blue card, each with the same number, and tearing an identical portion from each card and attaching it to the other card, so that you have new cards that are part red and part blue. You can do this in an almost infinite number of ways and all before you start to deal them as before!

that result with those shown in Figure 1. If there is no recombination by crossing over only two different types of cell are produced. However, if recombination does occur, four different cell types are produced. Crossing over therefore increases genetic variety even further.

Possible chromosome combinations following meiosis

Homologous pairs of chromosomes line up at the equator of a cell during meiosis I. Either one of a pair can pass into each daughter cell (independent segregation) and so there are a large number of possible combinations of chromosomes in any daughter cell. It is possible to make a mathematical calculation based on the number of chromosomes in an organism to determine the number of possible combinations of chromosomes for each daughter cell. The formula is:

$$2^n \text{ where } n = \text{the number of pairs of homologous chromosomes.}$$

So an organism with 4 homologous pairs of chromosomes can produce 2^4 or 16 possible different combinations of chromosomes of maternal and paternal origin in its daughter cells as a result of meiosis.

We have also seen that variety is further increased through the random pairing of male and female gametes. Where the gametes come from different parents two different genetic complements with different alleles are combined, providing yet more variety. Again, we can calculate this mathematically using the formula:

$$(2^n)^2 \text{ where } n = \text{the number of pairs of homologous chromosomes.}$$

Using our example of an organism with 4 homologous pairs of chromosomes, there are $(2^n)^2$ or 256 different combinations of chromosomes in the offspring produced as the result of sexual reproduction.

These calculations are based on chromosomes staying intact throughout meiosis. In practice we know that crossing over between chromatids during meiosis I exchanges sections of chromosomes between homologous pairs in the process called recombination. As recombination occurs each time gametes are made, it will greatly increase the number of possible chromosome combinations in the gametes.

Maths link √x̄

MS 0.5 and 1.4, see Chapter 11.

Summary questions

1 A cell is examined and found to have 27 chromosomes. Is it likely to be haploid or diploid? Explain your answer.

2 State **two** ways in which meiosis leads to an increase in genetic variety.

3 Study Figure 2. Imagine that both alleles of the gene on the smaller chromosome are for blood group A (rather than blood groups A and B). List all the different combinations of alleles in the gametes.

4 A mule is a cross between a horse (64 chromosomes) and a donkey (62 chromosomes). Mules therefore have 63 chromosomes. From your knowledge of meiosis, suggest why mules are sterile.

5 √x̄ Calculate the number of possible chromosome combinations produced from the fertilisation of two gametes, each of which contains five chromosomes (assume there is no crossing over).

Worked example

Calculate the number of possible chromosome combinations produced from the fertilisation of two gametes from separate individuals whose diploid number is 12 (assume no crossing over).

As you are looking at the possible chromosomes in the offspring (after fertilisation), you must use the formula:

$(2^n)^2$, where n = the number of pairs of homologous chromosomes.

First you need to find the value of n.

You are told the diploid number is 12.

Therefore the number of *pairs* of homologous chromosomes is $12 \div 2 = 6$.

Substituting in the formula you get: $(2^6)^2 = 4096$.

Organisms are varied. Around 1.8 million species of organisms on Earth have been identified and named. Many more are unnamed or undiscovered. Estimates of the total number of species on this planet today range from 5 million to 100 million. All of these species are different.

Even between members of the same species there are a multitude of differences. Almost every one of the 7.3 billion people alive in 2014 are similar enough to be recognised as humans and yet different enough to be distinguished from one another. What makes us and other species similar and yet different?

Genetic diversity

We saw Topic 8.1, that it is DNA which determines the considerable variety of proteins that make up each organism. Therefore genetic similarities and differences between organisms may be defined in terms of variation in DNA. Hence it is differences in DNA that lead to the vast genetic diversity we find on Earth.

We also saw in Topic 8.1 that a section of DNA that codes for one polypeptide is called a gene. All members of the same species have the same genes. For example, all humans have a gene for blood group, just as all snapdragons (*Antirrhinum majus*) have a gene for petal colour. Which blood group humans have depends on which two alleles of the gene they possess. Likewise, the colour of a snapdragon's petals depends on which two alleles for petal colour it possesses. Organisms of the same species differ in their combination of alleles, not their genes.

Genetic diversity is described as the total number of different alleles in a population. A population is a group of individuals of the same species that live in the same place and can interbreed. A species consists of one, or more, populations. The greater the number of different alleles that all members of a species possess, the greater the genetic diversity of that species. Genetic diversity is reduced when a species has fewer different alleles. The greater the genetic diversity, the more likely that some individuals in a population will survive an environmental change. This is because of a wider range of alleles and therefore a wider range of characteristics. This gives a greater probability that some individual will possess a characteristic that suits it to the new environmental conditions. Genetic diversity is a factor that enables natural selection to occur.

Natural selection in the evolution of populations

Not all alleles of a population are equally likely to be passed to the next generation. This is because only certain individuals are reproductively successful and so pass on their alleles.

Reproductive success and allele frequency

Differences between the reproductive success of individuals affects allele frequency in populations. The process works like this:

- Within any population of a species there will be a gene pool containing a wide variety of alleles.

▲ **Figure 1** *Examples of genetic diversity (from top to bottom): anemone; lichens; mountain goat; fritillary butterfly*

> **Hint**
>
> Remember that an allele is one alternative form of a gene and, as such, is a length of DNA on one chromosome of a homologous pair.

- Random mutation of alleles within this gene pool may result in a new allele of a gene which in most cases will be harmful.
- However in certain environments, the new allele of a gene might give its possessor an advantage over other individuals in the population.
- These individuals will be better adapted and therefore more likely to survive in their competition with others.
- These individuals are more likely to obtain the available resources and so grow more rapidly and live longer. As a result, they will have a better chance of breeding successfully and producing more offspring.
- Only those individuals that reproduce successfully will pass on their alleles to the next generation.
- Therefore it is the new allele that gave the parents an advantage in the competition for survival that is most likely to be passed on to the next generation.
- As these new individuals also have the new, 'advantageous' allele, they in turn are more likely to survive, and so reproduce successfully.
- Over many generations, the number of individuals with the new, 'advantageous' allele will increase at the expense of the individuals with the 'less advantageous' alleles.
- Over time, the frequency of the new, 'advantageous' allele in the population increases while that of the 'non-advantageous' ones decreases.

It must be stressed that what is 'advantageous' depends upon the environmental conditions at any one time. For example, alleles for black body colour may be 'advantageous' as camouflage against a smoke-blackened wall, but 'non-advantageous' against a snowy landscape.

Summary questions

1 State whether each of the following is likely to *increase* or *decrease* genetic diversity:

   a   increasing the variety of alleles within a population

   b   breeding together closely related cats to develop varieties with longer fur

   c   mutation (permanent change to the DNA) of an allele.

2 Explain how a difference in its DNA might lead to an organism having a different appearance and hence the species showing greater genetic diversity.

Natural selection in action

The peppered moth, *Biston betularia* normally has a light colour that camouflages it against the light background of the lichen-covered trees on which it rests. In 1848 a dark (melanic) form of the peppered moth appeared in Manchester. At this time, most buildings, walls and trees in the city were blackened by the soot from 50 years of industrial development. Both types of the moth are shown in Figure 4 against this blackened background. By 1895, 98% of Manchester's population of the moth was of the black type.

▲ **Figure 2** *Light and dark varieties of the peppered moth*

1   Suggest an explanation that accounts for the change from the light to dark variety of the moth.

Selection is the process by which organisms that are better adapted to their environment tend to survive and breed, while those that are less well adapted tend not to. Every organism is subjected to a process of selection, based on its suitability for surviving the conditions that exist at the time. Different environmental conditions favour different characteristics in the population. Depending on which characteristics are favoured, selection will produce a number of different results.

- Selection may favour individuals that vary in one direction from the mean of the population. This is called **directional selection** and changes the characteristics of the population.
- Selection may favour average individuals. This is called **stabilising selection** and preserves the characteristics of a population.

Most characteristics are influenced by more than one gene (**polygenes**). These types of characteristics are more influenced by the environment than ones determined by a single gene. The effect of the environment on polygenes produces individuals in a population that vary about the mean. When you plot this variation on a graph you get a normal distribution curve. The next part of this topic looks at how these two types of selection affect this curve.

Learning objectives

→ Describe what selection is
→ Describe the environmental factors which exert selection pressure
→ Explain what stabilising and directional selection are.

Specification reference: 3.4.4

Maths link $\sqrt{x}$

MS 3.1, see Chapter 11.

Directional selection

If the environmental conditions change, the phenotypes (the observable physical and biochemical characteristics of an organism) that are best suited to the new conditions are most likely to survive. Some individuals, which fall to either the left or right of the mean, will possess a phenotype more suited to the new conditions. These individuals will be more likely to survive and breed. They will therefore contribute more offspring (and the alleles these offspring possess) to the next generation than other individuals. Over time, the mean will then move in the direction of these individuals. To explain, let us take the example of the development of antibiotic resistance in bacteria.

Shortly after the discovery of antibiotics it became apparent that the effectiveness of some antibiotics at killing bacteria was reduced. It was found that these populations of bacteria had developed resistance to antibiotics such as penicillin. The resistance was not due to the development of tolerance to the antibiotic, but rather a chance mutation within the bacteria. We saw in Topic 9.1 that a mutation is a change in DNA that results in different characteristics, usually due to a change to some protein. As an example of directional selection, let us look at the case of resistance to penicillin:

▲ **Figure 1** *One in every six prescriptions issued by doctors in the UK is for an antibiotic*

- A spontaneous mutation occurred in the allele of a gene in a bacterium that enabled it to make a new protein. The new protein was an enzyme that broke down the antibiotic penicillin before it was able to kill the bacterium. The enzyme was given the name penicillinase.
- The bacterium happened, by chance, to be in a situation where penicillin was being used to treat an individual. In these circumstances, the mutation gave the bacterium an advantage in

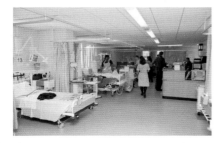
being able to use penicillinase to break down the antibiotic and so survive while the rest of the population of bacteria were killed by it.

- The bacterium that survived was able to divide by binary fission to build up a small population of penicillin-resistant bacteria.

- Members of this small penicillin-resistant population were more able to survive, and therefore multiply, in the presence of penicillin than members of the non-resistant population.

- The population of penicillin-resistant bacteria increased at the expense of the non-resistant population. Consequently the frequency of the allele that enabled the production of penicillinase increased in the population.

- The population's normal distribution curve shifted in the direction of a population having greater resistance to penicillin (see Figure 2).

While our example illustrates penicillin resistance, the process applies equally to any antibiotic.

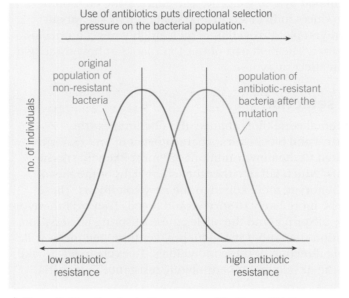

▲ **Figure 2** *Directional selection as exemplified by antibiotic resistance in bacteria*

It must be stressed that bacteria do not mutate because of the presence of antibiotics. Mutations occur randomly and are very rare. However as there are so many bacteria around, the total number of mutations is large. Many of these mutations will be of no advantage to a bacterium. Indeed most will be harmful, in which case the bacterium will probably die. Very occasionally a mutation will be advantageous. Even then it depends upon the circumstances. For example, a mutation that leads to the production of penicillinase is only an advantage when the bacterium is in the presence of penicillin. With continued use of antibiotics, there is a greater chance that the mutant population will out-compete, and replace, the original population.

▲ **Figure 3** *The use of many antibiotics in hospitals increases the chance of multiple antibiotic resistance developing in bacteria*

Directional selection therefore results in phenotypes at one extreme of the population being selected for and those at the other extreme being selected against.

Stabilising selection

If environmental conditions remain stable, it is the individuals with phenotypes closest to the mean that are favoured. These individuals are more likely to pass their alleles on to the next generation. Those individuals with phenotypes at the extremes are less likely to pass on their alleles. Stabilising selection therefore tends to eliminate the phenotypes at the extremes.

Let us look at the example of human birth weights.

Stabilising selection results in phenotypes around the mean of the population being selected for and those at both extremes being selected against. These events are summarised in Figure 4.

The body mass at birth of babies born at a hospital was measured over a 12-year period. In the graph in Figure 5 the percentage of births in the population (*y*-axis on the left) is plotted against birth mass of the infants as a histogram.

Over the same period, the infant mortality (death) rate was also recorded. The infant mortality rate is measured on a logarithmic scale (*y*-axis on the right) and plotted against infant body mass at birth as a line graph.

Looking at the histogram in Figure 4, you will see that the body mass of the babies at birth is within a relatively narrow range – mostly between 2.5 and 4.0 kg. The likely explanation for this can be found from looking at the line for infant mortality. This line climbs steeply where the birth weight is below 2.5 kg and again where it is above 4.0 kg. In other words, there is a much greater risk of infant death when the birth weight is outside the range 2.5–4.0 kg. Bigger isn't always better.

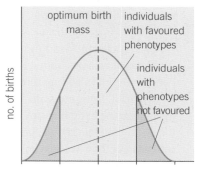

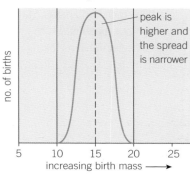

▲ **Figure 4** *Stabilising selection*

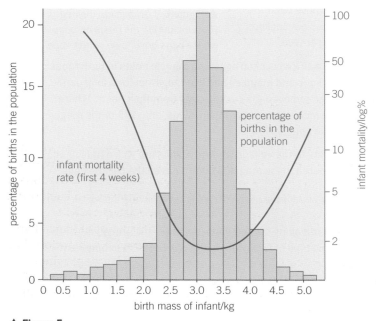

▲ **Figure 5**

Summary questions

1 What is selection?

2 Distinguish between directional selection and stabilising selection.

3 A severe cold spell in 1996 killed over 50% of swallows living on cliffs in Nebraska. Biologists collected nearly 2000 dead swallows from beneath the cliffs and captured around 1000 living ones. By measuring the body mass of the birds, they found that birds with a larger than average body mass survived the cold spell better than ones with a smaller than average body mass. State, giving your reasons, which type of selection was taking place here.

This illustrates stabilising selection because the mortality rate is greater at the two extremes. The infants with the highest and lowest birth masses are more likely to die (are being selected against) while those around the mean are less likely to die (are being selected for/favoured). The population's characteristics are being preserved rather than changed.

Stabilising selection therefore results in phenotypes around the mean of the population being selected for and those at both extremes being selected against. These events are summarised in Figure 5.

Natural selection results in species that are better adapted to the environment that they live in. These adaptations may be:

- **Anatomical**, such as shorter ears and thicker fur in arctic foxes compared to foxes in warmer climates.

- **Physiological**, for example oxidising of fat rather than carbohydrate in kangaroo rats to produce additional water in a dry desert environment.

- **Behavioural**, such as the autumn migration of swallows from the UK to Africa to avoid food shortages in the UK winter.

They must be cuckoo!

Cuckoos lay their eggs in the nests of other birds. The host birds will often raise these parasite chicks alongside their own.

In many valleys in southern Spain, great cuckoos and common magpies have lived together for hundreds of years. In some valleys, however, magpies have been around for centuries but cuckoos have only recently arrived.

▲ **Figure 6** *Cuckoos lay their eggs (bottom row) in the nests of magpies whose eggs are shown on the upper row*

Scientists placed artificial cuckoo eggs into magpie nests in both types of valley. Where cuckoos and magpies had lived together for a long period, 78% of the magpies removed the cuckoo eggs from their nests. Where cuckoos had only recently colonised the valleys, only 14% of the magpies removed the cuckoo eggs.

It would appear that, in the valleys where cuckoos are well established, selection has favoured those magpies that removed the cuckoo eggs.

1 Suggest **one** advantage to the magpies of removing cuckoo eggs from their nest.

2 Explain how removing cuckoo eggs increases the probability of the alleles for this type of behaviour being passed on to subsequent generations.

3 Suggest why this form of behaviour is not shown by magpies in those valleys where cuckoos have only recently arrived.

4 State, with your reasons, which type of selection is taking place here.

5 Explain how the different behaviour of the two groups of magpies might lead to selection that produces a change within the magpie population.

1   Phenylketonuria is a disease caused by mutations of the gene coding for the enzyme PAH. The table shows part of the DNA base sequence coding for PAH. It also shows a mutation of this sequence which leads to the production of non-functioning PAH.

DNA base sequence coding for PAH	C	A	G	T	T	C	G	C	T	A	C	G
DNA base sequence coding for non-functioning PAH	C	A	G	T	T	C	C	C	T	A	C	G

(a) (i)   What is the maximum number of amino acids for which this base sequence could code? *(1 mark)*

(ii) Explain how this mutation leads to the formation of non-functioning PAH. *(3 marks)*

PAH catalyses a reaction at the start of two enzyme-controlled pathways.

The diagram shows these pathways.

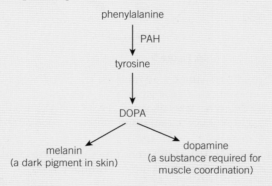

phenylalanine

↓ PAH

tyrosine

↓

DOPA

melanin (a dark pigment in skin)   dopamine (a substance required for muscle coordination)

(b) Use the information in the diagram to give two symptoms you might expect to be visible in a person who produces non-functioning PAH. *(2 marks)*

(c) One mutation causing phenylketonuria was originally only found in one population in central Asia. It is now found in many different populations across Asia. Suggest how the spread of this mutation may have occurred. *(1 mark)*

AQA Jan 2012

2   **Figure 3** shows a pair of chromosomes at the start of meiosis. The letters represent alleles.

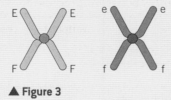

▲ **Figure 3**

(a) What is an allele? *(1 mark)*
(b) Explain the appearance of one of the chromosomes in **Figure 3**. *(2 marks)*
(c) The cell containing this pair of chromosomes divided by meiosis. **Figure 4** shows the distribution of chromosomes from this pair in four of the gametes produced.

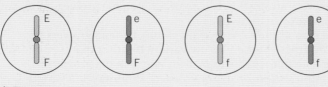

▲ **Figure 4**

(i)   Some of the gametes formed during meiosis have new combinations of alleles. Explain how the gametes with the combinations of alleles Ef and eF have been produced.   (2 marks)

(ii) Only a few gametes have the new combination of alleles Ef and eF. Most gametes have the combination of alleles EF and ef. Suggest why only a few gametes have the new combination of alleles, Ef and eF. (1 mark)

(d) **Figure 5** shows a cell with six chromosomes.

▲ Figure 5

(i)   This cell produces gametes by meiosis. Draw a diagram to show the chromosomes in one of the gametes.   (2 marks)

(ii) How many different types of gametes could be produced from this cell as a result of different combinations of maternal and paternal chromosomes? (1 mark)

AQA June 2010

(e) (i)   Calculate the number of different types of gametes that can be produced in a species with a diploid number of 24.   (1 mark)

(ii) Assuming random fertilisation, calculate the number of different combinations of maternal and paternal chromosomes in the zygotes of this species. (1 mark)

3   (a) Explain what is meant by genetic diversity.   (1 mark)

(b) Apart from genetic factors what other type of factor causes variation within a species?   (1 mark)

(c) The spotted owl is a bird. Numbers of spotted owls have decreased over the past 50 years. Explain how this decrease may affect genetic diversity. (2 marks)

AQA June 2011

4   The table shows some differences between three varieties of banana plant.

	Variety **A**	Variety **B**	Variety **C**
Number of chromosomes in a leaf cell	22	33	44
Growth rate of fruit / cm^3 week^{-1}	2.9	6.9	7.2
Breaking strength of leaf / arbitrary units	10.8	9.4	7.8

(a) (i)   How many chromosomes are there in a male gamete from variety **C**? (1 mark)

(ii) Variety **B** cannot produce fertile gametes. Use information in the table to explain why.   (2 marks)

In some countries very strong winds may occur. Banana growers in these countries choose to grow variety **B**.

(b) (i)   Use the data in the table to explain why banana growers in these countries choose to grow variety **B** rather than variety **A**. (1 mark)

(ii) Use the data in the table to explain why banana growers in these countries choose to grow variety **B** rather than variety **C**. (1 mark)

(c) Banana growers can only grow new variety **B** plants from suckers. Suckers grow from cells at the base of the stem of the parent plant.

Use your knowledge of cell division to explain how growing variety **B** on a large scale will affect the genetic diversity of bananas. (2 marks)

AQA Jan 2011

Scientists have identified and named around 1.8 million different living organisms. No one knows how many types remain to be identified. Estimates for the total number of species on Earth vary from 10 million to 100 million. The figure is likely to be around 14 million. These represent only the species that exist today. Some scientists have estimated that 99% of the species that have existed on Earth are now extinct, and almost all of them have left no fossil record. With such a vast number of organisms it is clearly important for scientists to name them and sort them into groups.

Classification is the organisation of living organisms into groups. This process is not random but is based on a number of accepted principles. Before we examine how organisms are grouped according to these principles, consider how scientists distinguish one type of organism from another.

The concept of a species

A species is the basic unit of classification. A definition of a species is not easy, but members of a single species have one main thing in common:

- **They are capable of breeding to produce living, fertile offspring.** They are therefore able to produce more offspring. This means that, when a species reproduces sexually, any of the genes of its individuals can, in theory, be combined with any other.

Naming species – the binomial system

At one time scientists often gave new organisms a name that described their features, for example blackbird, rainbow trout. This practice resulted in the same names being used in different parts of the world for very different species. Therefore, it was difficult for scientists to be sure they were referring to the same organism. Over 200 years ago the Swedish botanist Linnaeus overcame this problem by devising a common system of naming organisms. This system is still in use today.

Organisms are identified by two names and hence the system is called the **binomial system**. Its features are as follows:

- It is a universal system based upon Latin or Greek names.
- The first name, called the **generic name**, denotes the genus to which the organism belongs. This is equivalent to the surname used to identify people and shared by their close relatives.
- The second name, called the **specific name**, denotes the species to which the organism belongs. This is equivalent to the first (or given) name used to identify people. However, unlike in humans, it is never shared by other species within the genus.

There are a number of rules that are applied to the use of the binomial system in scientific writing:

- The names are printed in italics or, if handwritten, they are underlined to indicate that they are scientific names.

Learning objectives

→ Explain the concept of a species is.

→ Outline how species are named.

→ Explain how courtship is a precursor to mating.

→ Explain the principles of classification.

→ Explain how classification is related to evolution.

Specification reference: 3.4.5

Study tip

A common error is to state that members of the same species are capable of breeding to produce viable offspring rather than fertile offspring. Viable simply means alive not fertile.

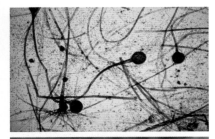

▲ **Figure 1** *(From top to bottom) the fungus* Mucor mucedo *(bread mould); the plant* Lathyrus odoratus *(sweet pea); the animal* Felix tigris *(tiger). The classification of these organisms is shown in Table 1 on the next page*

- The first letter of the generic name is in upper case (capitals), but the specific name is in lower case (small letters).
- If the specific name is not known, it can be written as '*sp.*', for example, *Felix sp.*

The naming of organisms is in a constant state of change. Current names reflect the present state of scientific knowledge and understanding. In the same way, the classification of species is regularly changing as our knowledge of their evolution, physical features, biochemistry and behaviour increases.

Courtship behaviour

Members of the same species have similar, or have the same genes and therefore resemble one another physically and biochemically. This helps them to distinguish members of their own species from those of other species. The same is true of behaviour. The behaviour of members of the same species is more alike than that of members of different species. Individuals can therefore recognise members of their own species by the way they act. Like the physical and biochemical features of a species, the ability to display a behaviour is genetically determined. It too has evolved and it influences the chances of survival. When it comes to survival of the species (as opposed to the individuals), courtship and mating are essential.

No individual lives forever. Reproduction is therefore the means by which a species can survive over time. Each individual has adaptations that help to ensure that their DNA is passed on, through the reproductive process, to the next generation. The females of most species only produce eggs at specific times, often as little as once a year. It is therefore important to ensure that mating is successful and that the offspring have the maximum chance of survival. Courtship behaviour helps to achieve this by enabling individuals to:

- **recognise members of their own species** to ensure that mating only takes place between members of the same species because only members of the same species can produce fertile offspring
- **identify a mate that is capable of breeding** because both partners need to be sexually mature, fertile and receptive to mating
- **form a pair bond** that will lead to successful mating and raising of offspring
- **synchronise mating** so that it takes place when there is the maximum probability of the sperm and egg meeting.
- **become able to breed** by bringing a member of the opposite sex into a physiological state that allows breeding to occur.

The females of many species undergo a cycle of sexual activity in which they can only conceive during a very short time. They are often only receptive to mating for a period around the time when they produce eggs. Courtship behaviour is used by males to determine whether the female is at this receptive stage. If she responds with the appropriate behavioural response, courtship continues and is likely to result in the production of offspring. If she is not receptive, she exhibits a different pattern of behaviour and the male ceases to court her, turning his attentions elsewhere.

During courtship, animals use signals to communicate with a potential mate and with members of their own sex. Typically there is a chain of actions between a male and female. The chain of actions is the same for all members of a species but differs for members of different species. In this way both individuals recognise that their partner is of the same species and that they may be prepared to mate.

Grouping species together – the principles of classification

With so many species, past and present, it makes sense to organise them into manageable groups. This allows better communication between scientists and avoids confusion. The grouping of organisms is known as **classification**, while the theory and practice of biological classification is called **taxonomy**.

There are two main forms of biological classification, each used for a different purpose.

- **Artificial classification** divides organisms according to differences that are useful at the time. Such features may include colour, size, number of legs, leaf shape, etc. These are described as analogous characteristics where they have the same function but do not have the same evolutionary origins. For example, the wings of butterflies and birds are both used for flight but they originated in different ways.

- **Phylogenetic classification:**

  a is based upon the evolutionary relationships between organisms and their ancestors

  b classifies species into groups using shared features derived from their ancestors

  c arranges the groups into a hierarchy, in which the groups are contained within larger composite groups with no overlap.

Relationships in a phylogenetic classification are partly based on homologous characteristics. Homologous characteristics have similar evolutionary origins regardless of their functions in the adult of a species. For example, the wing of a bird, the arm of a human and the front leg of a horse all have the same basic structure and evolutionary origins and are therefore homologous.

Organising the groups of species – taxonomy

Each group within a phylogenetic biological classification is called a taxon (plural taxa). Taxonomy is the study of these groups and their positions in a hierarchical order, where they are known as taxonomic ranks. These are based upon the evolutionary line of descent of the group members. A **domain** is the highest taxonomic rank and three are recognised: **Bacteria**, **Archaea** (a group of prokaryotes) and **Eukarya**.

▲ **Figure 1** *Courtship of great crested grebes – the weed presentation dance*

Study tip

When considering animal behaviour always remember that animals do not think like humans.

Hint

A useful mnemonic for remembering the order of the taxonomic ranks is 'Delicious King Prawn Curry Or Fat Greasy Sausages'.

Bacteria are a group of single-celled prokaryotes with the following features:

- the absence of membrane-bounded organelles such as nuclei or mitochondria
- unicellular, although cells may occur in chains or clusters
- ribosomes are smaller (70S) than in eukaryotic cells
- cell walls are present and made of murein (but never chitin or cellulose)
- single loop of naked DNA made up of nucleic acids but no histones

Archaea are a group of single-celled prokaryotes that were originally classified as bacteria which they resemble in appearance.

They differ from bacteria because:

- their genes and protein synthesis are more similar to eukaryotes
- their membranes contain fatty acid chains attached to glycerol by **ether** linkages
- there is no murein in their cell walls
- they have a more complex form of RNA polymerase.

Eukarya are a group of organisms made up of one or more eukaryotic cells. Their features are:

- their cells possess membrane-bounded organelles such as mitochondria and chloroplasts
- they have membranes containing fatty acid chains attached to glycerol by **ester** linkages
- not all possess cells with a cell wall, but where they do it contains no murein
- ribosomes are larger (80S) than in Bacteria and Archaea.

The Eukarya domain is divided into four kingdoms: **Protoctista**, **Fungi**, **Plantae** and **Animalia**. Within each kingdom the largest groups are known as **phyla**. Organisms in each phylum have a body plan radically different from organisms in any other phylum. Diversity within each phylum allows it to be divided into **classes**. Each class is divided into **orders** of organisms that have additional features in common. Each order is divided into **families** and at this level the differences are less obvious. Each family is divided into **genera** and each genus (singular) into **species**. As examples of how the system works (rather than names to be learnt), the classification of three organisms is given in Table 1.

▼ **Table 1** *Classification of three organisms from different kingdoms*

Rank	Pin mould	Sweet pea	Tiger
kingdom	Fungi	Plantae	Animalia
phylum	Zygomycota	Angiospermophyta	Chordata
class	Zygomycetes	Dicotyledonae	Mammalia
order	Mucorales	Rosales	Carnivora
family	Mucoraceae	Fabaceae	Felidae
genus	*Mucor*	*Lathyrus*	*Felix*
species	*mucedo*	*odoratus*	*tigris*

Phylogeny

The hierarchical order of taxonomic ranks is based upon the supposed evolutionary line of descent of the group members. This evolutionary relationship between organisms is known as **phylogeny**. The term is derived from the word 'phylum', which, in classification, is a group of related or similar organisms. The phylogeny of an organism reflects the evolutionary branch that led up to it. The phylogenetic relationships of different species are usually represented by a tree-like diagram called a phylogenetic tree. In these diagrams, the oldest species is at the base of the tree while the most recent ones are represented by the ends of the branches. An example is shown in Figure 3.

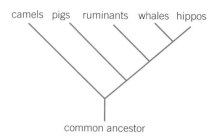

The closer the branches, the closer the evolutionary relationship. Hippos and whales are more closely related than hippos and ruminants.

▲ **Figure 3** *A phylogenetic tree showing the evolutionary relationship between certain mammals*

Summary questions

1 State the one thing that all members of a species share.

2 List the **three** features of a phylogenetic system of classification.

3 Explain why species recognition is important in courtship.

4 Suggest a way in which the courtship behaviour of one species might be used to determine which of two other species is most closely related to it.

5 *Rana temporaria* is the frog commonly found in Britain. Table 2, which is incomplete, shows part of its classification. State the most appropriate name for each of the blanks represented by the numbers 1–7.

▼ **Table 2**

Kingdom	Animalia
1	chordata
2	amphibia
3	anura
4	ranidae
genus	5
6	7

6 a State which group is the closest relative of the snakes.

b State whether dinosaurs are more closely related to crocodiles or birds.

c Suggest what C represents.

d Suggest a reason why dinosaurs are not shown along the time line like all the other groups.

Figure 4 shows a phylogenetic tree for birds and certain reptiles.

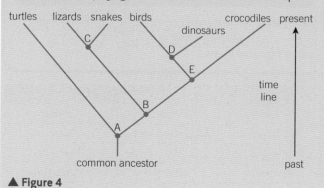

▲ **Figure 4**

▲ **Figure 5** *Turtles and crocodiles evolved from the same ancestor - but a very long time ago*

Application

The difficulties of defining species

A species may be defined in terms of observable similarities and the ability to produce fertile offspring. There are, however, certain difficulties with this definition. These include:

- Species are not fixed forever, but change and evolve over time. In time, some individuals may develop into a new species.
- Within a species there can be considerable variation among individuals. All dogs, for example, belong to the same species, but artificial selection has led to a variety of different breeds.
- Many species are extinct and most of these have left no fossil record.
- Some species rarely, if ever, reproduce sexually.
- Members of different populations of the same species may be isolated, for example by oceans, and so never meet and therefore never got the opportunity to interbreed.
- Populations of organisms that are isolated from one another may be classified as different species. These groups may turn out to be of the same species when their ability to interbreed is tested.
- Some types of organism are sterile (see below).

1 Even where groups of extinct organisms have left fossil records, it is very difficult to distinguish different species. Suggest **two** reasons why.
2 Suggest reasons why it is often difficult to classify organisms as distinct species.

A horse and a donkey (Figure 2) are capable of mating and producing offspring, which are known as mules. A horse and a donkey are, however, different species and the resulting mules are infertile, i.e. they almost never produce offspring when mated with each other. Why are mules infertile? It is all down to the number of chromosomes and the first stage of meiosis. A horse has 64 chromosomes (32 pairs) and a donkey has 62 chromosomes (31 pairs). The gametes of a horse and a donkey therefore have 32 and 31 chromosomes respectively. When the gametes of a horse and a donkey fuse, the offspring (the mule) has 63 chromosomes. Gametes are formed by meiosis but there is an odd number of chromosomes – they cannot pair up appropriately and so the gametes produced are not functional and so mules are infertile. However, mitosis can take place and therefore a mule grows and develops normally.

There have been occasional cases of a fertile female mule. This event is very rare, so much so that the Romans had a saying that meant 'when a mule foals', which was the equivalent of our modern 'once in a blue moon'.

3 Does the fact that fertile mules occasionally occur make a mule a distinct species? Give reasons for your answer.

▲ **Figure 2** A horse (right) and a donkey (left), although different species, are capable of mating and producing offspring called mules

10.2 Diversity within a community

Biodiversity is the general term used to describe variety in the living world. It refers to the number and variety of living organisms in a particular area and has three components:

- **Species diversity** refers to the number of different species and the number of individuals of each species within any one **community**.
- **Genetic diversity** refers to the variety of genes possessed by the individuals that make up a population of a species.
- **Ecosystem diversity** refers to the range of different **habitats**, from a small local habitat to the whole of the Earth.

One measure of species diversity is **species richness**. This is the number of different species in a particular area at a given time (community). Two communities may have the same number of species but the proportions of the community made up of each species may differ markedly. For example, a natural meadow and a field of wheat may both have 25 species. However, in the meadow, all 25 species might be equally abundant, whereas, in the wheat field, over 95% of the plants may be a single species of wheat.

Measuring the index of diversity

Consider the data shown in Table 1 about two different **habitats**. It does not tell us much about the differences between the two habitats because, in both cases, the total number of species and the total number of individuals are identical. However, if we measure the species diversity, we get a different picture.

▼ **Table 1** *Number and types of species found in two different habitats within the same ecosystem*

Species found	Numbers found in habitat X	Numbers found in habitat Y
A	10	3
B	10	5
C	10	2
D	10	36
E	10	4
no. of species	5	5
no. of individuals	50	50

One way of measuring species diversity is to use an index that is calculated as follows:

$$d = \frac{N(N-1)}{\Sigma n(n-1)}$$

Where:

d = **index of diversity**

N = **total number of organisms of all species**

n = **total number of organisms of each species**

Σ = **the sum of**

Learning objectives

→ Describe what we understand by species diversity.

→ Explain how a diversity index is used as a measure of species diversity.

Specification reference: 3.4.6

▲ **Figure 1** *In a tropical rainforest there is high species diversity*

Maths link √x̄

MS 2.3, see Chapter 11.

▲ **Figure 2** *In the sub-arctic tundra there is low species diversity*

Maths link

MS 2.3, see Chapter 11.

Worked example

Use the index to calculate the species diversity of the two habitats.

You must first calculate $n(n-1)$ for each species in each habitat. You can then calculate the sum of $n(n-1)$ for each species. These calculations are shown in Table 2.

▼ **Table 2** *Calculation of $n(n-1)$ and $\sum n(n-1)$ for habitats X and Y*

Species	Numbers (n) found in habitat X	$n(n-1)$	Numbers (n) found in habitat Y	$n(n-1)$
A	10	10(9) = **90**	3	3(2) = **6**
B	10	10(9) = **90**	5	5(4) = **20**
C	10	10(9) = **90**	2	2(1) = **2**
D	10	10(9) = **90**	36	36(35) = **1260**
E	10	10(9) = **90**	4	4(3) = **12**
	$\sum n(n-1)$	**450**	$\sum n(n-1)$	**1300**

You can now calculate the species diversity index for each habitat.

Habitat X: $\quad d = \dfrac{50(49)}{450} = \dfrac{2450}{450} = \mathbf{5.44}$

Habitat Y: $\quad d = \dfrac{50(49)}{1300} = \dfrac{2450}{1300} = \mathbf{1.88}$

The higher the value d, the greater is the species diversity. So, in this case, although the total number of species and the total number of individuals are the same in both habitats, the species diversity of habitat X is much greater.

Summary questions

1   Explain what is meant by species diversity.

2   Table 3 shows the numbers of each of six species of plant found in a salt-marsh community. Calculate the species diversity index for this salt-marsh community using the formula shown earlier. Show your working.

▼ **Table 3**

Species	Numbers in salt marsh
Salicornia maritima	24
Halimione portulacoides	20
Festuca rubra	7
Aster tripolium	3
Limonium humile	3
Suaeda maritima	1

3   Explain why it is more useful to calculate a species diversity index than just to record the number of species present.

Species diversity and ecosystems

Biodiversity reflects how well an ecosystem is likely to function. The higher the species diversity index, the more stable an ecosystem usually is and the less it is affected by change, for example, climate change. If there is a drought, a community with a high species diversity index is much more likely to have at least one species able to tolerate drought than a community with a low species diversity index. At least some members are therefore likely to survive the drought and maintain a community.

In extreme environments, such as hot deserts, only a few species have the necessary adaptations to survive the harsh conditions. The species diversity index is therefore normally low. This usually results in an unstable ecosystem in which communities are dominated by climatic factors rather than by the organisms within the community. In less hostile environments, the species diversity index is normally high. This usually results in a stable ecosystem in which communities are dominated by living organisms rather than climate.

1 Scientists believe that the production of greenhouse gases by human activities is contributing to climate change. Explain why an increase in greenhouse gases is more likely to result in damage to communities with a low species diversity index than communities with a high index.

2 The graph in Figure 3 shows the effect of environmental change on the stability and the functioning of ecosystems.

   a Describe the relationship between environmental change and the community with a low species diversity index.

   b Explain the different responses to environmental change in communities with a low and a high species diversity.

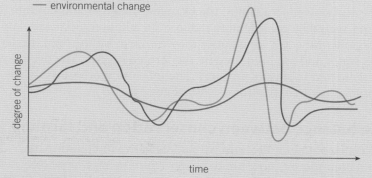

key
— community with low species diversity
— community with high species diversity
— environmental change

▲ Figure 3

▲ **Figure 4** *In harsh environments, like this hot desert, only a few species are adapted to survive the extreme conditions and therefore species diversity is low*

Maths link

MS 1.3, see Chapter 11.

Learning objectives

→ Describe the impact of agriculture on species diversity.

→ Explain the balance between conservation and farming.

Specification reference: 3.4.6

▲ **Figure 1** *High species diversity in a hay meadow*

▲ **Figure 2** *Low species diversity in a field grown for silage*

▲ **Figure 3** *Less intensively farmed agricultural land with hedgerows.*

In our efforts to provide enough food for the human population at a low cost, mankind has had a considerable impact on the natural world. This impact has led to a reduction in **biodiversity**. In this topic we will look at how agriculture reduces species diversity and the balance between conservation and farming.

Impact of agriculture

As natural ecosystems develop over time, they become complex **communities** with many individuals of a large number of different species. In other words, these communities have a high index of diversity. Agricultural **ecosystems** are controlled by humans and are different. Farmers often select species for particular qualities that make them more productive. As a result the number of species, and the genetic variety of **alleles** they possess, is reduced to the few that exhibit the desired features. To be economic, the number of individuals of these desirable species needs to be large. Any particular area can only support a certain amount of **biomass**. If most of the area is taken up by the one species that the farmer considers desirable, it follows that there is a smaller area available for all the other species. These many other species have to compete for what little space and resources are available. Many will not survive this competition. Even if species evolved to adapt to the changes, the population of the species would be considerably reduced. In addition, pesticides are used to exclude these species because they compete for the light, mineral ions, water and food required by the farmed species. The overall effect is a reduction in species diversity. The index of species diversity is therefore low in agricultural ecosystems.

The balance between conservation and farming

Food is essential for life, and with an ever-expanding human population there is pressure to produce it more and more intensively. In the UK, food production has doubled over the past 40 years. This has been achieved by the use of improved genetic varieties of plant and animal species, greater use of chemical fertilisers and pesticides, greater use of biotechnology and changes in farm practices, leading to larger farms and the conversion of land supporting natural communities into farmland. These changes have had many ecological impacts, but the overriding effect of intensive food production has been to diminish the variety of **habitats** within ecosystems and consequently reduce species diversity.

Certain practices have directly removed habitats and reduced species diversity. For example:

- removal of hedgerows and grubbing out woodland
- creating monocultures, for example replacing natural meadows with cereal crops or grass for silage
- filling in ponds and draining marsh and other wetland
- over-grazing of land, for example upland areas by sheep, thereby preventing regeneration of woodland.

Other practices have had a more indirect effect:

- use of pesticides and inorganic fertilisers
- escape of effluent from silage stores and slurry tanks into water courses
- absence of crop rotation and lack of **intercropping** or undersowing.

Despite the obvious conflicts between intensive food production and conservation, there are a number of management techniques that can be applied to increase species and habitat diversity, without unduly raising food costs or lowering yields. Examples of these conservation techniques include the following:

- Maintain existing hedgerows at the most beneficial height and shape. An A-shape provides better habitats than a rectangular one.
- Plant hedges rather than erect fences as field boundaries.
- Maintain existing ponds and where possible create new ones.
- Leave wet corners of fields rather than draining them.
- Plant native trees on land with a low species diversity rather than in species-rich areas.
- Reduce the use of pesticides – use biological control where possible or genetically modified organisms that are resistant to pests.
- Use organic, rather than inorganic, fertilisers.
- Use crop rotation that includes a **nitrogen-fixing** crop, rather than fertilisers, to improve soil fertility.
- Use intercropping rather than herbicides to control weeds and other pests.
- Create natural meadows and use hay rather than grasses for silage.
- Leave the cutting of verges and field edges until after flowering and when seeds have dispersed.
- Introduce conservation headlands – areas at the edges of fields where pesticides are used restrictively so that wild flowers and insects can breed.

It is recognised that these practices will make food slightly more expensive to produce, and therefore to encourage farmers there are a number of financial incentives from the Department for Environment, Food and Rural Affairs (DEFRA) and the European Union. Maintaining biodiversity is very important. If biodiversity is reduced the global living system becomes increasingly unstable and we all rely on the global system for food and other resources.

▲ **Figure 4** *Intensively farmed agricultural land with hedgerows removed.*

Summary questions

1   Explain how agriculture has reduced species diversity.

2   Explain why there is a reduction in species diversity when a forest is replaced by grassland for grazing sheep or cattle.

3   Suggest why the draining of ponds on agricultural land might have a greater effect on biodiversity than removing a hedgerow.

 Human activity and loss of species in the UK

The present rate of species extinction is thought to be between 100 and 1000 times greater than at any time in evolutionary history. The main cause of species loss is the clearance of land in order to grow crops and meet the demand for food from an ever-increasing human population. An area of rainforest roughly the size of the UK is cleared every year. Throughout the world habitats are being lost. Most of this habitat loss has entailed the replacement of natural communities of high species diversity with agricultural ones of low species diversity. The conservation agencies in the UK have made estimates of the percentage of various habitats that have been lost in the UK since 1900. These estimates are shown in Table 1.

▲ **Figure 5** *Deforestation (left) Heathland (middle) and mixed woodland (right)*

▼ **Table 1**

Habitat	Habitat loss since 1900/%	Main reason for habitat loss
hay meadow	95	conversion to highly productive grass and silage
chalk grassland	80	conversion to highly productive grass and silage
lowland fens and wetlands	50	drainage and reclamation of land for agriculture
limestone pavements in England	45	removal for sale as rockery stone
lowland heaths on acid soils	40	conversion to grasslands and commercial forests
lowland mixed woodland	40	conversion to commercial conifer plantations and farmland
hedgerows	30	to make larger fields to accommodate farm machinery

1 There are currently approximately 350 000 km of hedgerow in the UK. Calculate how many kilometres there were in 1900.

2 Some lowland mixed woodlands have been replaced by other woodland. Explain how this change might still result in a lower species diversity.

3 Suggest **one** benefit and **one** risk associated with the conversion of hay meadows and chalk grasslands to highly productive grass and silage.

4 Suggest in what ways the information in the table might be used to inform decision-making on preserving habitats and biodiversity.

5 The European Union gives grants to farmers to replant hedges. Explain how replanting hedges might affect the species diversity found on farms.

Hedge rows!

Hedgerows typify the conflict between conservation and farming.

They were originally created to mark the boundaries of fields and to contain livestock. Over the past 50 years there has been a farming revolution with an increase in the use of large farm machinery and larger farm sizes. Small fields are not suited to the new machinery and so hedgerows are removed to make it easier to manoeuvre large equipment. Hedgerows also take up land that could produce crops and so farmers removed them, often with grants that were once available to increase productive land area.

1 Suggest **three** ways in which the removal of hedges might benefit the farmer by increasing crop yields.

Hedges do, however, have a number of uses. They increase species diversity and act as corridors along which many species move to disperse themselves. They also produce food for both animals that live in the hedgerow as well as those that do not. Overall they add diversity and interest to the countryside.

2 Suggest **two** ways in which hedges could help farmers to increase crop yields in the long term.

In Topic 10.1, we saw that classification systems were originally based on features that could easily be observed. As science has developed it has become possible to use a wider range of evidence to determine the evolutionary relationships between organisms.

When organisms evolve it is not only their visible internal and external features that change, but also the molecules of which they are made. DNA determines the proteins of an organism, including enzymes and proteins determine the features of an organism. It follows that changes in the features of a species are due to changes in its DNA. Comparing the genetic diversity within, and between, species helps scientists to determine the evolutionary relationships between them. Let us look at the different ways that these comparisons can be made.

Comparison of observable characteristics

Traditionally, genetic diversity was measured by observing the characteristics of organisms. This method is based on the fact that each observable characteristic is determined by a gene or genes (with environmental influences). The variety within a characteristic depends on the number and variety of alleles of that gene (plus environmental influences).

Using observable characteristics has its limitations because a large number of them are coded for by more than one gene. They are polygenic. This means they are not discrete from one another but rather vary continuously. It is often difficult to distinguish one from another. Characteristics can also be modified by the environment. Differences may therefore be the result of different environmental conditions rather than different alleles. Height in humans for example is determined by a number of genes. However, environmental factors like diet can influence the actual height of an individual.

For these reasons, inferring DNA differences from observable characteristics has been replaced by directly observing DNA sequences themselves. This has been made possible through the advances in gene technology made over recent years.

Comparison of DNA base sequences

With the advent of gene technology, we can now read the base sequences of the DNA of any organism. Using various techniques, we can now accurately determine the exact order of nucleotides on DNA. DNA sequencing is now routinely done by automatic machines and the data produced analysed by computers. In these computerised systems, each nucleotide base can be tagged with a different coloured fluorescent dye – adenine (green), thymine (red), cytosine (blue) and guanine (yellow). This produces a series of coloured bands, each of which represents one of the four nucleotide bases as shown in Figure 1. We can measure the genetic diversity of a species by sampling the DNA of its members and sequencing it to produce a pattern of

Learning objectives

→ Explain the use of the following techniques in comparing genetic diversity within, and between, species:
  - observable characteristics
  - base sequence of DNA
  - base sequence of mRNA
  - amino acid sequence of proteins
→ Explain how immunological comparisons are used to investigate variations in proteins.

Specification reference: 3.4.7

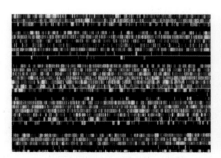

▲ **Figure 1** *Computer screen display of a DNA sequence. Each coloured band represents one of the four nucleotide bases*

Hint

As DNA determines the features of an organism, using the similarities in DNA as evidence for a close evolutionary relationship between species provides a direct record. However, some DNA is non-functional and does not code for proteins. Analysis of this DNA can provide new evidence of relationships between organisms.

coloured bands. Analysis of these patterns allows us to compare one species with another or one individual with another of the same species to determine how diverse they are. The process would be slow using the human eye and so the patterns are scanned by lasers and interpreted by computer software to give the DNA nucleotide base sequence in a fraction of the time. We can also use these techniques to determine the evolutionary relationships between species.

When one species gives rise to another species during evolution, the DNA of the new species will initially be very similar to that of the species that gave rise to it. Due to **mutations**, the sequences of nucleotide bases in the DNA of the new species will change. Consequently, over time, the new species will accumulate more and more differences in its DNA. As a result, we would expect species that are more closely related to show more similarity in their DNA base sequences than species that are more distantly related. As there are millions of base sequences in every organism, DNA contains a vast amount of information about the the genetic diversity and evolutionary history of all organisms.

Comparison of the base sequence of mRNA

We saw in Topic 8.4 that mRNA is coded for by DNA. The base sequences on mRNA are complementary to those of the strand of DNA from which they were made. It follows that we can measure DNA diversity, and therefore genetic diversity, by comparing the base sequence of mRNA.

Comparison of amino acid sequences in proteins

The sequence of amino acids in proteins is determined by mRNA which, in turn, is determined by DNA. Genetic diversity within, and between, species can therefore be measured by comparing the amino acid sequences of their proteins. The degree of similarity in the amino acid sequence of the same protein in two species will also reflect how closely related the two species are. Once the amino acid sequence for a chosen protein has been determined for two species, the two sequences are compared. This can be done by counting either the number of similarities or the number of differences in each sequence. An example is shown in Figure 2. Here there is a short sequence of seven amino acids of the same protein in six different species. The table on the right of the figure shows both the number of differences and the number of similarities.

Summary questions

1 Explain what causes the DNA sequences of genes to change over a period of time.

2 Using the information in Figure 2, state, with reasons, which two species are most closely related.

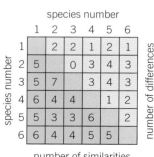

▲ **Figure 2** *Comparison of amino acid sequence in part of the same protein in six species*

Establishing relationships

The precise sequence of human evolution has long been a mystery. The evidence from different scientific techniques is often conflicting. Any conclusions drawn from the results of experiments are therefore tentative. Consequently scientists have been trying to refine their techniques in an attempt to clarify the evolutionary relationships of humans to other primates. As these new techniques have been adopted, our knowledge of primate evolution has changed as new evidence has provided a better explanation of the relationships between various primates. As a result the events in human evolution have been, and will continue to be, revised. Some of the

techniques and evidence that have led to these revisions are detailed below.

The proteins and DNA of organisms show differences between each species. These differences are thought to be due to changes that have occurred over long periods of time.

The sequences of amino acids in haemoglobin molecules have been used to clarify the evolution of primates. The amino acids found in seven specific positions in the haemoglobin molecules of six different primates were compared. The results of the study are shown in Table 1. Each amino acid is represented by a different letter.

▼ **Table 1** *Showing the sequence of amino acids in seven positions in the haemoglobin of six primates*

Primate	Position 1	Position 2	Position 3	Position 4	Position 5	Position 6	Position 7
Human	N	T	R	P	A	E	L
Gibbon	D	K	R	Q	T	D	H
Gorilla	N	T	K	P	A	D	L
Orang-utan	N	K	R	Q	T	D	L
Chimpanzee	N	T	R	P	A	E	L
Lemur	N	Q	T	A	T	E	H

1 Where the amino acid differs from that in human haemoglobin, the letter is shown in red. Use this information to list the evolutionary relationship of humans to the other primates shown. Start with the most closely related primate and end with the most distantly related one.

2 Deduce from figure 3 which two primates this immunological study suggests are the most closely related. Give reasons for your answer.

3 Deduce from Figure 3 which primate the study suggests is the nearest relative of the orang-utan. Give reasons for your answer.

4 The data in Figure 3 show the evolutionary relationships between humans and the five other primates. Outline in what **two** ways these relationships differ from that suggested by the haemoglobin study.

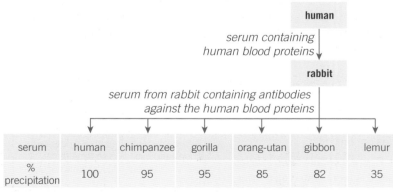

Another study compared the proteins found in a variety of primates using immunological techniques. The results of this study are illustrated in Figure 3.

human

serum containing human blood proteins ↓

rabbit

serum from rabbit containing antibodies against the human blood proteins

serum	human	chimpanzee	gorilla	orang-utan	gibbon	lemur
% precipitation	100	95	95	85	82	35

▲ **Figure 3**

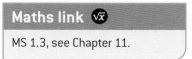

Maths link √x̄

MS 1.3, see Chapter 11.

A further study compared the number of base differences in the first 200 bases of a gene found in five species of primate. The results are shown in Table 2 below.

▼ **Table 2**

Human	0				
Gorilla	12	0			
Chimpanzee	15	15	0		
Orang-utan	29	33	26	0	
Lemur	48	49	49	50	0
	Human	Gorilla	Chimpanzee	Orang-utan	Lemur

Synoptic link

To help you follow this application and the following extension it would be useful to revise Topics 5.3, 5.4 and 5.5.

5 Evaluate what evidence there is from Table 2 to show that humans are more closely related to orang-utans than to lemurs.

6 Evaluate whether these data support the evolutionary relationships of these primates suggested by the other two studies. Explain your answer.

The conflicting evidence for the relationships between different primates illustrates the need to use a variety of evidence from different sources in drawing valid scientific conclusions.

Immunological comparisons of proteins

The proteins of different species can also be compared using immunological techniques. The principle behind this method is the fact that antibodies of one species will respond to specific antigens on proteins, such as albumin, in the blood serum of another. The process is carried out as follows:

- Serum albumin from species A is injected into species B.

- Species B produces antibodies specific to all the antigen sites on the albumin from species A.

- Serum is extracted from species B; this serum contains antibodies specific to the antigens on the albumin from species A.

- Serum from species B is mixed with serum from the blood of a third species C.

- The antibodies respond to their corresponding antigens on the albumin in the serum of species C.

- The response is the formation of a precipitate.

- The greater the number of similar antigens, the more precipitate is formed and the more closely the species are related.

- The fewer the number of similar antigens, the less precipitate is formed and the more distantly the species are related.

An example of this technique is illustrated in Figure 3. In this case, species A is a human, species B is a rabbit and species C is represented by a variety of other mammals.

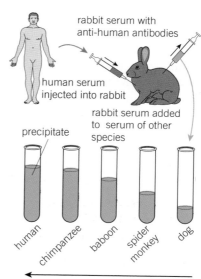

The results show that humans are very closely related to chimpanzees, less so to baboons and even less so to spider monkeys. They are only distantly related to dogs.

▲ **Figure 4** *Immunological comparisons of human serum with that of other species*

One look around us and it is clear that living things differ. If one species differs from another this is called **interspecific variation** (Figure 1). But members of the same species also differ from each other. This is called **intraspecific variation**. Every one of the billions of organisms on planet Earth is unique. Even identical twins, who are born with the same DNA, vary as a result of their different experiences. How then do we measure the differences between these characteristics?

Making measurements

All scientists measure things, but this is a particular problem for biologists. This is because they are usually measuring some aspect of living organisms and all living organisms are different. For this reason, biologists have to take many measurements of the same thing. They cannot reliably determine the height of buttercups or the number of red cells in 1 mm³ of human blood by taking a single measurement. Equally, they cannot measure every buttercup or human being in existence. What they do is take samples.

Random sampling

Sampling involves taking measurements of individuals, selected from the population of organisms which is being investigated. In theory, if these individuals are representative of the population as a whole, then the measurements can be relied upon. But are the measurements representative? There are several reasons why they might not be, including:

- **sampling bias.** The selection process may be biased. The investigators may be making unrepresentative choices, either deliberately or unwittingly. Are they as likely to take samples of buttercups from a muddy area as a dry one? Will they avoid areas covered in cow dung or rich in nettles?

- **chance.** Even if sampling bias is avoided, the individuals chosen may, by pure chance, not be representative. The 50 buttercup plants selected might just happen to be the 50 tallest in the population.

The best way to prevent sampling bias is to eliminate, as far as possible, any human involvement in choosing the samples. This can be achieved by carrying out **random sampling**. One method is:

1 Divide the study area into a grid of numbered lines, for example by stretching two long tape measures at right angles to each other.
2 Using random numbers, from a table or generated by a computer, obtain a series of coordinates.
3 Take samples at the intersection of each pair of coordinates.

We cannot completely remove chance from the sampling process but we can minimise its effect by:

- **using a large sample size.** The more individuals that are selected the smaller is the probability that chance will influence the result,

▲ **Figure 1** *Variation between species (interspecific variation)*

and the less influence anomalies will have. If we sample only five buttercups there is a high probability that they may all be taller than average. If we sample 500 there is a much lower probabilty that they will all be taller than average. The greater the sample size the more reliable the data will be.

- **analysis of the data collected.** Accepting that chance will play a part, the data collected can be analysed using statistical tests to determine the extent to which chance may have influenced the data. These tests allow us to decide whether any variation observed is the result of chance or is more likely to have some other cause.

The normal distribution curve

Figure 2 shows a normal distribution curve: its bell-shape is typical for a feature that shows continuous variation, for example height in humans. The graph is symmetrical about a central value. Occasionally the curve is shifted slightly to one side. This is called a skewed distribution and is illustrated in Figure 3. There are three terms associated with normal distribution curves. To illustrate these terms, consider the values given in Table 1 that compares the number of children in 11 different families.

The mean (arithmetic mean)

This the sum of the sampled values divided by the number of items. In our example, total the number of children in all families and divide by the number of families:

Total children = 0 + 1 + 1 + 1 + 2 + 2 + 3 + 4 + 6 + 6 + 7
$$= 33$$

Total number of families A–K = 11

Mean = $\dfrac{33}{11}$ = 3

The mode

This is the single value of a sample that occurs most often. In our example, more families have one child than any other number. The mode is therefore equal to 1.

The median

This is the central or middle value of a set of values. This requires arranging the values in ascending order. In our example, they are already arranged in ascending order of the number of children in each family. There are 11 families. The sixth family is therefore the middle family in the sample. This is family F. Family F has two children and so the median = 2.

Figure 2 shows a typical normal distribution curve in which the mean and mode (and often the median) have the same value. Figure 3 shows a skewed distribution in which the mean, mode and median all have different values.

▼ **Table 1**

Family	Number of children
A	0
B	1
C	1
D	1
E	2
F	2
G	3
H	4
I	6
J	6
K	7

Maths link

MS 1.2 and 1.6, see Chapter 11.

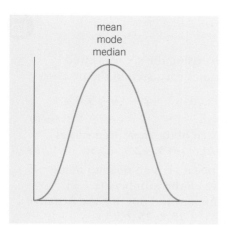

▲ **Figure 2** *A normal distribution curve where the mean, mode and median have the same value*

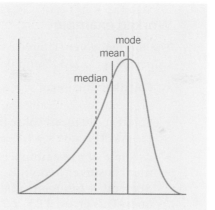

▲ **Figure 3** *A skewed distribution where the mean, mode and median have different values*

Mean and standard deviation

A normal distribution curve always has the same basic shape (Figure 2). It differs in two measurements: its maximum height and its width.

- The **mean** is the measurement at the maximum height of the curve. As we have seen, the mean of a sample of data provides an average value and is useful information when comparing one sample with another. It does not, however, provide any information about the range of values within the sample. For example, the mean number of children in a sample of eight families may be two. However, this could be made up of eight families each with two children or six families with no children and two families with eight children each.

- The **standard deviation** (*s*) is a measure of the width of the curve. It gives an indication of the range of values either side of the mean. A standard deviation is the distance from the mean to the point where the curve changes from being convex to concave (the point of inflexion). **68%** of all the measurements lie within ± 1.0 standard deviation. Increasing this width to almost ± 2.0 (actually ± 1.96) standard deviations takes in 95% of all measurements. These measurements are illustrated in Figure 4. To calculate the standard deviation with any accuracy there needs to be a minimum number of values.

Calculating standard deviation

At first sight, the formula for standard deviation can look complex:

$$\text{standard deviation} = \sqrt{\frac{\sum (x - \bar{x})^2}{n-1}}$$

Where:

$\sum$ = **the sum of**

x = **measured value (from the sample)**

$\bar{x}$ = **mean value**

n = **total number of values in the sample.**

However, it is straightforward to calculate and less frightening if you take it step by step.

Maths link

MS 1.2, see chapter 11.

Study tip

A large standard variation means a lot of variety, while a small standard deviation means little variety.

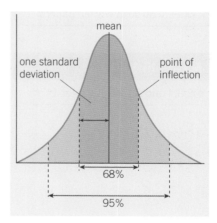

▲ **Figure 4** *The normal distribution curve showing values for standard deviation*

Hint

Remember to square all the numbers, not just the negative ones. It is also good practice not to round up figures too early in a calculation.

Maths link $\sqrt{x}$

MS 1.10, see Chapter 11.

Worked example

The following very simple example, using the six measured values (x) 4, 1, 2, 3, 5 and 0, illustrates each step in the process.

- Calculate the mean value ($\bar{x}$), i.e. $4 + 1 + 2 + 3 + 5 + 0 = 15$
$$15 \div 6 = 2.5$$

- Subtract the mean value from each of the measured values ($x - \bar{x}$). This gives: +1.5, −1.5, −0.5, +0.5, +2.5, −2.5.

- As some of these numbers are negative, you need to make them positive. To do this, square *all* the numbers $(x - \bar{x})^2$. Remember to square all the numbers and not just the negative ones. This gives: 2.25, 2.25, 0.25, 0.25, 6.25, 6.25.

- Add all these squared numbers together:

$$\sum(x - \bar{x})^2 = 17.5$$

- Divide this number by the original number of measurements less one, i.e. 5:

$$\frac{\sum(x - \bar{x})^2}{n - 1} = \frac{17.5}{5} = 3.5$$

- As all the numbers have been squared, the final step is to take the square root in order to get back to the same units as the mean:

$$\sqrt{\frac{\sum(x - \bar{x})^2}{n - 1}} = \sqrt{3.5} = 1.87$$

Study tip

You may need to calculate standard deviations as part of practical work but you will not be asked to do so in written papers.

Maths link $\sqrt{x}$

MS 0.4 and 1.1, see Chapter 11.

You will need to use a calculator to find out the value of standard deviations as it will considerably speed up your calculation. In doing so you will usually find the calculator provides a long figure running to many decimal places. In this example, the calculation $\sqrt{3.5}$ produces the answer 1.870828693. Clearly the significance of the latter digits is less than the earlier ones. It is normal to reduce these figures to a certain number of significant figures. In this case the answer has been rounded down to three significant figures, namely 1.87.

The number of significant figures to use can vary and you should always follow any instructions on how many significant figures to use in your answer. In the absence of specific instructions, it is best to use the same number of significant figures as the data you are given or are using. For example, in this case the square values were calculated to be 2.25, 0.25, 6.25, etc. As these had three significant figures, the same number was used in the final calculation. Remember, however, that where a numerical answer to one part of a question has to be used in a subsequent calculation, the answer to the first part should be carried forward without rounding it up or down.

Summary questions

1 List **two** reasons why a sample may not be representative of the population as a whole.

2 Explain how sampling bias may be prevented.

3 Distinguish between the terms 'mean', 'mode' and 'median'.

1 (a) What is a *species*?   (*2 marks*)
   (b) Scientists investigated the diversity of plants in a small area within a forest.
        The table shows their results.

Plant species	Number of individuals
Himalayan raspberry	20
Heartwing sorrel	15
Shala tree	09
Tussock grass	10
Red cedar	04
Asan tree	06
Spanish needle	8
Feverfew	8

The index of diversity can be calculated by the formula

$$d = \frac{N(N-1)}{\sum n(n-1)}$$

where
d = index of diversity
N = total number of organisms of all species
n = total number of organisms of each species

(i)   Use the formula to calculate the index of diversity of plants in the forest.
      Show your working.   (*2 marks*)
(ii) The forest was cleared to make more land available for agriculture.
      After the forest was cleared the species diversity of insects in the area
      decreased.
      Explain why.   (*3 marks*)

                                                                AQA June 2013

2 Organisms can be classified using a hierarchy of phylogenetic groups.
   (a) Explain what is meant by:
        (i)   a hierarchy   (*2 marks*)
        (ii) a phylogenetic group.   (*1 mark*)
   (b) Cytochrome c is a protein involved in respiration. Scientists determined
        the amino acid sequence of human cytochrome c. They then:
        • determined the amino acid sequences in cytochrome c from five other animals
        • compared these amino acid sequences with that of human cytochrome c
        • recorded the number of differences in the amino acid sequence compared with
          human cytochrome c.
        The table shows their results.

Animal	Number of differences in the amino acid sequence compared with human cytochrome c
A	1
B	12
C	12
D	15
E	21

(i)   Explain how these results suggest that animal **A** is the most closely
related to humans.   (*2 marks*)

(ii) A student who looked at these results concluded that animals **B** and **C**
are more closely related to each other than to any of the other animals.
Suggest **one** reason why this might **not** be a valid conclusion.   (*1 mark*)

(iii) Cytochrome c is more useful than haemoglobin for studying how
closely related different organisms are. Suggest **one** reason why.   (*1 mark*)

AQA June 2013

3 **(a)** What information is required to calculate an index of diversity for a
particular community?   (*1 mark*)

(b) Farmers clear tropical forest and grow crops instead. Explain how this
causes the diversity of insects in the area to decrease.   (*3 marks*)

(c) Farmers manage the ditches that drain water from their fields. If they do
not, the ditches will become blocked by plants. Biologists investigated the
effects of two different ways of managing ditches on farmland birds.
- Ditch **A** was cleared of plants on both banks
- Ditch **B** was cleared of plants on one bank.

The graph shows the number of breeding birds of all species along the two ditches,
before and after management.

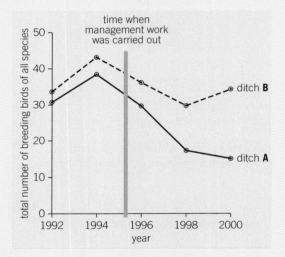

(i)   The points on the graph have been joined with straight lines rather than with a
smooth curve. Explain why they have been joined with straight lines.   (*1 mark*)

(ii) It would have been useful to have had a control ditch in this investigation.
Explain why.   (*1 mark*)

(d) A farmer who wanted to increase the diversity of birds on his land read about this
investigation.

He concluded that clearing the plants from one bank would not decrease diversity as
much as clearing the plants from both banks. Evaluate this conclusion.   (*3 marks*)

AQA Jan 2011

4 Costa Rica is a Central American country. It has a high level of species diversity.

(a) There are over 12 000 species of plants in Costa Rica. Explain how this has
resulted in a high species diversity of animals.   (*2 marks*)

(b) The number of species present is one way to measure biodiversity. Explain
why an index of diversity may be a more useful measure of biodiversity. (*2 marks*)

(c) Crops grown in Costa Rica are sprayed with pesticides. Pesticides are substances
that kill pests. Scientists think that pollution of water by pesticides has reduced the
number of species of frog.

(i)   Frogs lay their eggs in pools of water. These eggs are small. Use this information
to explain why frogs' eggs are very likely to be affected by pesticides in the
water.   (*2 marks*)

(ii)   An increase in temperature leads to evaporation of water. Suggest how evaporation may increase the effect of pesticides on frogs' eggs.   (*1 mark*)

AQA June 2011

5   To reduce the damage caused by insect pests, some farmers spray their fields of crop plants with pesticide. Many of these pesticides have been shown to cause environmental damage.

Bt plants have been genetically modified to produce a toxin that kills insect pests. The use of Bt crop plants has led to a reduction in the use of pesticides.

Scientists have found that some species of insect pest have become resistant to the toxin produced by the Bt crop plants.

Figure 6 shows information about the use of Bt crops and the number of species of insect pest resistant to the Bt toxin in one country.

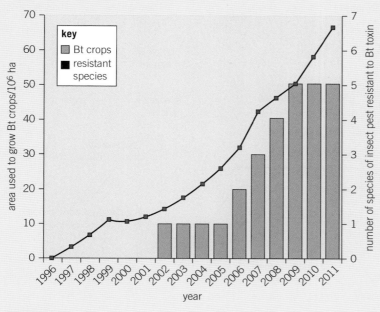

▲ **Figure 6**

(a)   Can you conclude that the insect pest resistant to Bt toxin found in the years 2002 to 2005 was the same insect species? Explain your answer.   (*1 mark*)

(b)   One farmer stated that the increase in the use of Bt crop plants had caused a mutation in one of the insect species and that this mutation had spread to other species of insect. Was he correct? Explain your answer.   (*4 marks*)

(c)   There was a time lag between the introduction of Bt crops and the appearance of the first insect species that was resistant to the Bt toxin. Explain why there was a time lag.   (*3 marks*)

AQA SAMS AS PAPER 2

(d)   Calculate the actual increase and the percentage increase in the area used to grow Bt crops between 2000 and 2010.   (*2 marks*)

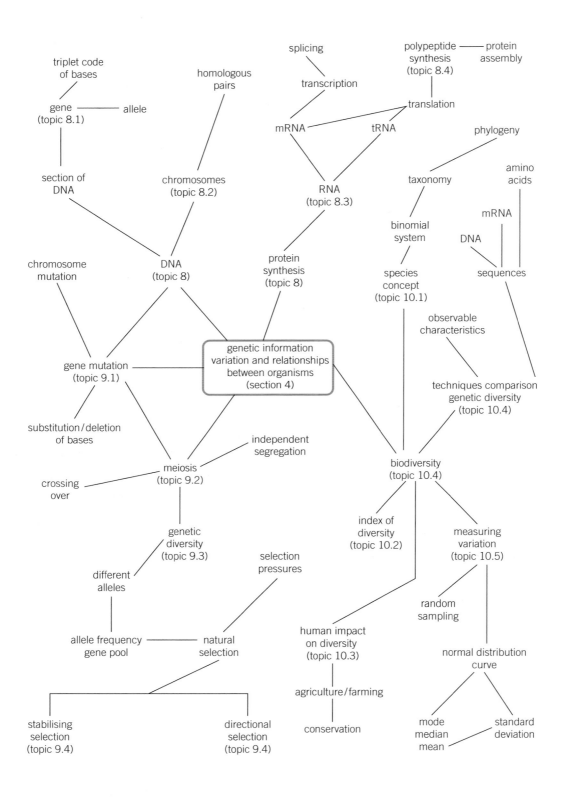

Practical skills

In this section you have met the following practical skills:

- Evaluating results and drawing conclusions
- Identifying variables and suitable controls
- Interpreting data, such as graphs, obtained from experiments
- How to use sampling techniques in fieldwork.

Maths skills

In this section you have met the following maths skills:

- Using the logarithmic function on a calculator
- Understanding simple probability
- Substituting values in algebraic equations like the Simpson's Index of Diversity
- Interpreting tables and histograms
- Using percentages
- Understanding and calculating mean, mode, median and standard deviation
- Using the appropriate number of significant figures
- Estimating results.

Extension task

Three factors that affect genetic diversity are:

- selective breeding
- the founder effect
- genetic bottlenecks.

Find out about these three factors in books, journals and the internet and then prepare a short talk to other students of AS Biology explaining how each affects genetic diversity.

Research by any appropriate means the topic of selective breeding in domesticated animals such as cattle. Using the information obtained, write an account that balances the advantages and disadvantages of selective breeding and discusses the ethical implications of the practice.

1 (a) Scientists can use protein structure to investigate the evolutionary relationships between different species. Explain why.   (*2 marks*)

   (b) Comparing the base sequence of genes provides more evolutionary information than comparing the structure of proteins. Explain why.   (*2 marks*)

   (c) The proteins of different species can be compared using immunological techniques. The protein albumin obtained from a human was injected into a rabbit. The rabbit produced antibodies against the human albumin. These antibodies were extracted from the rabbit and then added to samples of albumin obtained from four different animal species. The amount of precipitate produced in each sample was then measured. The results are shown in the table.

Species from which albumin was obtained	Amount of precipitate / arbitrary units
Rat	23
Chimpanzee	96
Marmoset	65
Trout	11

What do the results suggest about the evolutionary relationship between humans and the other species?   (*2 marks*)

AQA Jan 2012

2 Sugar beet is a crop grown for the sugar stored in its root. The sugar is produced by photosynthesis in the leaves of the plant. Plant breeders selected high-yielding wild beet plants. They used these plants to produce a strain of sugar beet to grow as a crop.

The drawings show a wild beet plant and a sugar beet plant. The drawings are to the same scale.

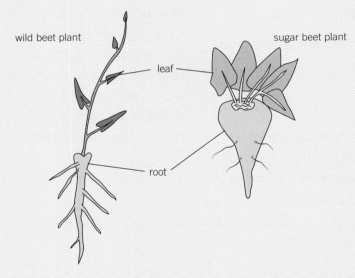

wild beet plant

sugar beet plant

leaf

root

   (a) Use the drawings to describe two ways in which a sugar beet plant is different from a wild beet plant.
   Explain how each of these differences would give an increased yield of sugar.   (*4 marks*)

   (b) Sugar beet plants have been selected for a faster rate of growth.
   Suggest how the faster rate of growth may increase profit for a farmer.   (*1 mark*)

   (c) Describe and explain how selection will have affected the genetic diversity of sugar beet.   (*2 marks*)

AQA June 2012

3 Ecologists investigated the size of an insect population on a small island. They used a mark-release-recapture method. To mark the insects they used a fluorescent powder. This powder glows bright red when exposed to ultraviolet (UV) light.

(a) The ecologists captured insects from a number of sites on the island. Suggest how they decided where to take their samples. *(2 marks)*

(b) Give **two** assumptions made when using the mark-release-recapture method. *(2 marks)*

(c) Suggest the advantage of using the fluorescent powder in this experiment. *(2 marks)*

The ecologists did **not** release any of the insects they captured 1–5 days after release of the marked insects.

Table 1 shows the ecologists' results.

▼ Table 1

Days after release	Number of marked insects remaining in population	Number of insects captured	Number of captured insects that were marked
1	1508	524	78
2	1430	421	30
3	1400	418	18
4	1382	284	2
5	1380	232	9

(d) Calculate the number of insects on this island 1 day after release of the marked insects. Show your working. *(2 marks)*

(e) The ecologists expected to obtain the same result from their calculations of the number of insects on this island on each day during the period 1–5 days after release. In fact, their estimated number increased after day 1.

During the same period, the number of insects they caught decreased. The method used by the ecologists might have caused these changes.

Use the information provided to suggest **one** way in which the method used by the ecologists might have caused the increase in their estimates of the size of the insect population. *(2 marks)*

AQA SAMS A LEVEL PAPER 3

4 **Table 1** shows the taxons and the names of the taxons used to classify one species of otter. They are not in the correct order.

▼ Table 1

Taxon	Name of taxon	
J	Family	Mustelidae
K	Kingdom	Animalia
L	Genus	Lutra
M	Class	Mammalia
N	Order	Carnivora
O	Phylum	Chordata
P	Domain	Eukarya
Q	Species	lutra

(a) Put letters from **Table 1** into the correct order. *(1 mark)*

(b) Give the scientific name of this otter. *(1 mark)*

Scientists investigated the effect of hunting on the genetic diversity of otters. Otters are animals that were killed in very large numbers for their fur in the past.

The scientists obtained DNA from otters alive today and otters that were alive before hunting started.

For each sample of DNA, they recorded the number of base pairs in alleles of the same gene. Mutations change the numbers of base pairs over time.

Figure 6 shows the scientists' results.

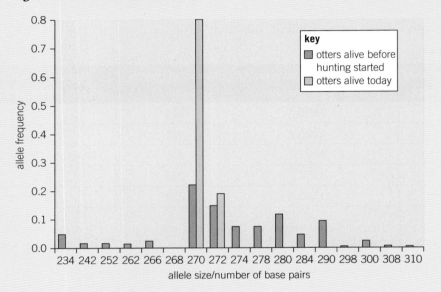

(c) The scientists obtained DNA from otters that were alive before hunting started. Suggest **one** source of this DNA. *(1 mark)*

(d) What can you conclude about the effect of hunting on genetic diversity in otters? Use data from **Figure 6** to support your answer. *(2 marks)*

(e) Some populations of animals that have never been hunted show very low levels of genetic diversity.

Other than hunting, suggest **two** reasons why populations might show very low levels of genetic diversity. *(2 marks)*

AQA SAMS AS PAPER 1

Section 5 Skills in A level Biology
Chapter 11 Mathematical skills

Biology students are often less comfortable with the application of mathematics compared with students such as physicists, for whom complex maths is a more obvious everyday tool. Nevertheless, it is important to realise that biology does require competent maths skills in many areas. It is important to practise these skills so you are familiar with them as part of your routine study of the subject.

Confidence with mental arithmetic is very helpful, but among the most important skills is that of taking care and checking calculations. We may not be required to understand the detailed theory of the maths we use, but we do need to be able to apply the skills accurately, whether simply calculating percentages or means, or substituting numbers into complex-looking algebraic equations, such as in statistical tests.

This chapter is designed to help with some of the regularly encountered mathematical problems in biology.

Working with the correct units

In biology it is very important to be secure in the use of correct units. These must always be written clearly in calculations.

Maths link $\sqrt{x}$

MS 0.1

Base units
The units we use are from the Système Internationale – the SI units. In biology we most commonly use the SI base units:

- metre (m) for length, height, distance
- kilogram (kg) for mass
- second (s) for time
- mole (mol) for the amount of a substance.

You should develop good habits right from the start, being careful to use the correct abbreviation for each unit used. For example, seconds should be abbreviated to s, not 'sec' or 'S'.

Derived units
Biologists also use SI derived units, such as:

- square metres (m^2) for area
- cubic metre (m^3) for volume
- cubic centimetre (cm^3), also written as millilitre (ml), for volume
- degree Celsius (°C) for temperature
- mole per litre (mol/L, mol dm^{-3}) is usually used for concentration of a substance in solutions (although the official SI derived unit is moles per cubic metre)
- joule (J) for energy
- pascal (Pa) for pressure
- volt (V) for electrical potential.

Non-SI units

Although examination boards use SI units, you may also encounter non-SI units elsewhere, for example:

- litre (cubic decimetre) (l, L, dm³) for volume;
- Minute (min) for time;
- hour (h) for time;
- svedberg (S) (for sedimentation rate), used for ribosome particle size.

Unit prefixes

To accommodate the huge range of dimensions in our measurements, they may be further modified using appropriate prefixes. For example, one thousandth of a second is a millisecond (ms). This is illustrated in the Table 1.

▼ Table 1

Division	Factor	Prefix	Length		Mass		Volume		Time	
one thousand millionth	10^{-9}	nano	nanometre	nm	nanogram	ng	nanolitre	nl	nanosecond	ns
one millionth	10^{-6}	micro	micrometre	μm	microgram	μg	microlitre	μl	microsecond	μs
one thousandth	10^{-3}	milli	millimetre	mm	milligram	mg	millilitre	ml/cm³	millisecond	ms
one hundredth	10^{-2}	centi	centimetre	cm						
whole unit			metre	m	gram	g	litre	l/L/dm³	second	s
one thousand times	10^{3}	kilo	kilometre	km	kilogram	kg				

Converting between units

You may need to convert between units in order to be able to scale and express numbers in sensible forms. For example, rather than refer to the width of a cell in metres you would use micrometres (μm). This allows your measurements to be understood within the relevant scale of the observation.

Divide by 1000 for each step to convert in this direction ⟹

nano-	micro-	milli-	whole unit	kilo-
e.g. nm	e.g. μm	e.g. mm	e.g. m	e.g. km

⟸ Multiply by 1000 for each step to convert in this direction

▲ Figure 1

Examples:

Convert 1 m to mm: 1 × 1000 = 1000 mm

Convert 1 m to μm: 1 × 1000 = 1000 mm, then 1000 × 1000 = 1 000 000 μm

Convert 1 l to cm³: 1 × 1000 = 1000 cm³

Convert 20 000 μm to mm: 20 000 ÷ 1000 = 20 mm

Converting between square or cube units requires a bit more care.

One m² = 1000 × 1000 = 1 000 000 mm², so your conversion factor becomes × or ÷ 1 000 000.

One m³ is 1000 × 1000 × 1000 = 1 000 000 000 mm³, so your conversion factor now becomes × or ÷ 1 000 000 000.

Examples:

Convert 20 m² to km²: 20 ÷ 1 000 000 = 0.000 02 km²

Convert 1 m² to mm²: 1 × 1 000 000 = 1 000 000 mm²

Convert 5 000 000 mm³ to m³: 5 000 000 ÷ 1 000 000 000 = 0.005 m³

Convert 0.000 000 7 m³ to mm³: 0.000 000 7 × 1 000 000 000 = 70 mm³

Decimals and standard form

When you are using numbers that are very small, such as dimensions of molecules and organelles, it is useful to use **standard form** to express them more easily. Standard form is also commonly called **scientific notation**.

Standard form is essentially expressing numbers in powers of ten. For example, 10 raised to the power 10 means 10 × 10, i.e. 100. This may be written down as 10×10^1 or 1×10^2. To get to 1000 you use 10 × 10 × 10, which would be written as 1×10^3.

An easy way to look at this is to imagine the decimal point moving one place per power of ten. For example, to write down 58 900 000 000 as standard form, you would follow the steps below.

Step 1: write down the smallest number between 1 and 10 that can be derived from the number to be converted. In this case it would be 5.89.

Step 2: write the number of times the decimal place will have to shift to expand this to the original number as powers of ten. On paper this can be done by hopping the decimal over each number like this:

5.89 0 0 0 0 0 0 0 0 0 0

▲ Figure 2

until the end of the number is reached. In this example, that requires 10 shifts, so the standard form should be written as 5.89×10^{10}.

Going the other way, for example expressing 0.000 007 8 as standard form, write the number in terms of the number of places the decimal place would have to hop forward to make the smallest number between 1 and 10, so to get to 7.8 you would have to hop over six times, so this number is written as 7.8×10^{-6}.

Significant figures

There are some simple rules to use when working out significant figures.

Rule 1: All non-zero digits are significant.

For example, 78 has two significant figures, 9.543 has four significant figures and 340 has two significant figures.

Maths link √x̄

MS 0.2

Maths link √x̄

MS 1.1

Rule 2: Intermediate zeros are significant.

For example, 706 has three significant figures and 5.900 76 has six significant figures.

Rule 3: Any leading zeroes are not significant.

For example, 0.005 67 has three significant figures (5,6 and 7; ignore the leading zeroes)

Rule 4: Zeroes at the ends of numbers containing decimal places are significant.

For example, 45.60 has four significant figures and 330.00 has five significant figures.

Significant figures and rounding

Table 2 shows the effect of rounding numbers to decimal places compared with significant figures. Remember that in rounding, when the next number is 5 or more round up, while if it is 4 or less don't round up. For example, 4.35 rounds to 4.4 and 4.34 rounds to 4.3.

Table 2 shows examples of rounding the number 23.336 00 to decimal places and to significant figures.

▼ Table 2

Measurements expressed by rounding to decimal places	Number of decimal places	Measurements expressed by rounding to significant figures	Number of significant figures	Measured to the nearest
23.336 00	5	23.336	5	100 thousandth
23.336 0	4	23.34	4	Ten thousandth
23.336	3	23.3	3	Thousandth
23.34	2	23	2	Hundredth
23.3	1	20	1	Tenth
23	0	—	—	Whole number

Significant figures and standard form

In standard form only the significant figures are written as digits, for example 5.600×10^3 has four significant figures. If this were written as a straight number it would be 5600. But according to the rules above, 5600 only has two significant figures – what does this mean?

In a given number, the significant figures are defined as the ones that contribute to its precision. Writing the number as 5600 implies precision only to the nearest whole hundred. The zeroes in the number could mean that it has simply been rounded, e.g. 5600.44 or even 5633. But if this number were actually more precise, for example it had been measured with equipment genuinely sensitive to the nearest hundredth part (2 decimal places) then 5600.00 is actually very precise and the two zeros have significance because they tell us that the measurement is *exactly* 5600 with no tenths or hundredths at all. So using standard form allows this precision to remain clearly as part of the stated number, because all significant figures are written.

Averages

An average value is actually a measure of central tendency. The most familiar measurement is the arithmetic mean (mean for short), but median or modal values are sometimes more appropriate to the data.

The arithmetic mean

Usually referred to simply as the mean, this is a measure of central tendency that takes into account the number of times each measurement occurs together with the range of the measurements. When repeated measurements are averaged, the mean will approach the true value, which will lie somewhere in the middle of the observed range, more accurately. This is why it is important to repeat experimental measurements, especially in biology where the natural unpredictability of living systems leads to inevitable fluctuations.

The mean is determined by adding together all the observed values and then dividing by the number of measurements made.

For a range of values of x, the mean $\bar{x} = \dfrac{\sum x}{n}$.

$\bar{x}$ is the mean value.

$\sum x$ is the sum of all values of x.

n is the number of values of x.

For example, five mice were weighed, giving masses of 6.2 g, 7.7 g, 6.7 g, 7.1 g and 6.3 g.

The mean mass is $(6.2 + 7.7 + 6.7 + 7.1 + 6.3) \div 5 = 6.8$ g

Be careful with your decimal places when calculating mean values. Your mean should normally have the same level of precision as the original measurements and therefore the same number of decimal places, otherwise you may be implying that the averaged measurements are more precise than they really are. For example, masses in whole grams would not average to a mass with one or more decimal places. Similarly averaging the numbers of whole objects should result in a whole number; if counts of bubbles in a pondweed experiment were averaged to a decimal place it implies you counted a fraction of one bubble, which is impossible!

The median

The median value in a set of data is calculated by placing the values in numerical order then finding the middle value in the range.

For example, the data set 12, 15, 10, 17, 9, 13, 13, 19, 10, 11 rearranges as 9, 10, 10, 11, 12, 13, 13, 15, 17, 19.

The middle of this range is 12.5.

The median value is very useful when data sets have a few values (outliers) at the extremes, which if included in an arithmetic mean could skew the data. It also allows comparison of data sets with similar means but a clear lack of overlap, skewed data and when there are too few measurements to calculate a reliable mean value.

For example, in the data set 1, 3, 3, 11, 12, 12, 12, 13, 14, 15, the median value is 12, a sensible looking mid point, but the mean would be 9.6, skewed to the left by the numbers at the lower extreme.

Maths link $\sqrt{x}$

MS 1.2 and 1.6

The mode

The modal value is the most frequent value in a set of data. It is very useful when interpreting data that is qualitative or in situations where the distribution has more than one peak (bimodal).

For example, in the data set 9, 10, 11, 11, 12, 13, 13, 13, 14, 17, 18, 19, the modal value is 13.

In biology, caution should be used because the sets of data are usually small and can introduce confusion. For example, in the data set 9, 10, 11, 11, 12, 13, 13, 14, 17, 18 there are apparently two modal values, 11 and 13, while in the set 11, 12, 13, 14, 17 there is no most frequent number and the mode is effectively every number and therefore of no value at all.

The modal value is not used very often, but it can be usefully applied when data is collected in categories, for example, numbers of moths attracted to lights of different colours.

Percentages

A percentage is simply expressing a fraction as a decimal. It is important to be confident with calculating percentages, which although straightforward are commonly calculated incorrectly.

Percentages as proportions and fractions

For example, two shapes of primrose flowers exist depending on stigma length; 'pin eyed' and 'thrum eyed'. In a survey of two areas of grassland, one area had 323 pin and 467 thrum (total 790 plants), the other had 667 pin and 321 thrum (total 988 plants). The percentage of pin eyed plants in each area is calculated as follows:

Area 1: $fraction = \frac{323}{790}$ which gives $decimal$ 0.41, which multiplied by 100 gives $percentage$ 41%.

The percentage of pin eyed flowers in Area 2 is $\frac{667}{988} \times 100 = 67.5\%$.

Percentages as chance

In genetics the likelihood of different offspring phenotypes should always be expressed as a percentage. For example, in a simple genetic cross between two heterozygous parents carrying the cystic fibrosis allele, one out of every four possible children could potentially be affected by the disorder. The chance of a cystic fibrosis child from these parents is therefore $\frac{1}{4} \times 100 = 25\%$.

Percentage change

This often comes up in osmosis experiments where samples (usually of potato tissue) gain and lose mass in different bathing solutions.

For example, a sample weighed 18.50 g at the start and at the end it weighed 11.72 g.

The actual loss in mass = 18.50 − 11.72 g = 6.78 g

The percentage change $= \frac{\text{mass change}}{\text{starting mass}} \times 100 = \frac{6.78}{18.50} \times 100 = -36.7\%$

Note the use of the minus sign to indicate that this is a loss.

Maths link

MS 0.3

Equations

Substituting into equations

There are several equations (mathematical formulae) that you will need to be able to use in advanced level biology. You do not need to learn the theoretical maths from which they are derived, but you do need to be able to put known numerical values in the right place (this is *substituting into the equation*) and then calculate the result of the equation by performing the different steps in the right order (*this is solving the equation*).

An example that you will encounter during ecology studies is called the Simpson's Index of Diversity, which has the formula $= \dfrac{N(N-1)}{\Sigma n(n-1)}$.

Each symbol (*term*) in the equation has a specific meaning. In this example:

N means the total number of all individual organisms in a survey.

n means the total number of each different species.

Σ means 'the total of' and requires you to add together all the indicated values.

Brackets indicate sub-calculations that must be done, for example $N-1$ means the total of all species found -1.

The figures in brackets need to be multiplied by the figures outside them, e.g. $N(N-1)$ means $N \times (N-1)$.

An example of the data to use could be counts of the plant species found in a certain area. To make life easy, use a table like Table 3.

Maths link

MS 2.2, 2.4 and 2.3

▼ **Table 3**

Plant species	Number of plants of each species found (n)	($n-1$)	$n(n-1)$
A	22	22 − 1 = 21	22 × 21 = 462
B	30	30 − 1 = 29	30 × 29 = 870
C	25	25 − 1 = 24	25 × 24 = 600
D	23	23 − 1 = 22	23 × 22 = 506
Totals of all plants = N	$N = 100$ $N - 1 = 99$		$\Sigma n(n-1) = 2438$

The brackets in equations always need to be solved first.

Begin by finding $n-1$ for each plant (see column 3 in Table 3) and $N-1$ (at the bottom of column 2).

Next work out $n(n-1)$ (column 4 in Table 3).

Now find $\Sigma n(n-1)$ by totalling the figures in column 4.

Substituting the known values into the equation works like this:

$D = \dfrac{N(N-1)}{\Sigma n(n-1)}$ becomes $D = \dfrac{100(99)}{2438}$ which calculates to $D = \dfrac{9900}{2438}$, which gives the result D = 4.1.

Maths link $\sqrt{x}$

MS 1.8

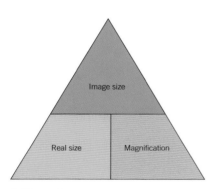

▲ Figure 3

Rearranging equations

The individual parts or *terms* in equations are all related, but sometimes you might know all the values of the terms except one. The equation can be re-written so that the unknown term can be calculated. This is called rearranging or *changing the subject of* an equation. A very useful example of this arises during the study of microscopy and magnification.

The different terms are magnification, size of image and actual size of the object being observed. The equation that relates them together is:

$$magnification = \frac{size\ of\ image}{size\ of\ real\ object}.$$

You can use the equation to calculate magnification factors quite simply. For example, if you had a photograph of your pet dog, the magnification of the image would be the height of the image of that dog divided by its real height.

Be very careful to use the same units for each measurement! If the dog is 9 cm tall in the photograph and the real dog is 0.4 m tall you would have to convert the units before starting. 0.4 m is 40 cm, so the sum would be 9 ÷ 40 = 0.23. You picture's magnification is ×0.23.

Suppose you only had the photo and the magnification. How would you find out how big the real object was? You may need to do this type of calculation on photomicrographs of cells or parts of cells.

For example, a photograph shows a mitochondrion which is 41 mm long in the picture and is taken at magnification ×34 000. How long is the original mitochondrion?

To find out, rearrange the equation. You might use an equation triangle to help.

On Figure 3 the horizontal line means divide and the vertical line means multiply.

So $magnification = \frac{size\ of\ image}{size\ of\ real\ object}$

rearranges as *size of image = magnification × real size*

and $real\ size = \frac{size\ of\ image}{magnification}.$

You need to find the real size of the mitochondrion, so your sum will be:

$real\ size = \frac{41}{34000} = 0.0012\ mm.$

At this point you need to check that your units are sensible. A mitochondrion is so small that the appropriate unit of measurement is a micrometre (µm). The question may even ask you to use this unit. Earlier in this chapter you saw that 1 µm is 1/1000th of a mm, so to convert you need to multiply by 1000. The real mitochondrion is 0.0012 × 1000 = 1.2 µm long.

Gathering data and making measurements

Estimating results

When measuring and recording data it is useful to be able to make an estimate of the number you should be getting. This will allow you to judge whether the results you actually record seem believable. This

Maths link $\sqrt{x}$

MS 0.4

is especially important when using a calculator, because it is easy to mis-type an entry and get a wrong answer. An estimate is really a sensible guess. It is a good idea to practise this skill, for example when collecting data from practical work in class.

Uncertainties in measurements

When making measurements, even using good quality instruments such as rulers and thermometers, there is a certain level of doubt in the precision of the measurement obtained. This is the *uncertainty of measurement*. The uncertainty can be stated, in which case a margin of error is identified. For example, measurements made using a good mm scale ruler a measurement may be reported as +/- 0.5mm, which is the maximum error likely when using the ruler carefully.

Percentage error is a way of using the maximum error to calculate the possible total error in a given measurement. Some types of instrument have maximum errors written on them, for example a balance may state +/- 0.01g. Other devices such as rulers and thermometers may rely on common sense, e.g. +/- 0.5mm or +/- 0.5 °C when recorded by eye. To find percentage error use the formula:

$$\% \ error = \frac{maximum \ error}{measured \ value \ recorded} \times 100$$

For example, with a ruler the maximum likely error is usually 0.5mm. If an object is measured at 6mm with the ruler, the percentage error = $\frac{0.5}{6} \times 100 = 8.3\%$.

A larger object will have a smaller % error because the +/- 0.5 mm is a lesser part of the total recorded, e.g. an object measured at 87mm has a % error = $\frac{0.5}{87} \times 100 = 0.6\%$.

Working with graphs and charts

Choosing the right type of graph or chart

During your course you will most commonly use line graphs, bar charts and histograms. You need to be able to choose the right one to suit the data and also to be able to draw graphs accurately.

The first part of your decision depends on the type of independent variable that you have measured. When you have used an independent variable that has specific values on a continuous scale, such as temperature, you should use a line graph, e.g. oxygen volume consumed by woodlice in a respirometer at a variety of temperatures. Alternatively your data might be in discrete categories, for example the number of left-handed or right-handed people. For this data a bar chart should be used with a space between each bar. When your categoric data is in groups that can be arranged on a continuous scale, e.g. height categories of plants such as 0 to 1 cm, >1 to 2 cm, >2 to 3 cm and so on, a histogram should be chosen, in which the bars are not separated by gaps.

Plotting the graph or chart

The rules when plotting the graph are:

- Ensure that the graph occupies the majority of the space available (this means more than half the space).

- Mark axes using a ruler and divide them clearly and equidistantly (i.e. 10, 20, 30, 40 not 10, 15, 20, 30, 45.

Maths link $\sqrt{x}$

MS 1.11

Maths link $\sqrt{x}$

MS 3.1, 3.2, 3.4, 3.5, and 1.3

- Ensure that the dependent variable that you measured is on the *y* axis and the independent variable that you varied is on the *x* axis.
- Ensure that both axes have full titles and units clearly labelled, e.g. pH of solution, not just 'pH'; mean height/m, not just 'height'.
- Plot the points accurately using a sharp pencil and '×' marks so the exact position of the point is obvious and is not obscured when you plot a trend line.
- Draw a neat best-fit line, either a smooth curve or a ruled line. It does not have to pass through all the points. Alternatively use a point to point ruled line, which is often used in biology where observed patterns do not necessarily follow mathematically predictable trends!
- Confine your line to the range of the points. Never extrapolate the line beyond the range within which you measured. Extrapolation is conjecture! A common mistake is to try and force the plotted line to go through the origin.
- Distinguish separate plotted trend lines using a key.
- Add a clear concise title.
- Where data ranges fall a long way from zero, a broken axis will save space. For example, if the first value on the *y* axis is 36 it may be sensible to start the axis from 34 rather than zero. This will avoid leaving large areas of your graph blank.

You will be expected to follow these conventions. If you do, then questions that involve drawing a graph become easy.

Adding range bars and error bars to your plotted points

The position of the point on a graph is always subject to uncertainty. It may be a mean value, which will depend on the values averaged or whether you include or exclude any possible anomalies. A way of indicating the level of certainty in the positioning of your points is to use a range bar or error bar. These are ways of pictorially indicating the possible range of positions of the point and reflect the spread or variability in the original measurements that were averaged. The more spread the measurements, the less certain the position of the mean when plotted.

Maths link $\sqrt{x}$

MS 1.10

Table 4 shows some example data from an experiment on gas production by a photosynthesising plant at different temperatures, with which the different styles can be demonstrated.

▼ Table 4

Temperature (°C)	Time taken to collect 10 cm³ of gas (s)								
	1	2	3	4	5	mean	*s*	*mean +s*	*mean −s*
15	87	95	102	121	117	104	14.4	118.4	89.6
20	67	78	61	90	86	76	12.3	88.3	63.7
25	57	59	48	66	51	56	7.0	63	49
30	47	45	39	42	21	39	10.4	49.4	28.6
35	118	123	145	136	132	131	10.7	141.7	120.3

Range bars

A range bar is the simplest way of showing the showing the spread in the data. Look for the maximum and minimum values in each set of repeats; they are picked out in bold in the table. After plotting the point on your graph mark the positions of the maximum and minimum values above and below the point using a small bar. Join the two extremes with a neat ruler line running vertically through the plotted point (Figure 4).

Notice that the range bars are not always symmetrical above and below the plotted points. The tops and bottoms just show the largest or smallest values among the measurements made.

Error bars

To plot error bars you use the standard deviation (a calculated measure of the spread of the data), to indicate your ranges. This is better than using range bars because it reduces the effect of any extreme values in the dataset. In the table the values of standard deviation are shown in the column headed *s*.

Plot the bars by marking the top and lower limits exactly plus and minus one standard deviation above and below the point. These values are also included in the table. The result is shown in Figure 5.

Notice that the error bars are symmetrical above and below the points, which are now indicated with a range of ± one standard deviation. The length of the bars now indicates not the maximum/minimum values but the mathematical spread in the data. The more the data spread out around the mean, the longer the error bar becomes and the less certain you are that the mean is really accurate.

Calculating rates from graphs

When data have been plotted on a line graph relating measured values on the y axis to time on the x axis, it is possible to calculate a rate of change for the y variable. There are two common graph forms that you will encounter, shown in Figure 6.

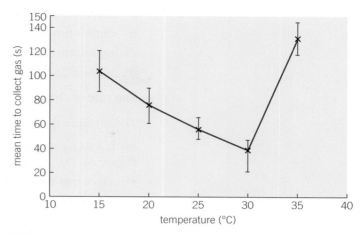

▲ **Figure 4** *Range bars*

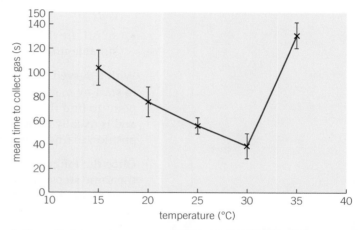

▲ **Figure 5** *Error bars*

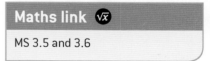

Maths link $\sqrt{x}$

MS 3.5 and 3.6

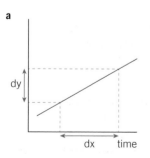

a

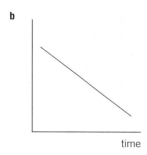

b

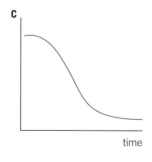

c

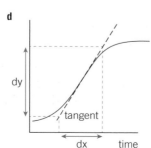

d

HAMPTON SCHOOL
BIOLOGY DEPARTMENT

a might represent oxygen production by a photosynthesising plant

b might represent oxygen consumption by a respiring organism

c might represent pH change during a lipid digestion experiment

d might represent growth of a bacterial population in a fermenter

The rate of change is simply the gradient of the graph.

The formula that is used is $\frac{change\ in\ y}{change\ in\ x}$ or $\frac{dy}{dx}$

With a straight line graph follow these steps, which are marked on Figure 6a

- select any two points on the plotted line
- use a ruler to mark construction lines from the two points to the x and y axes
- measure the difference between the two points on the y axis, this is dy
- measure the time difference between the points on the x axis, this is dx
- substitute the values into the equation and remember to quote suitable units for the result, e.g. cm^3 O_2 per minute.

With a curved line the procedure is the same except that you need to start by marking a tangent against the curve, usually at its steepest point to find the maximum rate of change. This takes a bit of practise and is usually done by eye, although it is possible to calculate a position mathematically.

Once the tangent is drawn, select two points on it and proceed using the same steps that were applied to the straight line. Figure 6d has been marked with a tangent as an example.

Scatter diagrams

A scatter diagram is a method of plotting two variables in order to try and identify a correlation between them. The dependent variable is plotted on the y axis and the independent along the x axis. Once the points are plotted a trend line can be added to show a possible relationship. An example might be plotting incidence of lung cancer against number of cigarettes smoked per day. Once such a plot has been made the relationship can be tested using a statistical test, for example the correlation coefficient, r.

Probability

When data appears to show a pattern, it is possible to determine whether the pattern is simply due to chance or whether it has an underlying cause. For example, 36 throws of die should give near enough six of each possible number. Throwing 7 4's and 5 6's is likely to be a fluke, but if you threw 23 6's then this is definitely against the rules of probability!

Probability is assessed using a statistical test, for example chi-squared to test how closely observed measurements fit with expectation, such as in genetic cross results, or student t, which compares the means of sets of data to assess whether they differ, e.g. leaf width of sun versus shade grown ivy plants.

Such tests produce a calculated value that may be found in a table of probability. In these tables, it is the *probability that the data observed differ*

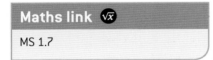

Maths link √x̄

MS 1.7

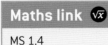

Maths link √x̄

MS 1.4

by chance alone that is being found. In biology we accept any probability greater than 5% as likely to be just chance or fluke, but probabilities of 5% or below show us that the data do differ significantly and there must be a cause influencing the outcome.

Maths link

MS 0.3, 1.5, 1.8, 1.9, 1.10, 2.2, 2.3, 2.4 and 4.1.

Table 5 Formulae commonly used in biology

Circumference of a circle	$\pi \times d$	d = diameter
Surface area of a cube or cuboid	$2(ab) + 2(ac) + 2(bc)$	a, b and c are side lengths
Surface area of a sphere	$4\pi r^2$	r is the radius
Surface area of a cylinder	$2\pi r^2 + (\pi d \times h)$	h is the length or height of a cylinder
Volume of a cube or cuboid	$a \times b \times c$	a, b and c are side lengths
Volume of a sphere	$\frac{4}{3}\pi r^3$	r is the radius
Volume of a cylinder	$\pi r^2 h$	r is the radius
magnification	$magnification = \dfrac{image\ size}{real\ size}$	Rearrange to find the other quantities
pH	$pH = -log_{10}[H^+]$	$[H^+]$ is the concentration of the hydrogen ion in moles per litre
Pulmonary ventilation rate	$PVR = tidal\ volume \times breathing\ rate$	
Cardiac output	$CO = stroke\ volume \times heart\ rate$	
Species diversity index	$D = \dfrac{N(N-1)}{\sum n(n-1)}$	N is the grand total of all species sampled n is the number of each individual species sampled
Efficiency of energy transfer	$\dfrac{energy\ transferred}{energy\ intake} \times 100\%$	
Standard deviation, s Used to assess spread or dispersion in a set of data	$s = \sqrt{\dfrac{\sum x^2 - \dfrac{(\sum x)^2}{n}}{n-1}}$	x refers to the values of the measurements taken n = the number of measurements $\sum$ means "the sum of"
An alternative sum for standard deviation	$s = \sqrt{\dfrac{\sum(x - \bar{x})^2}{n-1}}$	$\bar{x}$ is the mean value of the values of x
chi² (χ^2) test, used to compare agreement n = between sample and expectation	$x^2 = \sum \dfrac{(0 - E)^2}{E}$	O are the values you actually measure E are the values you expected to see
Student t test, used to compare the means of two sets of data	$t = \dfrac{\bar{x}_1 - \bar{x}_2}{\sqrt{\dfrac{s_1^2}{n_1} + \dfrac{s_2^2}{n_2}}}$	s^2 is the variance of a set of data Subscript denotes the data being compared, e.g. n_1 is the number of values of the first set s_2^2 is the variance of the second set $\bar{x}$ is the mean value of the values of x
Variance s² , which is also a measure of dispersion in a set of data	$s^2 = \dfrac{\sum(x - \bar{x})^2}{n-1}$	

Practical skills are at the heart of Biology. In the AS specification, the assessment of practical skills is found only in the written exams and there is no practical endorsement. 15% of the marks in the AS papers will relate to practical work.

By undertaking the set practical activities in this course, it will not only develop your manipulative skills with specific apparatus and techniques but will also help you to gain a deeper understanding into the processes of scientific investigations. Skills such as researching, planning, implementing by making and processing measurements safely, analysing and evaluating results will be reinforced and enhanced.

It is advantageous for you to answer practical questions when you have completed the practical – any exam questions on practical skills will have been written with the expectation that you will have carried out the practical activities. Having undertaken the practical, this helps with the teaching and learning of concepts in the specification. A richer practical experience will be gained if you do more practicals than the following six set practical activities in Table 1. For each activity, Table 1 references the relevant topic(s) in this book.

AS required practical activities

▼ **Table 1**

	Practical	Topic
1	Investigation into the effect of a named variable on the rate of an enzyme-controlled reaction	1.8 Factors affecting enzyme action
2	Preparation of prepared squashes of cells from plant root tips; set-up and use of an optical microscope to identify the stages of mitosis in these stained squashes and calculation of a mitotic index	3.1 Methods of studying cells 3.7 Mitosis
3	Production of a dilution series of a solute to produce a calibration curve with which to identify the water potential of plant tissue	4.3 Osmosis
4	Investigation into the effect of a named variable on the permeability of cell-surface membranes	4.1 Structure of the cell-surface membrane
5	Dissection of animal or plant gas exchange systems, a mass transport system or of an organ within such a system	6.2 Gas exchange in insects 6.3 Gas exchange in fish 6.4 Gas exchange in the leaf of a plant 6.6 Mammalian lungs
6	Use of aseptic technique to investigate the effect of antimicrobial substances on microbial growth	9 Genetic diversity and adaptation

Practical questions

The following questions are designed to give you some practice at this practical style of question. If you haven't completed the practical yet, just think of similar practicals you have done or when you have used the apparatus and this will help you.

Practical 1 – The effect of pH on catalases

A celery extract was liquidised and prepared by the technician as a source of the enzyme catalase. A burette had been filled up to the 50 cm³ mark with hydrogen peroxide. 10 cm³ of celery extract was added and the height of the upper level of the frothing liquid was recorded. The class was asked to repeat the procedure adding the following to the H_2O_2:

- Add 2 drops HCl / 2 drops distilled water
- Add 4 drops HCl
- Add 2 drops NaOH / 2 drops distilled water
- Add 4 drops NaOH

The pH of each solution was tested before starting the experiment.

1   (a)   Sketch a graph of your expected results. Remember to label your axes.
    (b)   List all variables that need to be controlled and how you would control them.

Describe how you could change the method to make it:

2   (a)   more reliable
    (b)   more valid

Practical 2 – The mitotic index

A student performed a root tip squash on tissue from a garlic root tip using acetic orcein stain. She counted the number of cells she could see in one of the stages of mitosis. In total there were 150 cells and 12 cells were undergoing mitosis.

1   (a)   Calculate the mitotic index for these cells. Show your working.   (*2 marks*)
    (b)   Another student didn't follow the exact procedure and as a result did not see any cells undergoing mitosis.
          Suggest two reasons why she did not see any cells in any stages of mitosis.   (*2 marks*)

In a further investigation to see the effect of cells environment on cell division, cells were taken from varying distances from the root tip and the number of cells undergoing mitosis was counted. To make the results quantitative, the student calculated the mitotic index for each sample and plotted it on graph 1.

Graph 1

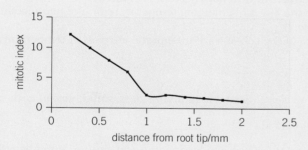

2   (a)   Using graph 1, describe the results.   (*2 marks*)
    (b)   Suggest one reason why this relationship exists?   (*1 mark*)

Practical 3 – Water potential in plant tissue

A practical was carried out to estimate the water potential in plant tissue. Six solid cylinders of potato were prepared each with identical dimensions. The students were given 100 cm³ of a stock 0.5 mol dm⁻³ solution sucrose. They were instructed to make up a series of six different dilutions. The mass (g) of the potato cylinder was measured before being submerged in the solution for 1 hour and then measured again after 1 hour.

Concentration of sucrose ($mol\ dm^{-3}$)	Mass before submerging in solution (g)	Mass after submerging in solution (g)	Percentage change in mass of potato tissue (%)
0.0	4.5	5.0	
0.1	3.9	4.3	
0.2	4.3	4.5	
0.3	4.1	4.2	
0.4	4.4	3.7	
0.5	4.4	3.6	

1   Construct a table to show how to make up 20 cm³ of each of the six dilutions required.

2   (a) Calculate the % change in mass for all results in the results table. Show your working.
     (b) Why is it important to calculate a % change of mass in this experiment?

3   (a) Plot a graph of sucrose concentration against % change in mass.
     (b) Use this graph to find the concentration of sucrose.
     (c) Describe how you can estimate the water potential with this value.

Practical 4 - Effect of temperature on beetroot cell-surface membranes

Core samples of beetroot were washed and put into tubes containing distilled water. Each tube was left in a different temperature for 20 minutes. The distilled water in the tube became coloured and was transferred to a colorimeter and a reading was taken.

Graph 2

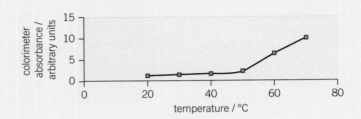

1   (a) Explain why the distilled water in the tubes becomes coloured?
     (b) Using the graph, describe the effect of changing temperature on the permeability of the cell-surface membrane.
     (c) Explain how the structure of the cell-surface membrane changes at temperatures above 50°C?

Practical 5 – Dissection

A live locust was being examined in class. Using a magnifying glass, tiny holes on each side of the segments were visible.

1   (a) Name these small holes?
     (b) Further observation showed these holes to be opening and closing. What benefit does this give the locust?
     (c) Some other insects have hairs around these holes.
          What environmental condition does this help them survive in and how does it aid their survival?

The teacher dissected a locust and exposed the inside of the body cavity. The teacher located the tracheal tubes and mounted a small sample of tissue onto a slide. The tracheal tubes were seen to be highly branched.

2   (a) Give one advantage of the tubes being highly branched.
     (b) What substances cause the ring thickening of the tracheae?
     (c) The tracheae branch into tubes of a much smaller diameter with little thickening. Suggest a reason for this structure.

Practical 6 – Aseptic technique

The technician set up a petri dish with nutrient agar jelly using equipment that had been in an autoclave. He inoculated it with non-pathogenic bacteria and left it to incubate for 48 hours.

1   (a) Why did he use equipment that had been in an autoclave?
     (b) Describe the steps he took to transfer the bacteria from the bottle to the petri dish. Only make reference to steps concerning keeping conditions aseptic.
     (c) Explain why he disinfected the work surface and washed his hands after the experiment was finished?

Additional practice questions

1 (a) Gas exchange in fish takes place in gills. Explain how two features of gills allow efficient gas exchange. *(2 marks)*

(b) A zoologist investigated the relationship between body mass and rate of oxygen uptake in four species of mammal. The results are shown in the graph.

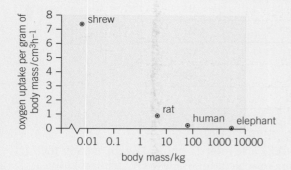

(i) The scale for plotting body mass is a logarithmic scale. Explain why a logarithmic scale was used to plot body mass. *(1 mark)*

(ii) Describe the relationship between body mass and oxygen uptake. *(1 mark)*

(iii) The zoologist measured oxygen uptake per gram of body mass. Explain why he measured oxygen uptake per gram of body mass. *(2 marks)*

(iv) Heat from respiration helps mammals to maintain a constant body temperature. Use this information to explain the relationship between body mass and oxygen uptake shown in the graph. *(3 marks)*

AQA, Jan 2010

2 (a) Describe how DNA is replicated. *(6 marks)*

(b) The graph shows information about the movement of chromatids in a cell that has just started metaphase of mitosis.

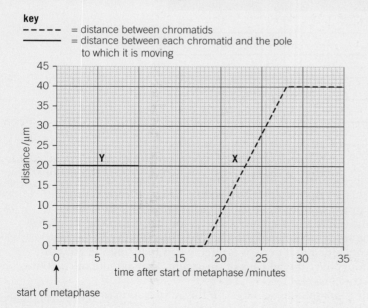

(i) What was the duration of metaphase in this cell? *(1 mark)*

(ii) Use line **X** to calculate the duration of anaphase in this cell. *(1 mark)*

(iii) Complete line **Y** on the graph. *(2 marks)*

(c) A doctor investigated the number of cells in different stages of the cell cycle in two tissue samples, **C** and **D**. One tissue sample was taken from a cancerous tumour. The other was taken from non-cancerous tissue. The table shows his results.

Stage of the cell cycle	Percentage of cells in each stage of the cell cycle	
	Tissue sample C	Tissue sample D
Interphase	82	45
Prophase	4	16
Metaphase	5	18
Anaphase	5	12
Telophase	4	9

   (i)   In tissue sample **C**, one cell cycle took 24 hours. Use the data in the table to calculate the time in which these cells were in interphase during one cell cycle. Show your working. *(2 marks)*

  (ii) Explain how the doctor could have recognised which cells were in interphase when looking at the tissue samples. *(1 mark)*

 (iii) Which tissue sample, **C** or **D**, was taken from a cancerous tumour? Use information in the table to explain your answer. *(2 marks)*

AQA Jan 2013

3   Taxol is a drug used to treat cancer. Research scientists investigated the effect of injecting taxol on the growth of tumours in mice. Some of the results are shown in **Figure 3**.

Number of days of treatment	Mean volume of tumour / mm³	
	Control group	Group injected with taxol in saline
1	1	1
10	7	2
20	21	11
30	43	20
40	114	48
50	372	87

▲ Figure 3

(a) Suggest how the scientists should have treated the control group. *(2 marks)*

(b) Suggest and explain **two** factors which should be considered when deciding the number of mice to be used in this investigation. *(2 marks)*

(c) The scientists measured the volume of the tumours. Explain the advantage of using volume rather than length to measure the growth of tumours. *(1 mark)*

(d) The scientists concluded that taxol was effective in reducing the growth rate of the tumours over the 50 days of treatment. Use suitable calculations to support this conclusion. *(2 marks)*

(e) In cells, taxol disrupts spindle activity. Use this information to explain the results in the group that has been treated with taxol. *(3 marks)*

(f) The research scientists then investigated the effect of a drug called OGF on the growth of tumours in mice. OGF and taxol were injected into different mice as separate treatments or as a combined treatment. **Figure 4** and **Figure 5** show the results from this second investigation.

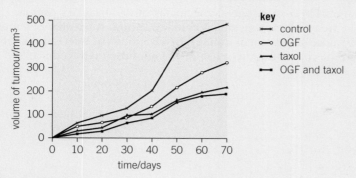

▲ Figure 4

Treatment	Mean volume of tumour following 70 days treatment / mm³ (± standard deviation)
OGF	322 (± 28.3)
Taxol	207 (± 22.5)
OGF and taxol	190 (± 25.7)
Control	488 (± 32.4)

▲ Figure 5

(i) What information does standard deviation give about the volume of the tumours in this investigation? *(1 mark)*

(ii) Use **Figure 4** and **Figure 5** to evaluate the effectiveness of the two drugs when they are used separately and as a combined treatment. *(4 marks)*

AQA Jan 2010

4 Read the following passage.

Gluten is a protein found in wheat. When gluten is digested in the small intestine, the products include peptides. Peptides are short chains of amino acids. These peptides cannot be absorbed by facilitated diffusion and leave the gut in faeces. Some people have coeliac disease. The epithelial cells of people with coeliac disease do not absorb the products of digestion very well. In these people, some of the peptides from gluten can pass between the epithelial cells lining the small intestine and enter the intestine wall. Here, the peptides cause an immune response that leads to the destruction of microvilli on the epithelial cells.

Scientists have identified a drug which might help people with coeliac disease. It reduces the movement of peptides between epithelial cells. They have carried out trials of the drug with patients with coeliac disease.

Use the information in the passage and your own knowledge to answer the following questions.

(a) Name the type of chemical reaction which produces amino acids from proteins. *(1 mark)*

(b) The peptides released when gluten is digested cannot be absorbed by facilitated diffusion (lines 2 – 3). Suggest why. *(3 marks)*

(c) The epithelial cells of people with coeliac disease do not absorb the products of digestion very well (lines 4 – 5). Explain why. *(3 marks)*

(d) Explain why the peptides cause an immune response (lines 7 – 8). *(1 mark)*

(e) Scientists have carried out trials of a drug to treat coeliac disease (lines 10 – 11). Suggest **two** factors that should be considered before the drug can be used on patients with the disease. *(2 marks)*

AQA June 2012

5 **(a)** Haemoglobin contains iron. One type of anaemia is caused by a lack of iron. This type of anaemia can be treated by taking tablets containing iron. A number of patients were given a daily dose of 120 mg of iron. **Figure 8** shows the effect of this treatment on the increase in the concentration of haemoglobin in their red blood cells.

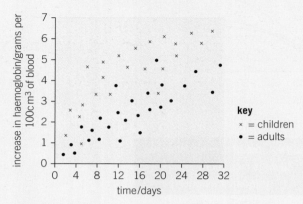

▲ **Figure 8**

(i) Give one difference in the response of adults and children to this treatment. *(1 mark)*

(ii) You could use the graph to predict the effect of this treatment on the increase in haemoglobin content of an adult after 40 days. Explain how. *(2 marks)*

(iii) Haemoglobin has a quaternary structure. Explain what is meant by a quaternary structure. *(1 mark)*

(b) (i) Pernicious anaemia is another type of anaemia. One method of identifying pernicious anaemia is to measure the diameter of the red blood cells in a sample of blood that has been diluted with an isotonic salt solution. Explain why an isotonic salt solution is used to dilute the blood sample. *(3 marks)*

(ii) A technician compared the red blood cells in two blood samples of equal volume.

One sample was from a patient with pernicious anaemia, the other was from a patient who did not have pernicious anaemia. **Figure 9** shows some of the results she obtained.

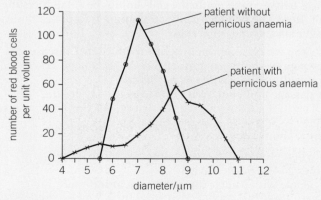

▲ **Figure 9**

Describe **two** differences between the blood samples. *(2 marks)*

(c) Scientists' analysis of blood proteins has indicated a lack of genetic diversity in populations of some organisms. Describe the processes that lead to a reduction in the genetic diversity of populations of organisms. *(6 marks)*

AQA June 2010

6 Students investigated the effect of different concentrations of sodium chloride solution on discs cut from an apple. They weighed each disc and then put one disc into each of a range of sodium chloride solutions of different concentrations. They left the discs in the solutions for 24 hours and then weighed them again. Their results are shown in the table.

Concentration of sodium chloride solution / mol dm^{-3}	Mass of disc at start / g	Mass of disc at end / g	Ratio of mass at start to mass at end
0.00	16.1	17.2	0.94
0.15	19.1	20.2	0.95
0.30	24.3	23.2	1.05
0.45	20.2	18.7	1.08
0.60	23.7	21.9	
0.75	14.9	13.7	1.09

(a) (i) Calculate the ratio of the mass at the start to the mass at the end for the disc placed in the 0.60 mol dm^{-3} sodium chloride solution. *(1 mark)*

(ii) The students gave their results as a ratio. What is the advantage of giving the results as a ratio? *(2 marks)*

(iii) The students were advised that they could improve the reliability of their results by taking additional readings at the same concentrations of sodium chloride. Explain how. *(2 marks)*

(b) (i) The students used a graph of their results to find the sodium chloride solution with the same water potential as the apple tissue. Describe how they did this. *(2 marks)*

(ii) The students were advised that they could improve their graph by taking additional readings. Explain how. *(2 marks)*

AQA Jan 2010

7 The diagram shows the structure of the cell-surface membrane of a cell.

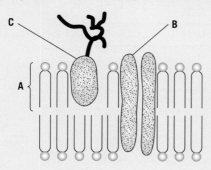

(a) Name **A** and **B**. *(2 marks)*

(b) (i) **C** is a protein with a carbohydrate attached to it. This carbohydrate is formed by joining monosaccharides together. Name the type of reaction that joins monosaccharides together. *(1 mark)*

(ii) Some cells lining the bronchi of the lungs secrete large amounts of mucus. Mucus contains protein.
Name **one** organelle that you would expect to find in large numbers in a mucus-secreting cell and describe its role in the production of mucus. *(2 marks)*

AQA June 2013

8 Students investigated the effect of removing leaves from a plant shoot on the rate of water uptake. Each student set up a potometer with a shoot that had eight leaves. All the shoots came from the same plant. The potometer they used is shown in the diagram.

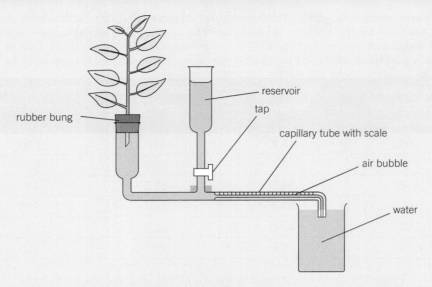

(a) Describe how the students would have returned the air bubble to the start of the capillary tube in this investigation. (*1 mark*)

(b) Give **two** precautions the students should have taken when setting up the potometer to obtain reliable measurements of water uptake by the plant shoot. (*2 marks*)

(c) A potometer measures the rate of water uptake rather than the rate of transpiration. Give two reasons why the potometer does not truly measure the rate of transpiration. (*2 marks*)

(d) The students' results are shown in the table.

Number of leaves removed from the plant shoot	Mean rate of water uptake / cm³ per minute
0	0.10
2	0.08
4	0.04
6	0.02
8	0.01

Explain the relationship between the number of leaves removed from the plant shoot and the mean rate of water uptake. (*3 marks*)

AQA Jan 2013

9 Read the following passage.

Some foods contain substances called flavenoids. Flavenoids lower blood cholesterol concentration and reduce the risk of developing coronary heart disease.

Some types of dark chocolate have a high concentration of flavenoids. One group of scientists investigated the effect of eating dark chocolate on the risk of developing coronary heart disease.

5

The scientists randomly divided healthy volunteers into two groups. Every day one group was given dark chocolate containing flavenoids to eat. The other group acted as a control.

The scientists measured the diameter of the lumen of the main artery in the arms of the volunteers every week. At the end of a month, the diameter of the lumen of the main artery in the arm of the volunteers who had eaten dark chocolate containing flavenoids had increased.

10

Use information from the passage and your own knowledge to answer the questions.

(a) High blood cholesterol concentration is a risk factor associated with coronary heart disease.
Give **two** other risk factors associated with coronary heart disease. *(2 marks)*

(b) (i) The scientists used healthy volunteers in this investigation (line 7).
Why was it important that the volunteers were healthy? *(1 mark)*

(ii) The scientists randomly divided the volunteers into two groups (line 7).
Explain why they divided them randomly. *(1 mark)*

(c) (i) Describe how the control group should have been treated. *(2 marks)*

(ii) Why was it important to have a control group in this investigation? *(1 mark)*

(d) Suggest why an increase in the diameter of the lumen of the main artery in the arm (lines 11–12) is associated with a reduced risk of coronary heart disease. *(3 marks)*

AQA June 2010

10 (a) What is intraspecific variation? *(1 mark)*

(b) Schizophrenia is a mental illness. Doctors investigated the relative effects of genetic and environmental factors on the development of schizophrenia. They used sets of identical twins and non-identical twins in their investigation. At least one twin in each set had developed schizophrenia.
 • Identical twins are genetically identical.
 • Non-identical twins are not genetically identical.
 • The members of each twin pair were raised together.

The table shows the percentage of cases where both twins had developed schizophrenia.

Type of twin	Percentage of cases where both twins had developed schizophrenia
Identical	50
Non-identical	15

(i) Explain why both types of twin were used in this investigation. *(2 marks)*

(ii) What do these data suggest about the relative effects of genetic and environmental factors on the development of schizophrenia? *(1 mark)*

(iii) Suggest two factors that the scientists should have taken into account when selecting the twins to be used in this study. *(2 marks)*

AQA Jan 2013

11 (a) Students measured the rate of transpiration of a plant growing in a pot under different environmental conditions. Their results are shown in the table.

Conditions		Transpiration rate/gh^{-1}
A Still air	15 °C	1.2
B Still air	15 °C	1.7
C Still air	25 °C	2.3

During transpiration, water diffuses from cells to the air surrounding a leaf.

(i) Suggest an explanation for the difference in transpiration rate between conditions **A** and **B**. *(2 marks)*

(ii) Suggest an explanation for the difference in transpiration rate between conditions **A** and **C**. *(2 marks)*

(b) Scientists investigated the rate of water movement through the xylem of a twig from a tree over 24 hours. The graph shows their results. It also shows the light intensity for the same period of time.

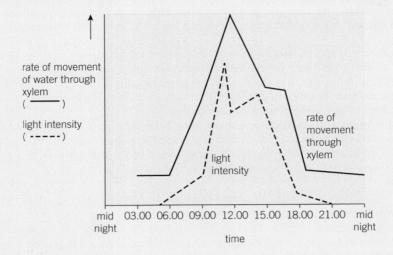

(i) Describe the relationship between the rate of water movement through the xylem and the light intensity. *(1 mark)*

(ii) Explain the change in the rate of water movement through the xylem between 06.00 and 12.00 hours. *(2 marks)*

(iii) The scientists also measured the diameter of the trunk of the tree on which the twig had been growing. The diameter was less at 12.00 than it was at 03.00 hours. Explain why the diameter was less at 12.00 hours. *(2 marks)*

AQA Jan 2011

12 Imatinib is a drug used to treat a type of cancer that affects white blood cells. Scientists investigated the rate of uptake of imatinib by white blood cells. They measured the rate of uptake at 4 °C and at 37 °C.

Their results are shown in the table.

Concentration of imatinib outside cells / mol dm^{-3}	Mean rate of uptake of imatinib into cells / g per million cells per hour	
	4 °C	37 °C
10.5	4.0	10.5
1.0	10.7	32.5
5.0	40.4	420.5
10.0	51.9	794.6
50.0	249.9	3156.1
100.0	606.9	3173.0

(a) The scientists measured the rate of uptake of imatinib in μg per million cells per hour. Explain the advantage of using this unit of rate in this investigation. *(2 marks)*

(b) Calculate the percentage increase in the mean rate of uptake of imatinib when the temperature is increased from 4 °C to 37 °C at a concentration of imatinib outside the cells of 1.0 μmol dm^{-3}. Give your answer to one decimal place. *(2 marks)*

(c) Imatinib is taken up by blood cells by active transport.

(i) Explain how the data for the two different temperatures support this statement. *(2 marks)*

(ii) Explain how the data for concentrations of imatinib outside the blood cells at 50 and 100 μmol dm^{-3} at 37 °C support the statement that imatinib is taken up by active transport. *(2 marks)*

AQA June 2013

Glossary

A

activation energy energy required to bring about a reaction. The activation energy is lowered by the presence of *enzymes*.

active immunity resistance to disease resulting from the activities of an individual's own immune system whereby an *antigen* induces plasma cells to produce *antibodies*.

active site a group of amino acids that makes up the region of an *enzyme* into which the *substrate* fits in order to catalyse a reaction.

active transport movement of a substance from a region where it is in a low concentration to a region where it is in a high concentration. The process requires the expenditure of *metabolic* energy.

aerobic connected with the presence of free oxygen. Aerobic respiration requires free oxygen to release energy from glucose. See also *anaerobic*.

allele one of a number of alternative forms of a *gene*. For example, the gene for the shape of pea seeds has two alleles: one for 'round' and one for 'wrinkled'.

allergen a normally harmless substance that causes the immune system to produce an immune response. See also *allergy*.

allergy the response of the immune system to an *allergen*. Examples include hay fever and *asthma*.

antibiotic a substance produced by living organisms that can destroy or inhibit the growth of microorganisms.

antibiotic resistance the development in microorganisms of mechanisms that prevent *antibiotics* from killing them.

antibody a protein produced by *lymphocytes* in response to the presence of the appropriate antigen.

anticodon a sequence of three adjacent *nucleotides* on a molecule of transfer RNA that is complementary to a particular *codon* on a messenger RNA molecule.

antigen a molecule that triggers an immune response by *lymphocytes*.

antioxidant chemical which reduces or prevents *oxidation*. Often used as an additive to prolong the shelf-life of certain foods.

apoplastic pathway route through the cell walls and intercellular spaces of plants by which water and dissolved substances are transported. See also *symplastic pathway*.

artificial selection breeding of organisms by human selection of parents/gametes in order to perpetuate certain characteristics and/or to eliminate others.

asthma a chronic illness in which there is resistance to air flow to the alveoli of the lungs as a result of the airways becoming inflamed due to an allergic response to an *allergen*.

atheroma fatty deposits in the walls of arteries, often associated with high *cholesterol* levels in the blood.

ATP (adenosine triphosphate) *nucleotide* found in all living organisms, which is produced during respiration and is important in the transfer of energy.

autosomes a chromosome which is not a sex chromosome.

B

B cell (B lymphocyte) type of white blood cell that is produced and matures within the bone marrow. B lymphocytes produce *antibodies* as part of their role in *immunity*. See also *T cell*.

Benedict's test a simple biochemical reaction to detect the presence of reducing sugars.

biodiversity the range and variety of genes, species and habitats within a particular region.

biomass the total mass of living material, normally measured in a specific area over a given period of time.

Biuret test a simple biochemical reaction to detect the presence of protein.

body mass index (BMI) a person's body mass in kilograms divided by the square of their height in metres.

C

cancer a disease, resulting from cells that break away from an original tumour to form secondary *tumours* elsewhere in the body.

carcinogen a chemical, a form of radiation, or other agent that causes *cancer*.

cardiac cycle a continuous series of events which make up a single heart beat.

cardiac output the total volume of blood that the heart can pump each minute. It is calculated as the volume of blood pumped at each beat (*stroke volume*) multiplied by the number of heart beats per minute (heart rate).

carrier molecule (carrier protein) a protein on the surface of a cell that helps to transport molecules and ions across plasma membranes.

Casparian strip a distinctive band of suberin around the endodermal cells of a plant root that prevents water passing into xylem via the cell walls. The water is forced through the living part (*protoplast*) of the endodermal cells.

centrifugation process of separating out particles of different sizes and densities by spinning them at high speed in a centrifuge.

cholesterol lipid that is an important component of cell-surface membranes. Excess in the blood can lead to *atheroma*.

chromatid one of the two strands of a *chromosome* that are joined together by a single centromere prior to cell division.

chromatin the material that makes up *chromosomes*. It consists of DNA and the protein histone.

chromosome a thread-like structure made of protein and DNA by which hereditary information is physically passed from one generation to the next.

clone a group of genetically identical cells or organisms formed from a single parent as the result of asexual reproduction or by artificial means.

codon a sequence of three adjacent *nucleotides* in mRNA that codes for one amino acid.

cohesion attraction between molecules of the same type. It is important in the movement of water up a plant.

collagen fibrous protein that is the main constituent of connective tissues such as tendons, cartilage and bone.

community all the living organisms present in an *ecosystem* at a given time.

complementary DNA DNA that is made from messenger RNA in a process that is the reverse of normal transcription.

condensation chemical process in which two molecules combine to form a more complex one with the elimination of a simple substance, usually water. Many biological *polymers*, such as polysaccharides and polypeptides, are formed by condensation. See also *hydrolysis*.

continuous variation variation in which organisms do not fall into distinct categories but show gradations from one extreme to the other.

coronary arteries arteries that supply blood to the cardiac muscle of the heart.

coronary heart disease (CHD) any condition, for example, *atheroma* and *thrombosis*, affecting the *coronary arteries* that supply heart muscle.

correlation when a change in one variable is reflected by a change in the second variable.

co-transport the transport of one substance coupled with the transport of another substance across a plasma membrane in the same direction through the same protein carrier.

countercurrent system a mechanism by which the efficiency of exchange between two substances is increased by having them flowing in opposite directions.

covalent bond type of chemical bond in which two atoms share a pair of *electrons*, one from each atom.

crossing over the process whereby a *chromatid* breaks during *meiosis* and rejoins to the chromatid of its *homologous chromosome* so that their *alleles* are exchanged.

D

denaturation permanent changes due to the unravelling of the three-dimensional structure of a protein as a result of factors such as changes in temperature or pH.

diastole the stage in the *cardiac cycle* when the heart muscle relaxes. See also *systole*.

differentiation the process by which cells become specialised for different functions.

diffusion the movement of molecules or ions from a region where they are in high concentration to one where their concentration is lower.

diploid a term applied to cells in which the nucleus contains two sets of *chromosomes*. See also *haploid*.

dipolar having a pair of equal and opposite electrical charges.

E

ecological niche describes how an organism fits into its environment. It describes what a species is like, where it occurs, how it behaves, its interactions with other species and how it responds to its environment.

ecosystem all the living and nonliving components of a particular area.

electron negatively charged subatomic particle that orbits the positively charged nucleus of all atoms.

emphysema a disease in which the walls of the alveoli break down, reducing the surface area for gaseous exchange, thereby causing breathlessness in the patient.

endocytosis the inward transport of large molecules through the cell-surface membrane.

enzyme a protein or RNA that acts as a catalyst and so alters the speed of a biochemical reaction.

epidemiology the study of the spread of disease and the factors that affect this spread.

eukaryotic cell a cell that has a membrane-bound nucleus and *chromosomes*. The cell also possesses a variety of other membranous organelles, such as mitochondria and endoplasmic reticulum. See also *prokaryotic cell*.

exocytosis the outward bulk transport of materials through the cell-surface membrane.

F

facilitated diffusion diffusion involving the presence of protein *carrier molecules* to allow the passive movement of substances across plasma membranes.

G

gamete reproductive (sex) cell that fuses with another gamete during fertilisation.

gene section of DNA on a *chromosome* coding for one or more polypeptides.

gene pool the total number of *alleles* in a particular population at a specific time.

glycolysis first part of cellular respiration in which glucose is broken down anaerobically in the cytoplasm to two molecules of pyruvate.

H

habitat the place where an organism normally lives and which is characterised by physical conditions and the types of other organisms present.

haemoglobin globular protein in blood that readily combines with oxygen to transport it around the body. It comprises four polypeptide chains around an iron-containing haem group.

haploid term referring to cells that contain only a single copy of each *chromosome*, e.g. the sex cells (*gametes*).

homologous chromosomes a pair of *chromosomes*, one maternal and one paternal, that have the same gene *loci* and therefore determine the same features. They are not necessarily identical, however, as individual *alleles* of the same *gene* may vary, e.g. one chromosome may carry the allele for blue eyes, the other the allele for brown eyes. Homologous chromosomes are capable of pairing during *meiosis*.

human genome the totality of the DNA sequences on the *chromosomes* of a single human cell.

hydrogen bond chemical bond formed between the positive charge on a hydrogen atom and the negative charge on another atom of an adjacent molecule, e.g. between the hydrogen atom of one water molecule and the oxygen atom of an adjacent water molecule.

hydrolysis the breaking down of large molecules into smaller ones by the addition of water molecules. See also *condensation*.

I

immunity the means by which the body protects itself from infection.

intercropping the practice of growing two or more crops in close proximity usually to produce a greater yield on a piece of land.

interspecific variation differences between organisms of different species.

intraspecific variation differences between organisms of the same species.

ion an atom or group of atoms that has lost or gained one or more *electrons*. Ions therefore have either a positive or negative charge.

ion channel a passage across a cell-surface membrane made up of a protein that spans the membrane and opens and closes to allow *ions* to pass in and out of the cell.

isotope variations of a chemical element that have the same number of protons and *electrons* but different numbers of neutrons. While their chemical properties are similar they differ in mass. One example is carbon which has a relative atomic mass of 12 and an isotope with a relative atomic mass of 14.

K

kinetic energy energy that an object possesses due to its motion.

L

latent heat of vaporisation heat taken in by a liquid in order to transform it into a vapour.

locus the position of a gene on a *chromosome*/DNA molecule.

lumen the hollow cavity inside a tubular structure such as the gut or a *xylem vessel*.

lymph a slightly milky fluid found in lymph vessels and made up of *tissue fluid*, fats and *lymphocytes*.

lymphocytes types of white blood cell responsible for the immune response. They become activated in the presence of *antigens*. There are two types: *B lymphocytes* and *T lymphocytes*.

M

meiosis the type of nuclear division in which the number of *chromosomes* is halved. See also *mitosis*.

mesophyll tissue found between the two layers of epidermis in a plant leaf comprising an upper layer of *palisade cells* and a lower layer of spongy cells.

metabolism all the chemical processes that take place in living organisms.

microvilli tiny finger-like projections from the cell-surface membrane of some animal cells.

middle lamella layer made up of pectins and other substances found between the walls of adjacent plant cells.

mitosis the type of nuclear division in which the daughter cells have the same number of *chromosomes* as the parent cell. See also *meiosis*.

monoclonal antibody an antibody produced by a single clone of *cells*.

monomer one of many small molecules that combine to form a larger one known as a *polymer*.

mono-unsaturated fatty acid fatty acid that possesses a carbon chain with a single double bond. See also *polyunsaturated fatty acid*.

mutation a sudden change in the amount or the arrangement of the genetic material in the cell.

myocardial infarction otherwise known as a heart attack, results from the interruption of the blood supply to the heart muscle, causing damage to an area of the heart with consequent disruption to its function.

N

nitrogen fixation the incorporation of atmospheric nitrogen gas into organic nitrogen-containing compounds.

normal distribution a bell-shaped curve produced when a certain distribution is plotted on a graph.

nucleotides complex chemicals made up of an organic base, a sugar and a phosphate. They are the basic units of which the nucleic acids DNA and RNA are made.

O

oral rehydration solution (ORS) means of treating dehydration involving giving, by mouth, a balanced solution of salts and glucose that stimulates the gut to reabsorb water.

osmosis the passage of water from a region of high *water potential* to a region where its *water potential* is lower, through a selectively permeable membrane.

oxidation chemical reaction involving the loss of *electrons*.

P

palisade cells long, narrow cells, packed with chloroplasts, that are found in the upper region of a leaf and which carry out photosynthesis.

passive immunity resistance to disease that is acquired from the introduction of *antibodies* from another individual, rather than an individual's own immune system, e.g. across the placenta or in the mother's milk. It is usually short-lived.

pathogen any microorganism that causes disease.

pentose sugar a sugar that possesses five carbon atoms. Two examples are ribose and deoxyribose.

peptide bond the chemical bond formed between two amino acids during *condensation*.

phagocytosis mechanism by which cells engulf particles to form a vesicle or a vacuole.

phospholipid triglyceride in which one of the three fatty acid molecules is replaced by a phosphate molecule. Phospholipids are important in the structure and functioning of plasma membranes.

photomicrograph photograph of an image produced by a microscope.

plasmid a small circular piece of DNA found in bacterial cells.

plasmodesmata fine strands of cytoplasm that extend through pores in adjacent plant cell walls and connect the cytoplasm of one cell with another.

plasmolysis the shrinkage of cytoplasm away from the cell wall that occurs as a plant cell loses water by *osmosis*.

polymer large molecule made up of repeating smaller molecules (*monomers*).

polymerases group of enzymes that catalyse the formation of long-chain molecules (*polymers*) from similar basic units (*monomers*).

polyunsaturated fatty acid (PUFA) fatty acid that possesses carbon chains with many double bonds. See also *mono-unsaturated fatty acid*.

primary structure of a protein the sequence of amino acids that makes up the polypeptides of a protein.

prokaryotic cell a cell of an organism belonging to the kingdom Prokaryotae that is characterised by lacking a nucleus and membrane-bound organelles. Examples include bacteria. See also *eukaryotic cell*.

protoplast the living portion of a plant cell, that is, the nucleus and cytoplasm along with the organelles it contains.

Q

quaternary structure of a protein a number of polypeptide chains linked together, and sometimes associated with non-protein groups, to form a protein.

R

reduction chemical process involving the gain of *electrons*.

S

saturated fatty acid a fatty acid in which there are no double bonds between the carbon atoms.

secondary structure of a protein the way in which the chain of amino acids of the polypeptides of a protein is folded.

selective breeding see *artificial selection*.

semi-conservative replication the means by which DNA makes exact copies of itself by unwinding the double helix so that each chain acts as a template for the next. The new copies therefore possess one original and one new strand of DNA.

serum clear liquid that is left after blood has clotted and the clot has been removed. It is therefore blood plasma without the clotting factors.

sinoatrial node (SAN) an area of heart muscle in the right atrium that controls and coordinates the contraction of the heart. Also known as the pacemaker.

species a group of similar organisms that can breed together to produce fertile offspring.

stoma (plural stomata) apore, mostly found in the lower epidermis of a leaf, through which gases diffuse in and out of the leaf.

stroke volume the volume of blood pumped at each ventricular contraction of the heart.

substrate a substance that is acted on or used by another substance or process. In microbiology, the nutrient medium used to grow microorganisms.

supernatant liquid the liquid portion of a mixture left at the top of the tube when suspended particles have been separated out at the bottom during *centrifugation*.

symplastic pathway route through the cytoplasm and *plasmodesmata* of plant cells by which water and dissolved substances are transported. See also *apoplastic pathway*.

systole the stage in the *cardiac cycle* in which the heart muscle contracts. It occurs in two stages: atrial systole when the atria contract and ventricular systole when the ventricles contact. See also *diastole*.

T

tertiary structure of a protein the folding of a whole polypeptide chain in a precise way, as determined by the amino acids of which it is composed.

thrombosis formation of a blood clot within a blood vessel that may lead to a blockage.

tidal volume the volume of air breathed in and out during a single breath when at rest.

tissue a group of similar cells organised into a structural unit that serves a particular function.

tissue fluid fluid that surrounds the cells of the body. Its composition is similar to that of blood plasma except that it lacks proteins. It supplies nutrients to the cells and removes waste products.

T cell (T lymphocyte) type of white blood cell that is produced in the bone marrow but matures in the thymus gland. T lymphocytes coordinate the immune response and kill infected cells. See also *B cell*.

transpiration evaporation of water from a plant.

triglyceride an individual lipid molecule made up of a glycerol molecule and three fatty acids.

tumour a swelling in an organism that is made up of cells that continue to divide in an abnormal way.

turgid a plant cell that contains the maximum volume of water it can. Additional entry of water is prevented by the cell wall stopping further expansion of the cell.

U

ultrafiltration filtration assisted by blood pressure, e.g. in the formation of *tissue fluid*.

unsaturated fatty acid a fatty acid in which there are one or more double bonds between the carbon atoms.

V

vaccination the introduction of a vaccine containing appropriate disease *antigens* into the body, by injection or mouth, in order to induce artificial *immunity*.

W

water potential the pressure created by water molecules. It is the measure of the extent to which a solution gives out water. The greater the number of water molecules present, the higher (less negative) the water potential. Pure water has a water potential of zero.

X

xerophyte a plant adapted to living in dry conditions.

xylem vessels dead, hollow, elongated tubes, with lignified side walls and no end walls, that transport water in most plants.

Answers to questions

Atoms, isotopes, and the formation of ions

1.1

1. a hydrogen

   b isotope

   c 100% (it is doubled as neutrons have the same mass as protons)

   d It is unchanged (the atomic number is the number of protons and this remains as one).

2. a hydrogen ion

   b It is not changed (mass number is the number of protons and neutrons)

1.2

1. Carbon atoms readily link to one another to form a chain.

2. polymer

3. Sugar donates electrons that reduce blue copper(II) sulfate to orange copper(I) oxide.

Semi-quantitative nature of Benedict's test

1. B, E, A, D, C

2. Dry the precipitate in each sample and weigh it. The heavier the precipitate the more reducing sugar is present.

3. Once all the copper(II) sulfate has been reduced to copper(I) oxide, further amounts of reducing sugar cannot make a difference.

1.3

1. a glucose + galactose; b glucose + fructose; c α glucose only

2. $C_{12}H_{22}O_{11}$ ($C_6H_{12}O_6$ + $C_6H_{12}O_6$ − H_2O)

3. Enzymes are denatured at higher temperatures and this prevents them functioning / enzymes lower the activation energy required.

1.4

1. starch

2. glycogen

3. α-glucose, β-glucose, starch, cellulose

4. starch, cellulose, glycogen

5. α-glucose

6. cellulose

7. starch, cellulose, glycogen

8. α-glucose, β-glucose

1.5

1. a triglycerides; b glycerol; c polyunsaturated; d two; e hydrophobic

2. triglyceride: 3 fatty acids / no phosphate group / nonpolar; phospholipid: 2 fatty acids / 1 phosphate group / hydrophilic 'head' and hydrophobic 'tail

3. Lipids provide more than twice as much energy as carbohydrate when they are oxidised. If fat is stored, the same amount of energy can be provided for less than half the mass. It is therefore a lighter storage product – a major advantage if the organism is motile.

1.6

1. peptide bond

2. condensation reaction

3. amino group (—NH_2), carboxyl group (—COOH), hydrogen atom (—H), R group

Protein shape and function

1. It has three polypeptide chains wound together to form a strong, rope-like structure that has strength in the direction of pull of a tendon.

2. prevents the individual polypeptide chains from sliding past one another and so they gain strength because they act as a single unit

3. The junctions between adjacent collagen molecules are points of weakness. If they all occurred at the same point in a fibre, this would be a major weak point at which the fibre might break.

1.7

1. a substance that alters the rate of a chemical reaction without undergoing permanent change

2. They are not used up in the reaction and so can be used repeatedly.

3. The changed amino acid may no longer bind to the substrate, which will then not be positioned correctly, if at all, in the active site.

4. The changed amino acid may be one that forms hydrogen bonds with other amino acids. If the new amino acid does not form hydrogen bonds the tertiary structure of the enzyme will change, including the active site, so that the substrate may no longer fit.

Lock and key model of enzyme action

1. It more clearly matches current observations such as enzyme activity being changed when molecules bind at sites other than the active site. This suggests

that enzyme molecules change shape when other molecules bind to them.

1.8

1 To function, enzymes must physically collide with their substrate. Lower temperatures decrease the kinetic energy of both enzyme and substrate molecules, which then move around less quickly. They hence collide less often and therefore react less frequently.

2 The heat causes hydrogen and other bonds in the enzyme molecule to break. The tertiary structure of the enzyme molecule changes, as does the active site. The substrate no longer fits the active site.

3 a High temperatures denature the enzymes and so they cannot spoil the food;

   b Vinegar is very acidic and the very low pH will denature the enzymes and so preserve the food

4 pH = 4

Enzyme action

1 a Enzyme X is not very specific as it acts on a number of different proteins.

      Enzyme Y is highly specific as it acts on a single protein.

   b Enzyme X could be used in biological washing powders to digest/remove stains from clothes.

      Enzyme Y could be used in making yoghurt/cheese from milk.

   c Milk is the only food in the diet of young mammals. The enzyme coagulates the milk causing it to remain in the stomach for longer. This gives time for enzymes there to act on it so that it can be broken down into products which can then be absorbed. If it had remained liquid, it would pass through the stomach more quickly and only be partially digested.

   d Enzyme X functions at much higher temperatures than enzyme Y and so must have a much more stable tertiary structure to prevent it becoming denatured. The bonds holding the polypeptide chain in its precise 3-D arrangement that makes up the active site must therefore be less easily broken than those of enzyme Y. It is therefore likely that the bonds of enzyme X are mostly, or entirely, disulfide bonds as these are not easily broken by heat. Enzyme Y is likely to have more of the heat-sensitive ionic and hydrogen bonds.

2 a

Temperature / °C	Time / min for hydrolysis of protein	Rate of reaction / $\frac{1}{time}$
15	5.8	0.17
25	3.4	0.29
35	1.7	0.59
45	0.7	1.43
55	0.6	1.67
65	0.9	1.11
75	7.1	0.14

b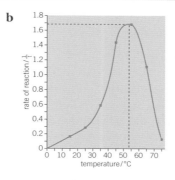

c The optimum temperature is found by dropping a vertical line from the highest point on the curve and reading the temperature where it transects the temperature (x) axis. The value is in the range 50–55 °C.

d Carry out the experiment in exactly the same way but use narrower temperature intervals (e.g. 1 °C) over the range 45–60 °C.

1.9

1 Competitive inhibitors occupy the active site of an enzyme while non-competitive inhibitors attach to the enzyme at a site other than the active site.

2 Increase the substrate concentration. If the degree of inhibition is reduced, it is a competitive inhibitor; if it stays the same, it is a non-competitive inhibitor.

Control of metabolic pathways

1 pH / substrate concentration (not temperature)

2 In a metabolic pathway, the product of one reaction acts as the substrate for the next reaction. By having the enzymes in appropriate sequence there is a greater chance of each enzyme coming into contact with its substrate than if the enzymes are floating freely in the organelle. This is a more efficient means of producing the end product.

3 a it would increase;   b it would be unchanged

4 Advantage – the level of the end product does not fluctuate with changes in the level of substrate.

Explanation – Non-competitive inhibition occurs at a site on the enzyme other than the active site. Hence it is not affected by the substrate concentration. Therefore, in non-competitive inhibition, changes in the level of substrate do not affect the inhibition of the enzyme, nor the normal level of the end product.

Competitive inhibition involves competition for active sites. In this case the end product needs to compete with the substrate for the active sites of enzyme A. A change in the level of substrate would therefore affect how many end product molecules combine with the active sites. As a result the degree of inhibition would fluctuate and so would the level of the end product.

2.1

1 pentose(sugar), phosphate group, organic base

2 The bases are linked by hydrogen bonds. The molecular structures could be such that hydrogen bonds do not form between adenine and cytosine and between guanine and thymine.

3 ACCTCTGA

4 30.1%. If 19.9% is guanine then, as guanine always pairs with cytosine, it also makes up 19.9% of the bases in DNA, so together they make up 39.8%. This means the remaining 60.2% of DNA must be adenine and thymine and, as these also pair, each must make up half of this, i.e. 30.1%.

Unravelling the role of DNA

1 Alternative theories can be explored and investigated. As a result, new facts may emerge and so a new theory is put forward or the existing one is modified. In this way, scientific progress can be made.

2 A suggested explanation of something based on some logical scientific reasoning or idea.

3 The harmful bacteria in the sample could be tested to ensure they were dead, e.g. by seeing if they multiply when grown in ideal conditions. Dead bacteria cannot multiply.

4 The probability of the mutation happening once is very small. The probability of the same mutation occurring each time the experiment is repeated is so minute that it can be discounted.

5 Society will probably be affected by new discoveries and so is entitled to say how they can or cannot be used.

A prime location

1 It means 'in life'. In other words the synthesis of DNA by a living organism rather than in a laboratory.

2 Enzymes are very specific. They have active sites that are of a specific shape that fits their substrate. The shape of the 3' end of the molecule with its hydroxyl group fits the active site of DNA polymerase whereas the shape of the 5' end does not.

2.2

1 TACGATGC

2 because half the original DNA is built into the new DNA strand

3 The linking together of the new nucleotides could not take place. While the nucleotides would match up to their complementary nucleotides on the original DNA strand, they would not join together to form a new strand.

Evidence for semi-conservative replication

1 the organic bases (adenine, guanine, cytosine and thymine)

2 Each DNA molecule is made up of one strand containing ^{15}N (the original strand) and one strand containing ^{14}N (the new strand). In other words, replication is semi-conservative.

3 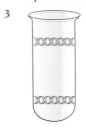   4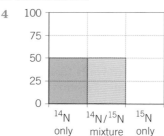

5 75%

2.3

1 ATP releases its energy very rapidly. This energy is released in a single step and is transferred directly to the reaction requiring it. It is too unstable to be a long term store.

2 ATP provides a phosphate that can attach to another molecule, making it more reactive and so lowering its activation energy. As enzymes work by lowering activation energy they have less 'work' to do and so function more readily.

3 Any 3 from: building up macromolecules (or named example of macromolecule) / active transport / secretions (formation of lysosomes) / activation of molecules.

2.4

a dipolar

b electrons

c hydrogen bonds

d surface tension

e hydrolysis

f photosynthesis

3.1

1 Magnification is how many times bigger the image is compared to the real object. Resolution is the minimum distance apart that two objects can be in order for them to appear as separate items.

2 200 times

3 10 mm

4 500 nm (0.5 μm)

5 Keep the plant cells in a cold, buffered solution with the same water potential as the cells. Break up the cells using a mortar and pestle / homogeniser. Filter the homogenate to remove cell debris. Centrifuge the homogenate at 1000 times gravity and remove the supernatant liquid (leaving nuclei behind in the sediment). Then centrifuge the supernatant liquid at 2000–3000 times gravity. The sediment produced will be rich in chloroplasts.

6 a 1.5 μm; b 19 nm

3.2

1 The EM uses a beam of electrons that has a much smaller wavelength than light.

2 Electrons are absorbed by the molecules in air and, if present, this would prevent the electrons reaching the specimen.

3 a plant cell and bacteria; b all of them; c plant cell, bacterium and virus

4 The preparation of the specimens may not be good enough. A higher energy electron beam is required and this may destroy the specimen

5 The organelle measures 25 mm (= 25 000 μm) long and represents 5 μm. Magnification is therefore 25 000 μm ÷ 5 μm = 5000 times.

3.3

1	light	2	eyepiece
3	stage	4	calibrated
5	2	6	8
7	20	8	3000

3.4

1 protein synthesis

2 glucose, fructose, galactose

3 a mitochondrion; b nucleus;
   c Golgi apparatus; d lysosome

4 a mitochondria, nucleus; b Golgi apparatus, lysosomes; c rough endoplasmic reticulum / ribosomes, mitochondria / smooth endoplasmic reticulum

3.5

1 a collection of similar cells aggregated together to perform a specific function.

2 An artery is made up of more than one tissue (epithelial muscle connective), whereas a blood capillary is made up of only one tissue (epithelial).

3 a organ b tissue c organ d tissue

3.6

1 A = absent; B = present; C = present;
   D = sometimes; E = sometimes; F = sometimes; G = present; H = present; I = sometimes;
   J = absent; K = present; L = present; M = absent;
   N = present

2 6 μm = 6000 nm; $\frac{6000\,nm}{150\,nm}$ = 40 times

3.7

1 a interphase; b prophase; c spindle; d nuclear envelope; e nucleolus; f metaphase; g anaphase;

Importance of mitosis

1 **advantage** – as the genetic make up of the parent has enabled it to survive and reproduce, if the offspring have the same genetic make up, they are also likely to survive and reproduce.

   disadvantage – genetic variety is limited – if environmental conditions change the species may not have individuals with the necessary genes to survive in the new conditions. It could fail to adapt and become extinct.

Recognising the stages of mitosis

1 A = telophase – chromosomes in two sets, one at each pole; B = prophase – chromosomes visible but randomly arranged; C = interphase – no chromosomes visible; D = metaphase – chromosomes lined up on equator; E = anaphase – chromatids in two sets, each being drawn towards pole

2 24 minutes. Number of cells in metaphase ÷ total number of cells observed × time for one cycle (in minutes), i.e. 20 ÷ 1000 × 1200 = 24

3 11% chromosomes visible in prophase, metaphase, anaphase and telophase $(73 + 20 + 9 + 8) \div 1000 \times 100$

3.8

1 interphase, nuclear division and cell division

2 **a** 12 hours and 24 hours;
   b 6–9 hours and 18–21 hours

Cancer and its treatment

1 0.2 million / 200 000

2 50%

3 8.33 times $(0.5 \div 0.06)$

4 More cancer cells are killed because they divide more rapidly than healthy cells and so are more susceptible to the drug.

5 **a** If the frequency was increased the healthy body cells would not have time to increase their numbers to near normal again between treatments. Their numbers would decline rapidly after a few treatments and possibly kill the patient.

   b The increased dose would kill even more healthy cells each time and again their numbers would decline rapidly after a few treatments and possibly kill the patient.

4.1

1 to control the movement of substances in and out of the cell

2 hydrophobic tail

3 **a** phospholipid;   **b** protein (carrier or channel)

4 Any 2 from: lipid-soluble / small in size / have no electrical charge (or if it does, the charge should be opposite to that on the protein channels).

4.2

1 Any three from: concentration gradient / area over which diffusion takes place / thickness of exchange surface / temperature.

2 Facilitated diffusion only occurs at channels on the membrane where there are special protein carrier molecules.

3 There is no ATP from respiration used in the process. The only energy used is the in-built (kinetic) energy of the molecules themselves.

4 Only lipid-soluble substances diffuse across the phospholipid bilayer easily. Water-soluble substances like glucose diffuse only very slowly.

5 It could increase its surface area with microvilli and it could have more proteins with pores that span the phospholipid bilayer. (Note: the thickness of

the cell-surface membranes does not vary to any degree).

6 **a** increases two times / doubles;

   b no change;

   c decreases four times / it is one quarter;

   d increase two times / doubles (The CO_2 concentration is irrelevant).

4.3

1 a membrane that is permeable to water molecules (and a few other small molecules) but not to larger molecules

2 zero

3 C, D, A, B

Osmosis and plant cells

1 Both cells have a lower water potential than pure water and so water enters them by osmosis. The animal cell is surrounded only by a thin cell-surface membrane and so it swells until it bursts. The plant cell is surrounded by a rigid cellulose cell wall. Assuming the cell is turgid, water cannot enter as the cellulose cell wall prevents the cell expanding and hence it bursting.

2 A = turgid, B = incipient plasmolysis, C = plasmolysed, D = turgid

3 solutions A, B and D (all except C)

4.4

1 Similarity – both use carrier proteins in the plasma membrane. Difference – active transport requires energy (ATP) / occurs against a concentration gradient.

2 Active transport requires energy in the form of ATP. Mitochondria supply ATP in cells and therefore they are numerous in cells carrying out active transport.

3 Diffusion, at best, can only reabsorb 50% of the glucose lost from the blood. The other 50% will be lost from the body. Active transport can absorb all the glucose, leaving none to be lost from the body.

4.5

1 by increasing the concentration gradient either side of it / by increasing the surface area / by increasing the density of protein channels (carrier proteins)

2 because glucose molecules and sodium ions move into the cells coupled together

3 **a** active;   **b** passive;   **c** passive

Oral rehydration therapy

1 Glucose stimulates the uptake of sodium ions from the intestine and provides energy as it is a respiratory substrate.

2 The sodium ions replace those lost from the body and encourage the use of the sodium-glucose transporter proteins to absorb more sodium ions.

3 Boiling the water will kill any diarrhoeal pathogens that would otherwise make the patient's condition worse.

4 Potassium in the banana replaces the potassium ions that have been lost. It also stimulates the appetite and so aids recovery.

5 Banana improves the taste and so makes it easier for children to drink the mixture.

6 Too much glucose might lower the water potential within the intestine to a level below that within the epithelial cells. Water will then pass out of the cells by osmosis, increasing dehydration.

7 Starch is a large insoluble molecule that has no osmotic effects.

8 Partially digest the starch with amylase. The smaller and more soluble molecules that result produce a less viscous drink.

9 Each species has different physical and chemical features and therefore may respond differently to the same drug. What is safe for some other animal may be harmful to a human.

10 It acts like a control experiment. Changes in the patients taking the real drug can be compared with patients taking the placebo to see whether they are due to the drug or to some other factor.

11 There is no risk of any deliberate or unwitting bias by the patients. Those knowing they are on the real drug might wrongly attribute changes in their symptoms to the drug.

5.1

1 A specific mechanism distinguishes between different pathogens but responds more slowly than a non-specific mechanism. A non-specific mechanism treats all pathogens in the same way but responds more rapidly than a non-specific mechanism.

2 The lymphocytes that will finally control the pathogen need to build up their numbers and this takes time.

3 The body responds immediately by 'recognising' the pathogen (and by phagocytosis); the delay is in building up numbers of lymphocytes and therefore controlling the pathogen.

5.2

1 a phagocytosis **b** phagosome **c** lysozyme **d** lysosome

2 The protective covering of the eye, and especially the tear ducts, are potential entry points for pathogens. The eyes are vulnerable to infection

because the coverings are thin to allow light through. Lysozyme will break down the cell walls of any bacterial pathogens and so destroy them before they can cause harm.

5.3

1 An organism or substance, usually a protein, that is recognised as foreign by the immune system and therefore stimulates an immune response.

2 Any 2 from: both are types of white blood cell / have a role in immunity / are produced from stem cells

3 T cells mature in the thymus gland while B cells mature in the bone marrow; T cells are involved in cell-mediated immunity while B cells are involved in humoral immunity.

Bird flu

1 H5N1 infects the lungs, leading to a massive production of T cells. Accumulation of these cells may block the airways / fill the alveoli and cause suffocation.

2 Birds carry H5N1 virus. They can fly vast distances across the world in a very short space of time.

5.4

1 In the primary response, the antigens of the pathogen have to be ingested, processed and presented by B cells. Helper T cells need to link with the B cells that then clone, some of the cells developing into the plasma cells that produce antibodies. These processes occur consecutively and therefore take time. In the secondary response, memory cells are already present and the only processes are cloning and development into plasma cells that produce antibodies. Fewer processes means a quicker response.

2 Examples of differences include:

Cell-mediated immunity	Humoral immunity
Involves T cells	Involves mostly B cells
No antibodies	Antibodies produced
First stage of immune response	Second stage of immune response after cell-mediated stage
Effective through cells	Effective through body fluids

3 rough endoplasmic reticulum – to make and transport the proteins of the antibodies; Golgi apparatus – to sort, process and compile the proteins; mitochondria – to release the energy needed for such massive antibody production

5.5

1 There must be a massive variety of antibodies as each responds to a different antigen, of which there are millions. Only proteins have the diversity of molecular structure to produce millions of different types.

2 An antigen is a molecule that triggers an immune response by lymphocytes while an antibody is the molecule that has a complementary shape to the antigen and is produced in response to it.

3 Any accurate response that includes an argument in favour (e.g. removes the risk of healthy volunteers being harmed / terminally ill patients have most to gain and less to lose) and an argument against (e.g. response of terminally ill might be different from those in the early stages of the disease and results therefore could be unreliable / sample size likely to be smaller / not typical).

Producing monoclonal antibodies

1 Detergents affect the lipid component of membranes causing them to develop 'holes'. When the detergent is washed out, the membranes reform, sometimes in combination with those of other cells that are adjacent.

2 to ensure the B cells and tumour cells repeatedly come into physical contact – essential if they are to fuse

3 because B cells are short-lived and do not divide outside of the body. Tumour cells are long-lived and divide outside the body. Using both of them leads to long-lived B cells that can be grown outside the body.

4 B cells with B cells, and tumour cells with tumour cells

5 because monoclonal antibodies from mouse tissue will be recognised as foreign (non-self) and will be destroyed by human antibodies if not 'humanised'.

6 The introduction of antibodies into humans could cause a reaction / disease that could be dangerous. The antigen could stimulate an over-response of the immune system.

5.6

1 Active immunity – individuals are stimulated to produce their own antibodies. Immunity is normally long-lasting.

Passive immunity – antibodies are introduced from outside rather than being produced by the individual. Immunity is normally only short-lived.

2 The influenza virus displays antigen variability. Its antigens change frequently and so antibodies no longer recognize the virus. New vaccines are required to stimulate the antibodies that complement the new antigens.

MMR vaccine

1 the MMR vaccine is given at 12–15 months – the same time as autism symptoms appear

2 It might: present the findings in an incomplete / biased fashion, ignore unfavourable findings, fund only further research that seems likely to produce the evidence that its seeks rather than investigating all possible outcomes, withdraw funding for research that seems likely to produce unfavourable findings.

5.7

1 It possesses RNA and the enzyme reverse transcriptase which can make DNA from RNA – a reaction that is the reverse of that carried out by transcriptase.

2 HIV is a virus – the human immunodeficiency virus – while AIDS (acquired immune deficiency syndrome) describes the condition caused by infection with HIV.

3 People with impaired immune systems, such as those with AIDS, are far less able to protect themselves from TB infections and so are more likely to contract and spread TB to others. Widespread use of condoms helps prevent HIV infection and so can reduce the number of people with impaired immune systems who are consequently more likely to contract TB.

6.1

1 respiratory gases, nutrients, excretory products and heat

2 0.6

3 Any 3 from: surface area / thickness of cell-surface membrane / permeability of cell-surface membrane to the particular substance / concentration gradient of substance between inside and outside of cell / temperature

Significance of surface area to volume ratio in organisms

1 They are very small and so have a very large surface area to volume ratio.

2 The blue whale has a very small surface area to volume ratio and so loses less heat to the water than it would if it were small.

Calculating a surface area to volume ratio

In making the calculation it is important to note that the cylinder sits on the rectangular box. This means that one end does not form part of the external surface. At the same time the equivalent area of the rectangular box also does not form part of the external surface. The area of these two discs is equal to $2 \times \pi r^2$ and must be subtracted from our calculation.

Since the surface area of the cylinder is calculated as $2\pi rh + 2\pi r^2$, we can therefore ignore the $2\pi r^2$ because this is the same as the area that must be subtracted from our calculation. The surface area of the cylinder is therefore taken to be $2\pi rh$ or $2 \times 3.14 \times 2 \times 8 = 100.48\,cm^2$. The surface area of the rectangular box is $2 \times (6 \times 5) + 2 \times (5 \times 12) + 2 \times (6 \times 12) = 324\,cm^2$. The total surface area = $100.48 + 324 = \mathbf{424.48\,cm^2}$.

The volume of the cylinder is calculated using the surface area of the base (πr^2) multiplied by its height (h). This equals $3.14 \times (2 \times 2) \times 8 = 100.48\,cm^3$. The volume of the rectangular box = $12 \times 6 \times 5 = 360\,cm^3$. The total volume is therefore $360 + 100.48 = \mathbf{460.48\,cm^3}$.

The surface area to volume ratio is $424.48 \div 460.48 = \mathbf{0.92}$.

6.2

1 diffusion over the body surface

2 Gas exchange requires a thin permeable surface with a large area. Conserving water requires thick, waterproof surfaces with a small area.

3 because it relies on diffusion to bring oxygen to the respiring tissues. If insects were large it would take too long for oxygen to reach the tissues rapidly enough to supply their needs.

Spiracle movements

1 It falls steadily and then remains at the same level.

2 Cells use up oxygen during respiration and so it diffuses out of the tracheae and into these cells. With the spiracles closed, no oxygen can diffuse in from the outside to replace it. Ultimately, all the oxygen is used up and so the level ceases to fall.

3 the increasing level of carbon dioxide

4 It helps conserve water because the spiracles are not open continuously and therefore water does not diffuse out continuously.

5 It contained more oxygen.

6.3

1 the movement of water and blood in opposite directions across gill lamellae

2 because a steady diffusion gradient is maintained over the whole length of the gill lamellae. Therefore more oxygen diffuses from the water into the blood.

3 Mackerel have more gill lamellae / gill filaments / larger surface area compared to plaice.

4 Less energy is required because the flow does not have to be reversed (important as water is dense and difficult to move).

6.4

1 Any 2 from: no living cell is far from the external air / diffusion takes place in the gas phase / need to avoid excessive water loss / diffuse air through pores in their outer covering (can control the opening and closing of these pores).

2 Any 2 from: insects may create mass air flow – plants never do / insects have a smaller surface area to volume ratio than plants / insects have special structures (tracheae) along which gases can diffuse – plants do not / insects do not interchange gases between respiration and photosynthesis – plants do.

3 Helps to control water loss by evaporation / transpiration.

Exchange of carbon dioxide

1 respiration

2 photosynthesis

3 At this light intensity the volume of carbon dioxide taken in during photosynthesis is exactly the same as the volume of carbon dioxide given out during respiration.

4 increase is $160 - 115 = 45\,cm^3\,h^{-1}$. Percentage increase $= \dfrac{45}{115} \times 100 = 39.13$.

5 With stomata closed, there is little, if any, gas exchange with the environment. While there will still be some interchange of gases produced by respiration and photosynthesis, neither process can continue indefinitely by relying exclusively on gases produced by the other. Some gases must be obtained from the environment. In the absence of this supply, both photosynthesis and respiration will ultimately cease and the plant will die.

6 The rate of respiration (in the dark)

6.5

1 Efficient gas exchange requires a thin, permeable surface with a large area. On land these features can lead to a considerable loss of water by evaporation.

2 waterproof covering to the body / ability to close the openings of the gas-exchange system (stomata and spiracles)

3 Plants photosynthesise and therefore need a large surface area to capture light.

4 a Water evaporating from the leaf is trapped. The region within the rolled up leaf becomes saturated with water vapour. There is no water potential gradient between the inside and outside of the leaf and so water loss is considerably reduced.

b Almost all stomata are on the lower epidermis. This would be exposed to air currents that would reduce the water potential immediately outside the leaf. The water potential gradient would be increased and a lot of water vapour would be lost.

Not only desert plants have problems obtaining water

1 The rain rapidly drains through the sand out of reach of the roots. Sand dunes are usually in windy situations, which reduces water potential and so increases the water potential gradient, leading to increased water loss.

2 The soil solution is very salty, i.e. it has a very low water potential, making it difficult for root hairs to draw water in by osmosis.

3 because in winter the water in the soil is frozen and therefore cannot be absorbed by osmosis

4 Being enzyme-controlled, photosynthesis is influenced by temperature. In cold climates enzymes work slowly and this limits the rate of photosynthesis. Therefore there is a reduced need for light as photosynthesis is taking place only slowly. In warm climates, photosynthesis occurs rapidly and therefore a large leaf area is needed to capture sufficient light.

6.6

1 Any 2 from: humans are large / have a large volume of cells; humans have a high metabolic rate / high body temperature.

2 alveoli, bronchioles, bronchus, trachea, nose

3 The cells produce mucus that traps particles of dirt and bacteria in the air breathed in. The cilia on these cells move this debris up the trachea and into the stomach. The dirt / bacteria could damage / cause infection in the alveoli.

6.7

1 17.14 breaths min^{-1}. Measure the time interval between any two corresponding points on either graph that are at the same phase of the breathing cycle (e.g. two corresponding peaks on the volume graph or two corresponding troughs on the pressure graph). The interval is always 3.5 s. This is the time for one breath. The number of breaths in a minute (60 s) is therefore 60 s ÷ 3.5 s = 17.14.

2 It is essential to first convert all figures to the same units. For example 3000 cm^3 is equal to 3.0 dm^3. From the graph you can calculate that the exhaled volume is 0.48 dm^3 less than the maximum inhaled volume. The exhaled volume is therefore 3.0 − 0.48 = 2.52 dm^3. If working in cm^3, the answer is 2520 cm^3.

3 The muscles of the diaphragm contract, causing it to move downwards. The external intercostals muscles contract, moving the rib cage upwards and outwards. Both actions increase the volume of the lungs. Consequently the pressure in the alveoli of the lungs is reduced.

Pulmonary ventilation

1 17 min^{-1}

6.8

1 **a** The rate of diffusion is more rapid the shorter the distance across which the gases diffuse.

b There is a very large surface area in 600 million alveoli (2 lungs) and this makes diffusion more rapid.

c Diffusion is more rapid the greater the concentration gradient. Pumping of blood through capillaries removes oxygen as it diffuses from the alveoli into the blood. The supply of new carbon dioxide as it diffuses out of the blood into the alveoli helps to maintain a concentration gradient that would otherwise disappear as the concentrations equalised.

d Red blood cells are flattened against the walls of the capillaries to enable them to pass through. This slows them down, increasing the time for gas exchange and reducing the diffusion pathway, thereby increasing the rate of diffusion.

2 four times greater

Correlations and causal relationships

1 Correlation between the incidence of lung cancer in men and the number of cigarettes smoked per day.

2 There is no experimental evidence in the data provided to show that smoking causes cancer. Hence there is no causal link between the two variables.

Risk factors for lung disease

1 Any 4 from: smoking / air pollution / genetic make-up / infections / occupation.

2 Allow figures in the range 50–60%.

3 two times. Around 80% of non-smokers live to age 70 compared to around 40% of people who smoke more than 25 cigarettes a day.

4 In general terms, she will live longer. More specifically she has a 50% chance of living to be 65 years if she carries on smoking but a 50% chance of living to 80 years if she gives up. Her life expectancy could increase by 15 years.

5 The aim of all the following measures is to reduce the number of cigarettes or other tobacco products

3 **a** muscular wall of atrium; **b** diastole; **c** semi-lunar valve

4 Training builds up the muscles of the heart and so the stroke volume increases / more blood is pumped at each beat. This means that, if the cardiac output is the same, the heart rate / number of beats per minute decreases.

5 One complete cycle takes 0.8 s. Therefore the number of cycles in a minute = 60 ÷ 0.8 = **75.** As there is 1 beat per cycle then there are 75 beats in a minute.

6 $\dfrac{5.2}{0.065} = 80$ beats min^{-1}.

Electrocardiogram

1 A = normal – large peaks and small troughs repeated identically suggesting a regular rhythm;

B = heart attack – less pronounced peaks and smaller troughs repeated in a similar, but not identical, way – a disrupted rhythm;

C = fibrillation – highly irregular pattern with no discernible rhythm

7.6

1 **a** elastic tissue allows recoil and hence maintains blood pressure / smooth blood flow / constant blood flow

b muscle can contract, constricting the lumen of the arterioles and therefore controlling the flow of blood into capillaries;

c valves prevent flow of blood back to the tissues and so keep it moving towards the heart / keep blood at low pressure flowing in one direction

d The wall is very thin, making the diffusion pathway short and exchange of material rapid.

2 **a** C; **b** B; **c** E; **d** D; **e** A

3 hydrostatic pressure (due to pumping of the heart)

4 via the capillaries and via the lymphatic system

Blood flow in various blood vessels

1 Rate of blood flow decreases gradually in the aorta and then very rapidly in the large and small arteries. It remains relatively constant in the arterioles and capillaries before increasing, at an increasing rate, in the venules and veins and vena cava.

2 Contraction of the left ventricle of the heart causes distension of the aorta. The elastic layer in the aorta walls creates a recoil action. There is therefore a series of pulses of increased pressure, each one the result of ventricle contraction.

3 Because the total cross-sectional area is increasing / there is increased frictional resistance from the increasing area of blood vessel wall.

4 Blood flow is slower, allowing more time for metabolic materials to be exchanged.

5 Capillaries have a large surface area and very thin walls (single cell thick) and hence a short diffusion pathway.

7.7

1 **a** transpiration; **b** stomata; **c** lower / reduced / more negative; **d** osmosis; **e** cohesion; **f** increases

Hug a tree

1 at 12.00 hours because this is when water flow is at its maximum. As transpiration creates most of the water flow they are both at a maximum at the same time.

2 Rate of flow increases from a minimum at 00.00 hours to a maximum at 12.00 hours and then decreases to a minimum again at 24.00 hours.

3 As evaporation / transpiration from leaves increases during the morning (due to higher temperature / higher light intensity) it pulls water molecules through the xylem because water molecules are cohesive / stick together. This transpiration pull creates a negative pressure / tension. The greater the rate of transpiration, the greater the water flow. The reverse occurs as transpiration rate decreases during the afternoon and evening.

4 As transpiration increases up to 12.00 hours, so there is a higher tension (negative pressure) in the xylem. This reduces the diameter of the trunk. As transpiration rate decreases, from 12.00 hours to 24.00 hours, the tension in the xylem reduces and the trunk diameter increases again.

5 Transpiration pull is a passive process / does not require energy. Xylem is non-living and so cannot provide energy. Although root cortex and leaf mesophyll cells are living – the movement of water across them uses passive processes, e.g. osmosis, and so continues at least for a while, even though the cells have been killed.

Measurement of water using a potometer

1 **a** As xylem is under tension, cutting the shoot in air would lead to air being drawn into the stem, which would stop transport of water up the shoot. Cutting under water means water, rather than air, is drawn in and a continuous column of water is maintained.

b Sealing prevents air being drawn into the xylem and stopping water flow up it / Sealing prevents water leaking out which would produce an inaccurate result.

2 that all water taken up is transpired

3 Volume of water taken up in one minute:
$3.142 \times (0.5 \times 0.5) \times 15.28 = 12.00\,mm^3$. Volume of water taken up in 1 hour:
$12.00 \times 60 = 720\,mm^3$. Volume in cm^3 =
$720 \div 1000 = 0.72\,cm^3$.
Answer = $0.72\,cm^3\,h^{-1}$

4 their surface area / surface area of the leaves

5 An isolated shoot is much smaller than the whole plant / may not be representative of the whole plant / may be damaged when cut.

Conditions in the lab may be different from those in the wild, e.g. less air movement / greater humidity / more light (artificial lighting when dark).

Specialised plant cells:
1 thin cell wall, large surface area / long, hair-like extension

2 Osmosis is the passage of water from a region where it has a higher water potential to a region where it has a lower water potential, through a selectively permeable membrane.

3 The water potential of the soil solution is higher than that in the vacuole / cytoplasm of the root hair cell. Water therefore moves along a water potential gradient.

4 mitochondria because they release energy / make ATP during respiration and this energy / ATP is essential for active transport

5 They have thick walls to prevent the vessels collapsing.

6 hollow; elongated

7 Living cells have a cell-surface membrane and cytoplasm, and water movement would be slowed as it crossed this membrane / cytoplasm.

8 waterproofing

9 Any 3 from: allows the vessel to elongate as the plant grows / uses less material and therefore is less wasteful / uses less material and therefore the plant has lower mass / allows stems to be flexible.

7.8
a phloem b sources
c sinks d mass flow
e sieve tube f co-transport
g photosynthesising/chloroplast containing
h lower/more negative
i xylem j higher/less negative
k osmosis

7.9
1 a There would be a large swelling above the ring in summer but little, if any, swelling in winter.

b In summer the rate of photosynthesis, and therefore production of sugars, is greater due to higher temperatures, longer daylight and higher light intensity. The translocation of these sugars leads to their accumulation, and therefore a swelling, above the ring. In winter lower temperatures, shorter daylight and lower light intensity mean the rate of photosynthesis is less and any swelling is therefore smaller. In deciduous plants, the lack of leaves means there is no photosynthesis and therefore no swelling at all.

2 If the squirrel strips away the phloem around the whole circumference of the branch it may not have sufficient sugar for its respiration to release enough energy for survival as none can reach it from other parts of the plant.

3 If the branch has sufficient leaves to supply its own sugar needs from photosynthesis, rather than depending on supplies from elsewhere, it might survive for a while at least.

4 It is unlikely that squirrels would strip bark from around the whole circumference of a large tree trunk. Any intact phloem could still supply sufficient sugars to its roots to allow it to survive.

5 a It takes time for the sucrose from the leaves to be transported across the mesophyll of the leaf by diffusion and then to be actively transported into the phloem.

b The sucrose in the phloem is diluted with the water that enters it from the xylem. A little sucrose may be converted to glucose and used up by the leaves during respiration but this alone would not be sufficient to explain the reduction in concentration in the phloem.

Using radioactive tracers
1 The data suggest that the translocation of ^{42}K is almost entirely in the xylem with very little in the phloem.

2 The wax paper is 'impervious' which means that materials cannot pass across it. In the middle of the region where xylem and phloem are separated by wax paper 99% of the ^{42}K is in the xylem. Even at the beginning and ends of the separated regions at least 85% of the ^{42}K is in the xylem. Where the two are in contact (control) the levels of ^{42}K are much more equal.

3 In sections 1 and 5, xylem and phloem are not separated by wax paper and so lateral movement of ^{42}K can take place. The ^{42}K therefore diffuses from the xylem into the phloem until the concentrations in both are similar.

4 The xylem and phloem could have been separated over the 225mm portion of the control branch and then rejoined but without the wax paper. This is an improvement as it eliminates the physical disruption caused by separating xylem and phloem as the explanation for there being no lateral movement of ^{42}K.

8.1

1 a base sequence of DNA that codes for the amino acid sequence of a polypeptide or functional RNA

2 18

3 A different base might code for a different amino acid. The sequence of amino acids in the polypeptide produced will be different. This change to the primary structure of the protein might result in a different shaped tertiary structure. The enzyme shape will be different and may not fit the substrate. The enzyme–substrate complex cannot be formed and so the enzyme is non-functional.

4 a 5

b the first and last (5th) / the two coded for by the bases TAC

c because some amino acids have up to six different codes, while others have just one triplet

Interpreting the genetic code

1 trp – UGG and met – AUG

2 a leu

b lys

c asp

3 a try-ala-ile-pro-ser

b arg-phe-lys-gly-leu

8.2

1 In prokaryotic cells the DNA is smaller, circular and is not associated with proteins (i.e. does not have chromosomes). In a eukaryotic cell it is larger, linear and associated with proteins / histones to form chromosomes.

2 It fixes the DNA into position.

3 it is looped and coiled a number of times

4 a 50 mm (46 chromosomes in every cell);

b 2.3 m (all diploid cells have same quantity of DNA)

8.3

1 mRNA is larger, has a greater variety of types and is shaped as a long single helix. tRNA is smaller, has fewer types and is clover-leaf in shape.

2 Any 3 from: RNA is smaller than DNA / RNA is usually a single strand and DNA a double helix / the sugar in RNA is ribose while the sugar in DNA is deoxyribose / in RNA the base uracil replaces the base thymine found in DNA

3 A codon is the triplet of bases on messenger RNA that codes for an amino acid. An anticodon is the triplet of bases on a transfer RNA molecule that is complementary to a codon.

Comparison of DNA, mRNA and tRNA

1 a The amount of DNA in a gamete is half that in a body cell.

b It allows gametes to fuse during sexual reproduction without doubling the total amount of DNA at each generation. In so doing it increases genetic variety by allowing the genetic information of two parents to be combined in the offspring.

2 a DNA needs to be stable to enable it to be passed from generation to generation unchanged and thereby allow offspring to be very similar to their parents. Any change to the DNA is a mutation and is normally harmful.

b mRNA is produced to help manufacture a protein, e.g. an enzyme. It would be wasteful to produce the protein continuously when it is only needed periodically. mRNA therefore breaks down once it has been used and is produced again only when the protein is next required.

8.4

1 The enzyme RNA polymerase moves along the template DNA strand, causing the bases on this strand to join with the individual complementary nucleotides from the pool that is present in the nucleus. The RNA polymerase adds the nucleotides one at a time, to build a strand of pre-RNA until it reaches a particular sequence of bases on the DNA that it recognises as a 'stop' code.

2 DNA helicase – This acts on a specific region of the DNA molecule to break the hydrogen bonds between the bases, causing the two strands to separate and expose the nucleotide bases in that region.

3 Splicing is necessary because pre-mRNA has nucleotide sequences derived from introns in DNA. These introns are non-functional and, if left on the mRNA, would lead to the production of non-functional polypeptides or no polypeptides at all. Splicing removes these non-functional introns from pre-mRNA.

4 a UACGUUCAGGUC

b 4 amino acids (1 amino acid is coded for by 3 bases so 12 bases code for 4 amino acids)

5 Some of the base pairs in the genes are introns (non-functional DNA). These introns are spliced from pre-mRNA so the resulting mRNA has fewer nucleotides.

8.5

1 ribosome

2 **a** UAG on tRNA

   b TAG on DNA

3 A tRNA molecule attaches an amino acid at one end and has a sequence of 3 bases, called an anticodon, at the other end. The tRNA molecule is transferred to a ribosome on an mRNA molecule. The anticodon on tRNA pairs with the complementary codon sequence on mRNA. Further tRNA molecules, with amino acids attached, line up along the mRNA in the sequence determined by the mRNA bases. The amino acids are joined by peptide bonds. Therefore the tRNA helps to ensure the correct sequence of amino acids in the polypeptide.

4 One of the codons is a stop codon that indicates the end of polypeptide synthesis. Stop codons do not code for any amino acid so there is one less amino acid than there are codons.

Protein synthesis

1 X = ribosome;   Y = mRNA

2 amino group

3 AUG

4 Val-Thr-Arg-Asp-Ser

5 CAATGGGCT

6 The mutation changes CAG to UAG. UAG is a stop codon that signifies the end of an amino acid sequence at which point the polypeptide is complete and is 'cast off'. The polypeptide chain is therefore shorter than it should be and may not function as normal.

7 **a** Glutamine has two codes GAG and GAA. The reversal of GAG produces the same codon and so still translates as glutamine and hence the polypeptide that is formed is unchanged. Reversal of GAA changes the codon to AAG which translates to a different amino acid – Lys. As a result, the polypeptide has a different primary structure which may affect bonding within the molecule and so change its tertiary structure also.

   b Enzyme function depends on the substrate becoming loosely attached to an enzyme within its active site.

If the mutation has changed the amino acid from glutamine to Lys., then the primary structure of the polypeptide will be different. Hydrogen and ionic bonds between the amino acids of the polypeptide may not be formed in the same way as before and so its tertiary shape may be changed. This change may alter the shape of the active site of the enzyme of which the polypeptide is a part and so the substrate no longer fits.

Glutamine may have been one of the amino acids in the active site to which the substrate normally attaches. Its replacement by another amino acid may mean that, although the shape of the active site is unchanged, the substrate cannot attach normally.

Cracking the code

1 DNase is an enzyme that breaks down DNA. The DNA of the cell needs to be destroyed because it would produce its own mRNA and so make it be impossible to determine which of the many polypeptides produced was due to the synthetic DNA.

2 The codon UUU – because the very radioactive polypeptide (39 800 counts min^{-1}) was only produced from the mixture containing poly U. This polypeptide must be made up of phenylalanine because this is the only radioactive amino acid present. As the synthetic mRNA contains only the base sequence UUUUUU, etc., one codon for phenylalanine must be UUU.

3 As a control experiment to show that the radioactivity was due to the labelled phenylalanine rather than some other factor, e.g. background radiation.

4 It may not be possible to say whether the mRNA sequence starts with U and therefore reads UGU GUG UGU, etc. or starts with G and reads GUG UGU GUG. Equally it may not be possible to say whether the polypeptide sequence begins with cysteine or valine. It may therefore be impossible to relate a particular codon to a particular amino acid.

5 As most amino acids have more than one codon, the code is degenerate. However, any one triplet only codes for a single amino acid and so it is not ambiguous.

6 Because they do not code for any amino acid – they are 'stop' codons that mark the end of a polypeptide.

9.1

1 A deletion because the fifth nucleotide (A) has been lost. The sequence prior to and after this

is the same. (Note: The last base in the mutant version was previously the 13th in the sequence and therefore not shown in the normal version.)

2 In a deletion, all codons after the deletion are affected. Therefore most amino acids coded for by these codons will be different and the polypeptide will be significantly affected. In a substitution, only a single codon, and therefore a single amino acid, will be affected. The effect on the polypeptide is likely to be less severe.

3 The mutation may result from the substitution of one base in the mRNA with another. Although the codon affected will be different, as the genetic code is degenerate, the changed codon may still code for the same amino acid. The polypeptide will be unchanged and there will be no effect.

4 These errors may be inherited and may therefore have a permanent affect on the whole organism. Errors in transcription usually affect only specific cells, are temporary and are not inherited. They are therefore less damaging.

Hybridisation and polyploidy

1 The numbers of chromosomes in the hybrids do not allow them to form homologous pairs during prophase I of meiosis I therefore they cannot produce gametes. This would arise if the hybrid has an odd, rather than an even, number of chromosomes.

2 They are less likely to blow over in storms and are easier to harvest.

3 They cannot breed to produce fertile offspring.

9.2

1 haploid because 27 is an odd number. Diploid cells have 2 sets of chromosomes and so their total must be an even number.

2 independent segregation of homologous chromosome and recombination by crossing over.

3 roller and blood group A, non-roller and blood group A.

4 Gametes are produced by meiosis. In meiosis, homologous chromosomes pair up. With 63 chromosomes precise pairings are impossible. This prevents meiosis and hence gamete production, making them sterile.

5 1024 (the haploid number of 5 is the same as the number of homologous chromosomes)

9.3

1 a increase; b decrease; c increase

2 Different DNA – different codes for amino acids – different amino acids – different protein shape – different protein function (e.g. non-functional

enzyme) – change in a feature determined by that protein – altered appearance – greater genetic diversity

9.3 Natural Selection in action

1 Against the black background the dark form was less conspicuous than the light natural form. As a result, the light form was eaten by birds, or other predators, more frequently than the dark form. More black-coloured moths than light-coloured moths survived and successfully reproduced. Over many generations, the frequency of the 'advantageous' dark-colour allele increased at the expense of the 'less advantageous' light-colour allele.

9.4

1 Selection is the process by which organisms that are better adapted to their environment survive and breed, while those less well adapted fail to do so.

2

Directional selection	Stabilising selection
Favours / selects phenotypes at one extreme of a population	Favours / selects phenotypes around the mean of a population
Changes the characteristics of a population	Preserves the characteristics of a population
Distribution curve remains the same shape but the mean shifts to the left or right	Distribution curve becomes narrower and higher but the mean does not change

3 Directional selection – because birds to one side of the mean (heavier birds) were being selected for, while those to the other side of the mean (lighter birds) were being selected against. The population's characteristics are being changed, not preserved.

They must be cuckoo!

1 Removing cuckoo eggs means there will be more food for the magpie's own chicks. These chicks have a greater probability of being successfully raised to adulthood.

2 Alleles for this type of behaviour are obviously present in the adult birds. There is a high probability that some of the chicks will inherit these alleles. Removing cuckoo eggs increases the probability of more of these chicks surviving to breed and therefore passing on the alleles for this behaviour to subsequent generations.

3 Displaying this behaviour has previously been of no advantage to magpies and so no selection for this behaviour has taken place. Although cuckoos have now arrived, it will take many generations

for selection to operate and for allele frequencies to change.

4 Directional selection – because the population's characteristics are being changed, not preserved.

5 Magpies that do not remove cuckoo eggs will raise both cuckoo and magpie chicks. There will be less food/space for the magpie chicks compared to those raised in magpie nests and so less chance of them surviving to adulthood and breeding. The chicks of magpies that remove cuckoo eggs will be selected for in preference to the chicks of parents that do not. Over many generations the population will change to have a greater proportion of magpies that remove cuckoo eggs at the expense of those that do not.

10.1

1 They are capable of breeding to produce offspring which are themselves fertile

2 It is based on evolutionary relationships between organisms and their ancestors; it classifies species into groups using shared characteristics derived from their ancestors; it is arranged in a hierarchy in which groups are contained within larger composite groups with no overlap.

3 To ensure that mating only takes place between members of the same species as only they can produce fertile offspring.

4 The courtship display that most closely resembles that of the first species is likely to be the closest relative.

5 1 phylum, 2 class, 3 order, 4 family,
   5 *Rana*, 6 species, 7 *temporaria*

6 a lizards
   b birds
   c common ancestor of lizards and snakes
   d Dinosaurs are extinct but all the other groups are still living and so they are shown extending further along the time line – as far as 'present'.

The difficulties of defining species

1 Fossil records are normally incomplete and not all features can be observed (there is no biochemical record) and so comparisons between individuals are hard to make. Fossil records can never reveal whether individuals could successfully mate.

2 Species change and evolve over time, sometimes developing into different species. There is considerable variety within a species. Fossil records are incomplete / non-existent. Current classifications only reflect current scientific knowledge and, as this changes, so does the naming and classifying of organisms.

3 No, it does not. Only fertile female mules are known, so interbreeding (a feature of any species) is impossible. The event is so rare that it can be considered abnormal and it would be wrong to draw conclusions from it. If a mule were a species, it would mean that the parents were the same species – however, donkeys and horses are sufficiently different to be recognised as separate species.

10.2

1 the number of different species and the proportion of each species within a given area / community

2

Species	Numbers in salt marsh	$n(n-1)$
Salicornia maritima	24	$24(23) = 552$
Halimione portulacoides	20	$20(19) = 380$
Festuca rubra	7	$7(6) = 42$
Aster tripolium	3	$3(2) = 6$
Limonium humile	3	$3(2) = 6$
Suaeda maritima	1	$1(0) = 0$
$\Sigma n(n-1)$ 986		

$$D = \frac{58(57)}{986} = \frac{3306}{986} = \text{Answer } 3.35$$

3 It measures both the number of species and the number of individuals. It therefore takes account of species that are only present in small numbers.

Species diversity and ecosystems

1 Greenhouse gases lead to climate change. Communities with a high species diversity index are likely to include at least one species adapted to withstand the change and therefore survive. When the index is low, the community is less likely to include a species adapted to withstand the change and is therefore at greater risk of being damaged.

2 a The community fluctuates in line with environmental change – rising and falling in the same way but a little later in time.

   b Communities with a high species diversity are more stable because they have a greater variety of species and therefore are more likely to have species that are adapted to the changed environment. Those with a low species diversity are less stable because they have fewer species and are less likely to include a species adapted to the change.

10.3

1 The few species possessing desirable qualities are selected for and bred. Other species are excluded,

as far as possible, by culling or the use of pesticides. Many individuals of a few species = low species diversity.

2 Because forests, with their many layers, have many habitats with many different species, i.e. a high species diversity. Grasslands have a single layer, fewer habitats, fewer species and lower species diversity.

3 Ponds provide a habitat for a wide range of aquatic species that are unlikely to find alternative habitats as aquatic habitats are few and far between. Ponds may be a source of food and water for terrestrial species which may also not survive without them. Hedgerow species are likely to have a larger range of alternative habitats as most of the area around will be terrestrial with other sources of food and shelter and so fewer species are likely to be lost.

Human activity and loss of species in the UK

1 500 000 km (350 000 × 100 ÷ 70)

2 Mixed woodlands comprise many species while the commercial conifer plantations that replace them are largely of a single predominant species

3 Any 1 from each: benefit – cheaper grazing / fodder for animals and hence cheaper food / more efficient food production; risk – loss of species diversity / less stable ecosystem / more fertilizers and pesticides needed

4 It provides evidence to inform and support decision making. Data show where the most change has occurred and therefore the habitats most at risk. These can be prioritised and measures taken to conserve them, e.g. by giving them special protection. Funds can be directed towards reverting land to its former use, e.g. by grants to farmers to create hay meadows / convert set-aside to woodland / re-establish hedgerows. Helps decision-makers form appropriate rules / legislation to prevent habitat destruction, e.g. ban on drainage of certain sites / rock removal. Informs decisions on planning applications for planting forests / reclaiming land.

5 Hedges provide more habitats / niches / food sources and therefore more species can survive. Species diversity is therefore increased

Hedge rows!

1 There is more land on which to grow crops / as hedges harbour pests, diseases and weeds, especially in winter, their removal reduces the chances of these affecting crops which therefore produce greater yields / hedges compete with the crop for light, water and nutrients and therefore reduce yields.

2 Hedges are a habitat for a wide range of organisms, including some that are natural predators of crop pests and therefore provide a means of biological control / pollinating insects live in hedges and these are essential for the production of certain crops like fruits / hedges provide wind-breaks and so reduce soil erosion.

10.4

1 mutations

2 species 2 and 3 because their amino acid sequences are identical

Establishing relationships

1 chimpanzee, gorilla, orang-utan, lemur, gibbon

2 the chimpanzee and the gorilla because they both show the same % precipitation (95%)

3 the gibbon because it shows only a 3% difference (85–82) in precipitation between itself and the orang-utan. All the other primates show a greater difference.

4 These data suggest that the gibbon is much more closely related to humans than the lemur. The haemoglobin study suggested the lemur was a closer relative. The chimpanzee is shown to be most closely related to humans in the haemoglobin study, while in the gene bases study it is the gorilla.

5 There are fewer differences between the bases in the gene of a human and that of an orang-utan (29) than there are between the genes of a human and a lemur (48). This suggests that the evolution of humans and lemurs diverged earlier than that of humans and orang-utans, giving more time for the amino acid differences to occur.

6 No, it does not. This study suggests that gorillas (with fewer base differences) are more closely related to humans than chimpanzees. The other studies suggest chimpanzees are more closely related to humans. The position of the orang-utan is the same in all three studies. The position of the lemur is the same as in the immunological study but different from that in the haemoglobin study.

10.5

1 sampling bias; chance variation

2 by using random sampling – effectively using a computer to generate sampling sites

3 In a given set of values, the mean is the sum of a set of values divided by the number of items, the mode is the single value that occurs most often while the median is the central or middle value when the values are arranged in ascending order.

Practical 1

1 (a) Sketch a graph of your expected results.
       Remember to label your axes.

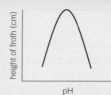

   (b) List all variables that need to be controlled and how you would control them.
       Temperature – Keep celery extract and H_2O_2 in a thermostatically controlled water
       bath at 30°C
       Enzyme concentration – use the same source of celery extract which has been
       mixed evenly
       Substrate concentration – use same volume and concentration of H_2O_2

2 (a) Repeat each pH at least twice and calculate a mean.
   (b) This method is very subjective to decide on the highest point of the froth.
       Change method to using a gas syringe to collect the O_2 gas released.
       Celery extract may contain varying concentrations of enzyme.
       Change method to use a pure source of a specific concentration of enzyme.

Practical 2

1 (a) $\frac{12}{150} \times 100 = 8\%$

   (b) No stain used / not root tip / cells not dividing in this small sample / more than one
       layer of cells as not squashed firmly enough.

2 (a) As distance increases from the root tip, the mitotic index decreases.
       Above 1 mm an increase in distance from root tip has little effect on the mitotic
       index / plateaus.
       Correctly quote paired set of data.

   (b) Meristem tissue only nearest the tip has the ability to divide and there is less
       meristem tissue as the distance increases from the tip.
       Nearest the tip gets more damage, therefore needs to do more cell division to repair
       the tissue.

Practical 3

1

Concentration of sucrose (mol dm⁻³)	Volume of distilled water (cm³)	Volume of 0.5 mol dm⁻³ stock solution sucrose (cm³)
0.0	20	0
0.1	16	4
0.2	12	8
0.3	8	12
0.4	4	16
0.5	0	20

2 (a) Calculate the % change in mass of potato tissue.

Concentration of sucrose (mol dm⁻³)	Mass before submerging in solution (g)	Mass after submerging in solution (g)	Mass change (g)	Percentage change in mass of potato tissue (%)
0.0	4.5	5.0	+0.5	+11.1
0.1	3.9	4.3	+0.4	+10.3
0.2	4.3	4.5	+0.2	+4.7
0.3	4.1	4.2	−0.1	−2.4
0.4	4.4	3.7	−0.7	−15.9
0.5	4.4	3.6	−0.8	−18.2

(b) There are different starting and finishing masses.
The mass change is very small; therefore a % change is easier to compare real differences.

3 (a) Correctly labelled axes with units;
Uniform axes;
plots taking up over $\frac{1}{2}$ space of graph;
accurate plots;
smooth line of best fit.

(b) Use this graph to find the concentration of sucrose (where curve crosses x-axis). Between 0.25 and 0.3 mol dm^{-3}

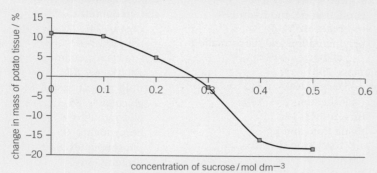

(c) Use a data resource with listed sucrose concentrations and water potentials to find the water potential for the sucrose solution read off the graph

Practical 4

1 (a) The red pigment is water soluble and held in the vacuole;
The cell-surface membrane is selectively permeable and some pigment diffuses out.

(b) As temperature increases from 20 to 40°C, there is a small increase in absorbance reflecting a small increase in the permeability of the cell-surface membrane.
Above 50°C there is a steep increase in the permeability of the cell-surface membrane.

(c) The proteins embedded in the cell-surface membrane become denatured.
The structure of the cell-surface membrane has been permanently disrupted so is now fully permeable and most of the pigment diffuses out.

Practical 5

1 (a) Spiracles.
(b) Control water vapour loss by closing spiracles if need to conserve water.
(c) High temperature environment causes more water to evaporate;
hairs trap water <u>vapour</u> and this reduces water potential gradient and therefore water vapour loss.

2 (a) Penetrate deep into muscle tissues;
delivers more air / oxygen to muscles.
(b) Chitin
(c) Smaller diameters are more permeable to gases and get closer to body cells for gaseous exchange by diffusion.

Practical 6

1 (a) To sterilise the equipment/ to kill any microbes on the equipment.
(b) 1 Washing hands / cleaning work surface with disinfectant
     2 Flame sterilising the inoculating loop
     3 Flaming the neck of the culture tube containing the bacteria
     4 Streaking the plate with the inoculating loop **quickly**
     5 Only lifting the lid of the petri dish a small amount
(c) To kill any harmful / pathogenic bacteria so they don't harm anyone.

Index

Acknowledgements

The authors wish to thank Graham Read for his invaluable help with the manuscript, Mitch Fitton for her meticulous editing, James Penny for the mathematical aspects, Louise Garcia for her work on the practice questions, Ellena Bale for her help with the practical elements and, of course, Alison Schrecker and Amy Johnson from OUP for their support, hard work and encouragement.

p30: Darin Burks/Shutterstock; **p09**: Martin Shields/Alamy; **p25**: J.C. Revy, ISM/Science Photo Library; **p62**: Biophoto Associates/Science Photo Library; **p95**: Steve Gschmeissner/Science Photo Library; **p97(L)**: Deepblue-photographer/Shutterstock; **p97(R)**: Andy Crawford/Getty Images; **p117**: Oksana Kuzmina/Shutterstock; p118: Saturn Stills/Science Photo Library; **p155(L)**: Manfred Kage/Science Photo Library; **p161**: Francis Leroy, Biocosmos/Science Photo Library; **p168**: Claudia Paulussen/Shutterstock; pg 184: Science Photo Library; **p202**: J.C. Revy, ISM/Science Photo Library; **p221**: Bon Appetit/Alamy; **p222(TL)**: WILDLIFE GmbH/Alamy; pg 222(BL): Blickwinkel/Alamy; **p222(R)**: Tim Gainey/Alamy; **p238(T)**: Claude Nuridsany & Marie Perennou/Science Photo Library; **p238(M)**: Chris2766/Shutterstock; **p238(B)**: Michal Ninger/Shutterstock; **p246(C)**: AC Rider/Shutterstock; **p247**: Oticki/Shutterstock; **p81**: Image Point Fr/Shutterstock; **p141**: Pawel Kazmierczak/Shutterstock; **p154**: Istetiana/Shutterstock; **p241**: Nancy Tripp Photography/Shutterstock; **p230**: Michael W. Tweedie/Science Photo Library; **p02-03** Background: Vitstudio/Shutterstock; **p53** Background: Vitstudio/Shutterstock; **p56-57** Background: Fusebulb/Shutterstock; **p125** Background: Fusebulb/Shutterstock; **p128-129** Background: Triff/Shutterstock; **p111**: Alfred Pasieka/Science Photo Library; **p200-201** Background: Aquapix/Shutterstock; **p261** Background: Aquapix/Shutterstock; **p239**: Duncan Usher/Alamy; **p209**: Alfred Pasieka/Science Photo Library; **p234**: David Hosking/Alamy; **p13**: Power and Syred/Science Photo Library; **p48**: Hermann Eisenbeiss/Science Photo Library; **p49**: Rido/Shutterstock; **p60**: Chris Priest/Science Photo Library; **p61**: Mauro Fermariello/Science Photo Library; **p63(T)**: David Mccarthy/Science Photo Library; **p63(B)**: Susumu Nishinaga/Science Photo Library; **p66**: Dr Gopal Murti/Science Photo Library; **p68**: CNRI/Science Photo Library; **p69**: Dr.Jeremy Burgess/Science Photo Library; **p70(T)**: Don Fawcett/Science Photo Library; **p70(B)**: Science Photo Library; **p76**: Dr Gopal Murti/Science Photo Library; **p79(A)**: Pr. G Gimenez-Martin/Science Photo Library; **p79(B)**: Pr. G Gimenez-Martin/Science Photo Library; **p79(C)**: Pr. G Gimenez-Martin/Science Photo Library; **p79(D)**: Pr. G Gimenez-Martin/Science Photo Library; **p82**: DR GOPAL MURTI/SCIENCE PHOTO LIBRARY; p91(L): Prof. P. Motta/Dept. Of Anatomy/University "La Sapienza", Rome/Science Photo Library; **p91(R)**: J.C. Revy, ISM/Science Photo Library; **p102**: Lowell Georgia/Science Photo Library; **p103**: K.R. Porter/Science Photo Library; **p104(T)**: Eye of Science/Science Photo Library; **p104(B)**: Science Photo Library; **p108**: Eye of Science/Science Photo Library; **p110**: Dr Gopal Murti/Science Photo Library; **p114**: Dr Jeremy Burgess/Science Photo Library; **p187**: Ed Reschke/Getty Images; **p113(B)**: Dr Rob Stepney/Science Photo Library; **p113(T)**: Author; **p115**: Volker Steger/Science Photo Library; **p120**: Dr. Hans Gelderblom, Visuals Unlimited / Science Photo Library; **p134**: Microfield Scientific Ltd/Science Photo Library; **p137**: Dr Jeremy Burgess/Science Photo Library; **p139(T)**: Author; **p139(B)**: Author; **p140**: Author; **p143(T)**: CNRI/Science Photo Library; **p143(B)**: Proff. Motta, Correr & Nottola/ University "La Sapienza", Rome/Science Photo Library; **p146**: Manfred Kage/Science Photo Library; **p155(R)**: Eye of Science/Science Photo Library; **p165(L)**: William J. Howes/FLPA/Alamy; **p165(R)**: Author; **p166(T)**: Ron Steiner/Alamy; **p166(B)**: Natural Visions/Alamy; **p175**: Proff. P. Motta/G. Macchiarelli/University "La Sapienza", Rome/Science Photo Library; **p179**: CNRI/Science Photo Library; **p180(T)**: Steve Gschmeissner/Science Photo Library; **p180(B)**: Susumu Nishinaga/Science Photo Library; **p183**: Power and Syred/Science Photo Library; **p189**: J.C. Revy, Ism/Science Photo Library; **p206**: Biophoto Associates/Science Photo Library; **p229(A)**: Author; **p229(B)**: Author; **p229(C)**: Author; **p229(D)**: Author; **p231**: David Levenson/Alamy; **p232**: WRPublishing/Alamy; **p242**: Mark J. Barrett/Alamy; **p243(T)**: Photodisc; **p243(B)**: SIMON FRASER/SCIENCE PHOTO LIBRARY; **p245**: Art Kowalsky/Alamy; **p246(A)**: Author; **p246(B)**: Author; **p248(L)**: Worldwide Picture Library/Alamy; **p248(M)**: Leslie J Borg/Science Photo Library; **p248(R)**: Photodisc; **p249**: James King-Holmes/Science Photo Library; **p253(T)**: George Ranalli/Science Photo Library; **p253(C)**: Tony Wood/Science Photo Library; **p253(B)**: Bob Gibbons/Science Photo Library; **p207**: CNRI/Science Photo Library; **p197**: Levente Gyori/Shutterstock.

Artwork by Q2A Media